D0776850

NOBLE-GAS COMPOUNDS

NOBLE-GAS COMPOUNDS

Edited with Introductions by

HERBERT H. HYMAN

THE UNIVERSITY OF CHICAGO PRESS

CHICAGO AND LONDON

Library of Congress Catalog Card Number: 63-20907

The University of Chicago Press, Chicago & London
The University of Toronto Press, Toronto 5, Canada

Composed and printed in the U.S.A.

PREFACE

In the summer of 1962 a number of the scientists whose work is described in this volume established the existence of a rich and complicated chemistry for xenon. Although neither experiments nor theory had ever demonstrated that compound formation was impossible for an "inert" gas, it is fair to say that many scientists were very surprised and even mildly incredulous when they first heard of this development.

It was clearly desirable that the rapidly appearing fragments of research results in this interesting area should be collected in a single volume as soon as an opportunity arose. In April, 1963, most of the scientists who had contributed to the study of noble-gas compounds assembled at Argonne National Laboratory, where much of the early work was done. At a two-day meeting they summarized and compared their observations. This volume has been distilled from the material made available through this conference.

The amount of research that has accumulated on noble-gas compounds in less than one year is impressive, and this collection should serve as a starting point for subsequent work in the same field. Moreover, as written material was accumulated, it became apparent to the editor that the resulting book might become a pedagogical tool with value well beyond that of a simple collection of current research.

The experimental techniques that have been employed in this short period in elucidating the behavior of xenon compounds cover a strikingly wide range. Virtually the entire battery of modern apparatus at the disposal of the physical chemist has been employed. In many cases, such as the structure determinations by X-ray and neutron diffraction, highly automated and elaborate equipment has been used to yield results on a level of precision not yet attained for many more familiar compounds. The most recently introduced techniques are included. The Mössbauer effect, in itself a startling innovation a few years ago, has been employed readily and effectively to study the nature of xenon compounds. Yet the traditional methods of physical chemistry are not neglected: approximate vapor-density determinations are used to establish molecular weight; thermochemical measurements, to fix thermodynamic stability or instability; conventional aqueous and non-aqueous solution–chemistry studies, to distinguish both ionic and undissociated species; titration curves, to establish valence; and, of course, the powerful tools of absorption and Raman spectroscopy, to identify molecular species. Here in a single volume the student will find how modern chemists study inorganic chemistry—the questions that are asked and the way answers are found.

As we leave the experimental realm and look for theoretical explanations and correlations of our observations, we find a fascinating story with many elements of a good mystery. There are as yet too few clues for even the distinguished battery of detectives that has applied itself to this challenge. It has not been easy to predict, or even explain, the chemistry of a family of elements that formerly had no chemistry.

The challenge of this mystery has been stimulating, indeed, as the reader of Part IX will readily note. No consensus about the best approach has been reached. It is clear, however, that the problems associated with integrating the family of "inert" gases with the other families of the periodic table involve fundamental questions of compound formation in general. The student of modern chemistry will find here an appraisal of the current state of theoretical knowledge, with both its strengths and its weaknesses exposed.

It is a pleasure at this point to acknowledge the assistance of those who made this volume possible. The research has been done in many laboratories, which have been supported in various ways. Some of these are acknowledged in the text. Most of the experimental work, however, has been done at laboratories supported by the United States Atomic Energy Commission. The following authors report work performed under the auspices of the AEC: in Part I, Claassen *et al.*; in Part II, Chernick, Smith, Studier and Sloth, Sheft and Hyman, MacKenzie and Wiswall, Weeks and Matheson, Chernick *et al.*, and Fields *et al.*; in Part III, Pomeroy, and Smith; in Part IV, Gunn and Williamson, Stein and Plurien, and Gunn; in Part V, Kilpatrick, Malm *et al.*, Gruen, and Appelman; in Part VI, Siegel and Gebert, Hamilton and Ibers, Templeton *et al.*, Burns *et al.*, Levy and Agron, Burns *et al.*, and Bohn *et al.*; in Part VII, Hindman and Svirmickas, Perlow *et al.*, Ruby, Smith, Claassen, and Hyman and Quarterman; and in Part VIII, Finkel *et al.*

Many of the articles include excerpts and figures and plates taken from earlier contributions by the authors to the periodical literature. We gratefully acknowledge permission to use such material from the *Journal of the American Chemical Society* (Part I, Claassen *et al.*; Part II, Streng *et al.*, Weeks and Matheson, and Fields *et al.*; Part VI, Templeton *et al.* [both papers] and Levy and Agron; and Part IX, Lohr and Lipscomb and Jortner *et al.*); *Science* (Part I, Claassen *et al.*; Part II, Smith, and Streng *et al.*; Part IV, Gunn and Williamson; Part V, Williamson and Koch; Part VI, Hamilton and Ibers and Burns *et al.*; and Part VII, Brown *et al.* and Claassen); the *Journal of Physical Chemistry* (Part II, Studier and Sloth); the *Journal of Chemical Physics* (Part II, Smith); *Inorganic Chemistry* (Part II, Dudley *et al.*); and the *Proceedings of the Chemical Society* (Part VII, Falconer and Morton).

Finally, the editor would like to acknowledge the assistance of many of his colleagues in the arrangements for the conference and preparation of this volume, particularly W. M. Manning (Director, Chemistry Division, Argonne National Laboratory) for consistently valuable advice and support, and C. L. Chernick, H. H. Claassen, and M. Kilpatrick, who covered important conference sessions and prepared appropriate résumés for this book. Gordon L. Goodman discussed the material covered in Part IX with each contributor and assumed editorial responsibility for that most important section. Conversations with J. H. Martens, of the Argonne Technical Publications Department, have been most helpful. Finally, the outstanding co-operation of the clerical staff at Argonne has substantially eased many editorial burdens.

HERBERT H. HYMAN

ARGONNE, ILLINOIS
July 15, 1963

CONTENTS

5. Aqueous Chemistry of Noble-Gas Compounds

6. Diffraction Studies and the Structure of Xenon Compounds

7. Studies of ESR, NMR, Mössbauer, IR, and Raman Spectra and Related Experiments

8. Physiological Properties of Noble-Gas Compounds

9. Theoretical Studies of Noble-Gas Compounds

[*Part 9 Is Edited with an Introduction by G. L. Goodman*]

PART 1 HISTORICAL AND INTRODUCTORY MATERIAL

The recent discovery of compounds of the noble gases and the rapid proliferation of related experimental and theoretical studies taught those of us involved an interesting lesson: The span of time necessary for one development in experimental chemistry to lead to another is very short. The excitement that surrounded the original discovery of a family of elements that apparently had no chemistry is a thing of the past for virtually all working scientists; yet this excitement occurred within the lifetime of many we have known as teachers if not as colleagues.

In the first paper of this volume, prepared as the dinner address for the 1963 conference on noble-gas compounds, Hiebert quotes from the addresses and papers which reported the discovery, in 1894–95, of a family of "inert" gases; he attempts to recapture the excitement that attended that discovery and sees as its counterpart the excitement now surrounding a discovery based on a conclusion opposite to that reported in 1895—that this family in the periodic table does indeed have a chemistry.

In the second paper, Yost describes some of the difficulties he encountered thirty years ago in his attempts to determine whether or not the noble gases were inert. In the next paper, Bartlett and Jha discuss their studies of the reaction between xenon and platinum hexafluoride—a continuation of Bartlett's studies that first demolished the "myth" of a family of completely inert gases.

In the last paper in this section, Claassen, Malm, and Selig describe the circumstances leading to their observations—observations which suggested not only that xenon had a chemistry but that the noble gases had a much richer and more complicated chemistry than anyone had anticipated.

HISTORICAL REMARKS ON THE DISCOVERY OF ARGON THE FIRST NOBLE GAS

Erwin N. Hiebert
University of Wisconsin

The so-called inert, rare, or noble gases form a group of related elements, the discovery of which constitutes one of the most exciting episodes in the history of the physical sciences. The history of the discovery and characterization of argon between mid-1894 and mid-1895 is not merely the story of the discovery of a new element. This discovery led to new interpretations in the kinetic theory of gases, spectral studies, and notions of atomicity. It also led to a new classificatory category for the chemical elements and eventually to fundamentally important developments in the theory of the structure of the atom. To understand the origins of that discovery we must go back briefly to the time of Henry Cavendish, chemist of the eighteenth century.

In 1785 in a paper addressed to the Royal Society of London, Cavendish expressed some doubts as to the homogeneity of what was then called "phlogisticated air," i.e., nitrogen [1]. In his experiments he had passed electric sparks from a hand-driven frictional electric machine through a short column of air confined with "soap-lees" (potash) over mercury in an inverted U-tube. He remarked that "when five parts of pure dephlogisticated air [oxygen] were mixed with three parts of common air, almost the whole of the air was made to disappear."

Cavendish thereupon designed "an experiment to determine whether the whole of a given portion of the phlogisticated air of the atmosphere could be reduced to nitrous acid, or whether there was not a part of a different nature to the rest, which would refuse to undergo that change." After several weeks of electrical sparking of air, in columns about 1 in. long and 0.1 in. in diameter, Cavendish and his assistant discovered that "a small bubble of air remained unabsorbed." He remarked: "If there is any part of the phlogisticated air of our atmosphere which differs from the rest and cannot be reduced to nitrous acid, we may safely conclude that it is not more than $\frac{1}{120}$ part of the whole."

Cavendish's discovery of an inert residue from atmospheric nitrogen apparently passed out of the recollection of most nineteenth-century chemists, who undertook to relate atmospheric oxygen and nitrogen analyses to locality, time, and weather conditions. This activity was initially stimulated by the desire to determine the "salubricity" of air as a factor in locating the optimum environment for healthy living.

In 1804 Gay-Lussac and Humboldt, using Volta's eudiometric explosion procedure with measured volumes of hydrogen, obtained values for atmospheric oxygen and nitrogen more precise than those attainable by Cavendish's cumbersome electric-spark method [2]. In 1846 Bunsen concluded that the atmospheric oxygen varied between 20.840 and 20.970 parts per hundred of air by volume [3]. After 1850 air analysis data for locations from various parts of the world were readily available [4, 5]. By the 1890's it was generally believed that the relative proportions of oxygen to nitrogen in air at sea level were almost constant, being comprised of about 79 volumes of nitrogen to 21 of oxygen, with small and variable amounts of carbon dioxide and water vapor, and trace amounts of substances like ammonia, hydrogen dioxide, and ozone.

The immediate motivations for the discovery in 1894 of argon as a new constituent of the atmosphere are related to investigations a decade earlier by Lord Rayleigh [6–9]. In him we discover a keen and unaffected mathematical and theoretical approach to physical problems, coupled with great patience and aptitude for designing and carrying out experiments. He demonstrated in this case his ability for framing significant experimental questions and making exciting discoveries in an area where everything seemed to be clear and secure.

Thomson wrote [10] that Rayleigh had the gift of seeing what was the vital point in his experimental work; that the part of his apparatus "which really affected the accuracy of the results was sure to be all right and carefully made, while the other parts might be made up of bits of sealing-wax, string and glass tubes." Rayleigh was a prolific writer, having published some 445 papers, and was, Thomson says, "very clear, so clear in fact that the reader may not realize how difficult the problem was unless he had attacked it himself before reading the paper."

When, in 1882, Lord Rayleigh—then forty years old—began his work on the relative densities of gases, he was Cambridge professor of experimental physics and head of the Cavendish Laboratory. In his presidential address to the British Association for the Advancement of Science that year [11] he alluded to experiments on the relative densities of hydrogen and oxygen. The object of this investigation was to determine whether the relative weights of the atoms of hydrogen and oxygen deviated from the simple integer number ratio of 1:16 demanded by Prout's law. He expressly referred, "though with diffidence" to this law, "according to which the atomic weights of the elements, or at any rate of many of them, stand in simple relation to that of hydrogen." He added:

> Some chemists have reprobated strongly the importation of *à priori* views into the consideration of the question, and maintain that the only numbers worthy of recognition are the immediate results of experiment. Others, more impressed by the argument

that the close approximations to simple numbers cannot be merely fortuitous, and more alive to the inevitable imperfections of our measurements, consider that the experimental evidence against the simple numbers is of a very slender character, balanced, if not out-weighed, by the *à priori* argument in favour of simplicity. The subject is eminently one for further experiment; and as it is now engaging the attention of chemists, we may look forward to the settlement of the question by the present generation. The time has perhaps come when a re-determination of the densities of the principal gases may be desirable—an undertaking for which I have made some preparations.

In 1888 Lord Rayleigh, then professor of natural philosophy at the Royal Institution in London and successor to Sir George Stokes in the secretaryship of the Royal Society, published a preliminary notice [12] on the relative densities of hydrogen and oxygen with a calculated ratio of atomic weights of 15.912. "In the hope of being able to prepare lighter hydrogen" (since impure hydrogen can only be too heavy), but with a result that was disappointing, Rayleigh the following year reported [13] a calculated ratio of atomic weights of 15.89. In 1892, after a host of unusually tedious experiments, he gave a result of 15.880, a value which he felt might be too low because of the presence of mercury vapor in his hydrogen [14]. His concerns at this point were obviously directed toward the deviation of almost 1 per cent from the value of 16 which would be required according to Prout's law.

To shed more light upon a discrepancy which he regarded with disgust and impatience [15], he published in *Nature*, a letter [16] inviting criticisms from chemists who might be interested in such questions. Saying that he had followed a suggestion from Professor Ramsay, he reported having prepared nitrogen by the passage of a mixture of dry air and ammonia over red-hot copper and had found it to be 0.1 per cent lighter than the nitrogen obtained from atmospheric air by removal of oxygen in the usual manner over red-hot copper. Rayleigh sought the source of the discrepancy and asked his fellow chemists: "Is it possible that the difference is independent of impurity, the nitrogen itself being to some extent in a different (dissociated) state?"

A reply from Ramsay stated that he had been sleeping over the nitrogen problem and had "resolved" various points in his dreams. Rayleigh wrote in his notebook that he had received "various useful suggestions, but none going to the root of the matter." Several wrote that the explanation might be sought in the partial dissociation of the nitrogen derived from ammonia—a suggestion which had been in his own letter to *Nature* [7, 17].

Rayleigh's next step was to try to exaggerate the discrepancy. Concerning this decision, he wrote two years later: "One's instinct at first is to try to get rid of a discrepancy, but I believe that experience shows such an endeavour to be a mistake. What one ought to do is to magnify a small discrepancy with a view to finding out the explanation" [15]. This he did by substituting pure oxygen for atmospheric air so that the whole, instead of only a part, of the nitrogen would come from the ammonia. The discrepancy was at once magnified five times, for the nitrogen so obtained proved to be about 0.5 per cent lighter than "atmospheric" nitrogen. The result stood out fairly sharply from the first, but Rayleigh went on to confirm it by comparison with nitrogen derived chemically in other ways.

A paper by Rayleigh [18] in March, 1893, reveals a shift of emphasis in the approach to his gas density researches, viz., from O/H to O/N density measurements. In this way he was able to avoid the large errors associated with hydrogen weighings. He also introduced some novel gas pressure–measurement features which gave him more reliability in relative-density measurements. His paper briefly touches again upon the problem of the anomalous density of nitrogen prepared from different sources.

> Although the subject is not yet ripe for discussion, I cannot omit to notice here that nitrogen prepared from ammonia, and expected to be pure, turned out to be decidedly lighter than the above. When the oxygen of air is burned by excess of ammonia, the deficiency is about 1/1000th part. When oxygen is substituted for air, so that all (instead of about one-seventh part) of the nitrogen is derived from ammonia, the deficiency of weight may amount to ½ per cent. It seems certain that the abnormal lightness cannot be explained by contamination with hydrogen, or with ammonia, or with water, and everything suggests that the explanation is to be sought in a dissociated state of the nitrogen itself. Until the questions arising out of these observations are thoroughly cleared up, the above number for nitrogen must be received with a certain reserve. But it has not been thought necessary, on this account, to delay the presentation of the present paper, more especially as the method employed in preparing the nitrogen for which the results are recorded is that used by previous experimenters.

It was at this stage of Rayleigh's research in 1893 that William Ramsay, professor of chemistry at University College, London, asked and received permission to undertake some experiments designed to explain if possible the anomalous behavior of atmospheric nitrogen. He did not publish anything on the subject for over a year.

Rayleigh's suspicions concerning the anomalous density of nitrogen prepared from ammonia were explored explicitly for the first time in a paper communicated to the Royal Society April 19, 1894 [19]. On the assumption that similar nitrogen should be obtained from the atmosphere and ammonia, he entertained an explanation of the discrepancy based either upon the atmospheric nitrogen being too heavy on account of imperfect removal of oxygen or of the nitrogen from ammonia being too light on account of contamination with gases lighter than nitrogen. Imperfect removal of oxygen was excluded on the basis of "the fact that the action of the copper . . . was pushed to great lengths," that in view of the small difference in weights between the two gases a large and detectable amount of oxygen would have to be present, and that the intentional introduction and subsequent removal of hydrogen over the hot oxide of copper revealed no difference in the densities.

The acceptance of the second alternative—contamination of the nitrogen from ammonia with lighter gases such as hydrogen, ammonia, or water vapor—

was hardly less formidable. The presence of water vapor and ammonia under the given experimental conditions was excluded at once. The existence in appreciable amounts of hydrogen or light combustible hydrocarbons such as CH_4 seemed unlikely in the presence of hot copper oxide, especially in view of the fact that hydrogen purposely introduced into the atmospheric nitrogen was readily oxidized and efficiently removed.

In a similar way Rayleigh discovered that nitrogen derived from ammonia, nitric oxide, nitrous oxide, and ammonium nitrite was materially lighter than atmospheric nitrogen obtained by removal of oxygen either in the cold with ferrous hydrate or by hot iron or copper. The difference amounted to "about $\frac{1}{200}$ part of the whole." To rule out the possibility of partial dissociation of N_2 molecules into detached atoms with the subsequent production of N_3 molecules (similar to the ozone produced from oxygen by silent discharge), he demonstrated that electrification had no appreciable effect upon the densities of the atmospheric or chemical nitrogen.

Rayleigh's experiment left no doubt that the atmosphere contained a new constituent, although whether this was a new element or a compound was still an open question. His paper was devoid of any expressed opinion as to the cause of the discrepancy. His reluctance to make even a preliminary announcement of any discovery at this time was possibly due to the fact that his residue gas was not always strictly proportional to the quantity of air treated. This was later shown to be a result of the appreciable absorption of argon by water. Apart from such factors, Rayleigh was unquestionably cautious in his publications. Schuster wrote [8] that Rayleigh "disliked premature publications of discoveries, not only because they did not satisfy his own high standard, but because 'scoffers'—as he expressed it—'would be encouraged.' "

The one person who seriously took up the challenge of Rayleigh's paper of April, 1894, was William Ramsay. His interest in the subject can be traced back to attempts several years earlier to cause nitrogen and hydrogen to combine directly by passing them over heated metals. Unsuccessful in this attempt, he had learned, instead, that red-hot magnesium turnings readily absorb nitrogen. After reading Rayleigh's paper, his plan was to absorb carefully purified atmospheric nitrogen in order to discover whether any portion of it was different from the rest [9, 17]. His first attempt in May, 1894, gave 40 cc. of residual gas weighing 0.050 gm. with a density 15, rather than 14, times greater than that of hydrogen. "The result," wrote Ramsay, "was encouraging, and led to the probability of the nitrogen being altered in some way, or of the presence of some new component of the atmosphere."

On May 24, 1894, Ramsay wrote to Rayleigh [9]:

> Has it occurred to you that there is room for gaseous elements at the end of the first column of the periodic table? Thus:—Li Be B C N O F X X X . . . etc. Such elements should have the density 20 or thereabouts, and 0.8 pc. (1/120th about) of the nitrogen of the air could so raise the density of nitrogen that it would stand to pure nitrogen in the ratio 230 ÷ 231.

By July, Ramsay was deeply involved personally in experiments on the absorption of nitrogen by magnesium, whereas prior to this time his laboratory assistant, Mr. Williams, had been doing the work. With additional precautions to remove CO_2 and water vapor, after a 12-day stint, a residue of 0.2190 gm. was left—having a density of 16.1. He remarked:

It was supposed that just as oxygen, when exposed to an electric discharge, undergoes a cleavage of its molecules, two-atom molecules become one-atom molecules for an instant, which then unite to form three-atom molecules, so the action of magnesium on the nitrogen might be to withdraw one atom of nitrogen from the two-atom molecule leaving a single uncombined atom, which might not improbably find two partners, each of its own kind, to form with them a three-atom molecule—a sort of nitrogen-ozone, in fact. Hence it was resolved to continue the absorption with fresh magnesium for a still longer time, in the hope of its being possible to isolate the three-atom nitrogen molecules.

Continuation of the experiment revealed that the bright metallic magnesium was not much further attacked but that the purified gas now corresponded to a density of 19.086 [17].

Resorting now to Cavendish's old electric-spark process, Ramsay found that sparking the residual gas with oxygen in the presence of caustic soda resulted in a contraction, after removal of excess oxygen with alkaline pyrogallate, which amounted to 15.4 per cent of the original volume, i.e., the remaining gas should now be 20 times as heavy as hydrogen.

A portion of the gas confined in a Plücker's tube at low pressure and connected with the secondary terminals of a Ruhmkorff's coil revealed a spectrum through a glass prism "exhibiting the bands of nitrogen . . . somewhat hazy bands, red, orange, yellow, and yellow-green in colour . . . [and] showed in addition certain groups of red and green lines which did not appear to belong to the spectrum of any known gas" [17].

In the meantime Rayleigh had been busy on the problem. By the beginning of July he was also in possession of the necessary induction coils to begin to isolate the new gas along the lines of Cavendish's sparking method. This work took him far into the night. Since the hammer break of the coil was liable to stick, he had a telephone so arranged that the hum of the induction coil could be transmitted to him as he dozed in his chair [7].

In August the two men joined forces. Up to that point their work had been more or less independent, though not uncommunicative. Now, letters passed back and forth almost daily between Rayleigh's laboratory at Essex and Ramsay's at University College. Ramsay wrote [9] to Rayleigh on August 4:

I have isolated the gas. Its density is 19.075, and it is not absorbed by magnesium.

I have been watching the density of X creep up as absorption proceeds; so you see this is no chance determination with a possible source of error.

I have filled the vacuum tubes with the gas. The results are very curious. My impression is that it gives no spectrum—no visible one. Perhaps a blue green line or band is due to it; the band is just visible in the spectrum of chemical nitrogen, but is bright in that of X. It appears still to contain a trace of nitrogen, for the lines of N_2 are still visible, though not very strong.

I am going to spark the gas with oxygen and with chlorine on Monday; also to try to liquefy it by pressure. [The lowest temperature Ramsay could reach was about −90° C., using liquid nitrous oxide.] I have arranged my critical-point apparatus and made a tube to fill.

I have also arranged to start with a big gas holder of nitrogen 89 litres. It is a long business and very dreary work absorbing nitrogen day after day; making finely divided magnesium etc, etc. However le jeu vaut la chandelle.

I should like to talk to you about this. Are you to be at Oxford? If so we will meet there? I didn't want to trespass on your preserves and yet I feel I have done so.

Rayleigh replied two days later:

I believe that I too have isolated the gas; though in miserably small quantities. . . .

I have concentrated X by diffusion, collecting at the end of a tobacco pipe $\frac{1}{30}$ of the gas which goes through the pores into a vacuum. The air so prepared contains twice as much X as ordinary air. I am preparing to develop this method further. My attempts to accumulate larger quantities than 1 c.c. have only partially succeeded, I think, because of the solubility of X in water. . . . However 1 c.c. was sufficient to allow of the spectrum being observed between platinum points. Like you, I could find no new line as I had hoped, but the nitrogen line was conspicuously absent, or extremely faint.

As to publication, I had thought of giving at Oxford some definite results of the work. . . . But it seems so mixed up with your work, as to be difficult or impossible to treat separately. My own feeling is that the only solution is joint publication. Doubtless your last results go further than mine, and are probably better established. But as you suggest, the whole is founded upon work which I carried up to a certain point, and was continuing. If this course is adopted, the question is whether anything had better be said as yet. If not, I would keep back my further results as to chemical nitrogen. I shall be at Oxford . . . and shall be glad to hear your views.

Ramsay replied:

I think that joint publication would be the best course, and I am much obliged to you for suggesting it. For I feel that a lucky chance has enabled me to get Q in quantity. (There are two other X's, so let us call it Q or Quid?)

Just prior to the meeting of the British Association at Oxford in August, 1894, "it was decided that the proof of the existence of a new constituent gas in air was sufficiently clear to render it advisable to make to the Association a short announcement of the discovery" [17].

On August 13, 1894, Rayleigh made a brief announcement at the Oxford meeting stating that he and Ramsay had found atmospheric nitrogen, carefully purified from every other known constituent of the air, to be contaminated to the extent of about 1 per cent with another gas even more inert than nitrogen. It was reported furthermore that its density lay between 18.9 and 20; that its spectrum at 8 mm. pressure was a definite and characteristic one with lines differing in position from those of nitrogen and without the flutings of the latter. Inertness was emphasized as the chief characteristic of the new gas. Nothing had appeared in the printed program, but the *Times* and *Chemical News* published brief notes stressing the interest which the announcement had awakened [20]. Ramsay wrote the following year:

> The statement was received with surprise and interest; chemists were naturally somewhat incredulous that air, a substance of which the composition had been so long and so carefully studied, should yield anything new. One of the audience inquired whether the name of this new substance had been discovered; as a matter of fact it was then under consideration [17].

In the discussion which followed the announcement, Emerson Reynolds, professor of chemistry of Trinity College, Dublin, noted the place which the new substance, if it proved to be an element, would occupy in Mendeléeff's table among the platinum metals. Roberts-Austen, an English metallurgist from the Royal School of Mines, suggested that this gas might be the one frequently found as a residue among the gases extracted from steel. James Dewar, Rayleigh's colleague and Fullerian professor of chemistry at the Royal Institution, noted that the observed turbidity in liquefied air might be due to the new gas [21].

A letter of August 15 to the editors of the *Times* on "the new element" came from James Dewar, calling attention once more to the well-known phenomenon of a white deposit found in purified liquid air. "Can this substance," asked Dewar, "which has so often been seen in the theatre of the Royal Institution, be in the main anything else than Rayleigh's new nitrogen in the solid form?" [22].

In a second letter to the *Times* the following day Dewar suggested that the new gas might be tri-atomic nitrogen "condensed molecularity into an allotropic form, having $1\frac{1}{2}$ times its normal density." He remarked that "electrical stimulation of nitrogen does produce two distinct spectra presumably due to different molecular modifications"; that oxygen is converted to ozone by electrical discharge; that red phosphorus is an inert body relative to yellow phosphorus; and that the theoretical density of 21 for N_3 was close to the experimentally reported value lying between 19 and 20. Dewar furthermore inferred "that the new substance is being manufactured by the respective experimenters, and not separated, as they imagine, from ordinary air." He concluded:

> It is not the first time that chemists and physicists have been tempted to believe in the production of an allotropic form of nitrogen, and to accept it as explaining certain curious phenomena, but hitherto the assumption has always broken down on more careful investigation. This time we may be permitted to hope that the elusive allotropic form has been fairly captured [22].

Other scientists besides Dewar regarded the discovery of argon with reserved skepticism, but none of these opinions seemed to weigh very heavily on Rayleigh or Ramsay. They knew they had discovered a new element. Rayleigh went to Scotland with his family for a vacation. Ramsay went back to his laboratory in London to work on the solubility of argon and a plan for a continuous process of its isolation. When Rayleigh returned, the two men spent the rest of the year purifying the gas to its "bitter end," attempting chemical combination, and making compressibility and velocity of sound measurements in the gas.

If the discovery of argon was accepted reluctantly by chemists, the most influential British physicist of the day was rooting for the discoverers. In his Anniversary Address [23] on November 30, 1894, the president of the Royal Society, Lord Kelvin, remarked:

> The greatest scientific event of the past year is, to my mind, undoubtedly the discovery of a new constituent of our atmosphere. If anything could add to the interest which we must all feel in this startling discovery, it is the consideration of the way by which it was found. . . . The arduous work . . . commenced in 1882, has been continued for twelve years, by Rayleigh, with unremitting perseverance.

Kelvin took this opportunity to point out that the discovery, resulting from the collaborative efforts of Rayleigh and Ramsay, provided "a fresh and most interesting verification" of a statement which he had made in his presidential address to the British Association 23 years earlier. He had said at that time:

> Accurate and minute measurements seem to the non-scientific imagination a less lofty and dignified work than looking for something new. But nearly all the grandest discoveries of science have been but the rewards of accurate measurement and patient long-continued labour in the minute sifting of numerical results.

He added:

> The investigation of the new gas is now being carried on vigorously, and has already led to the wonderful conclusion that the new gas does not combine with any other chemical substance which has hitherto been presented to it. We all wait with patience for further results of their work; we wish success to it, and we hope that it will give us, before the next anniversary meeting of the Royal Society, much knowledge of the properties, both physical and chemical, of the hitherto unknown and still anonymous fifth constituent of our atmosphere.

It was announced that a paper by Rayleigh and Ramsay on the new gas would be the subject of discussion at the next meeting on January 31, 1895. It was also stated that the name *argon*, symbol A, had been given provisionally to the new element [24]; that is, ἀ-εργόν from ἔργον (work; plus α, meaning not) [9].

By the end of 1894 the investigators had not published a word on the subject of argon, although their paper had been entered (along with 217 other contestants) in the Smithsonian Institution's Hodgkins Prize competition for the most important discovery with atmospheric air [25]. Ramsay wrote: "One of the conditions is that nothing shall be published before it is submitted to the committee, and hence we have not written a word, except what we sent to America as a preliminary account of the work" [9]. Needless to say, they won the prize for 1895. In the meantime tremendous curiosity about the whole subject was awakened as the time for the formal delivery of the paper arrived.

The new discoveries which were contained in the famous collaborative fifty-four-page paper, "Argon a new constituent of the atmosphere," were presented by Ramsay before an audience of at least eight hundred in the theater of the University of London on January 31, 1895, with Lord Kelvin in the presidential chair [26]. It was the largest meeting in the history of the Society. Professor Dewar was conspicuously absent. After Ramsay had given the paper, Rayleigh, being asked for comments, replied: "I have very little to add. . . . I am not without experience of experimental difficulties, but I have never encountered them in anything like so severe and aggravating a form as on this investigation." Two related papers were read by William Crookes and K. Olszewski.

The full text report in the *Philosophical Transactions* [27] was followed by abstracted versions of the paper in virtually every chemical journal of the day [28–33]. The first volume of *Chemical News* for 1895 (January through June) carried some twenty-five miscellaneous articles, notes, and letters on the subject.

Analysis of the paper itself is a rewarding experience. It provides a review of all the phenomenological evidence which formed the basis of the conclusions of the main investigators at this time, in addition to a detailed description of the experimental devices employed to isolate and characterize the new gas.

The simplest explanation of the anomalous density of atmospheric nitrogen and the existence of a residue "in proportion to the amount of air operated upon" the authors felt "was to admit the existence of a second ingredient in air from which oxygen, moisture, and carbonic anhydride had already been removed," but in accepting this explanation even provisionally, they felt that they had to face the improbability that a gas surrounding them on all sides "and present in enormous quantities, could have remained so long unsuspected."

The report contained a fairly detailed statement of Cavendish's experiments of 1785 with the comment:

> Attempts to repeat Cavendish's experiment in Cavendish's manner have only increased the admiration with which we regard this wonderful investigation. Working on almost microscopical quantities of material, and by operations extending over days and weeks, he thus established one of the most important facts in chemistry. And what is still more to the purpose, he raises as distinctly as we could do, and to a certain extent resolves, the question above suggested [9, 27, 34].

In the "atmolysis" experiments, relying upon Graham's method of separating a mixture of gaseous components of different densities by virtue of different rates of diffusion, the investigators showed that "atmospheric nitrogen" passing through a series of "churchwarden" tobacco pipes behaved as a mixture and not as a simple body, i.e., they demonstrated that a gas was obtained which was about 2 per cent heavier than "chemical nitrogen." The difference of weight in the best experiments was about $3\frac{1}{2}$ mg. They remarked that perhaps the heavier constituent of the atmosphere could be prepared in the pure state "if the diffusion apparatus could be set up on a large scale and be made thoroughly self-acting." The importance of this work was that the evidence of separation here could be based exclusively upon a physical process and was not open to the suspicion of new chemically manufactured substances.

There is also a lengthy discussion of the devices designed to separate argon on a large scale by means of magnesium. Residual nitrogen was converted into oxides by the electric spark, and excess oxygen was consumed by hydrogen. In one device, 921 cc. of argon was obtained from 100 liters of "atmospheric nitrogen," giving a yield of approximately 1 per cent. In another device an absorption of 3 liters of mixed gas per hour was attained—"about 3000 times the rate at which Cavendish could work." The major difficulties encountered were losses due to the relatively high solubility of argon in water, overheating of the vessels from the electric arc, and slow removal of the last traces of nitrogen.

The density of pure argon prepared by the Cavendish sparking method and the Ramsay magnesium method were, respectively, 19.7 and 19.9. For an indication of the extent of nitrogen removal, the testimony of the spectroscope was most helpful; the absence, even with a wide slit, of the characteristic yellow line of nitrogen was accepted as decisive.

Some of the difficulties and discrepancies which Rayleigh and Ramsay had encountered in the early stages of their investigations were understood after a study of the solubility of argon in water. It had been troublesome at first while trying to isolate argon to find that small quantities of gas seemed always to be disappearing. They found that about 4 cc. of argon dissolved in 100 cc. of water at room temperature, i.e., about 2½ times the solubility of nitrogen in water. This discovery led them to expect an increased proportion of argon over nitrogen in rain water. With a boiler constructed from an old oil can, they extracted the gases from cistern rain water and found it relatively twice as rich in argon as in nitrogen. However, an analysis of gas from a thermal spring revealed less argon than in air.

The two most important results communicated in the paper dealt with the experimental determination of the ratio of the specific heats of argon and the attempts to induce chemical combination with argon.

A great many experiments were made without any positive result. The chemical substances which failed to combine with argon under conditions ranging from direct contact at temperatures up to red-heat and by sparking were: oxygen, hydrogen, chlorine, phosphorus, sulfur, tellurium, sodium, caustic soda, soda lime, potassium nitrate, sodium peroxide, persulfides of sodium and calcium, nitrohydrochloric acid, bromine water, a mixture of potassium permanganate and hydrochloric acid, and platinum black. The investigators concluded:

> We do not claim to have exhausted the possible reagents. But this much is certain, that the gas deserves the name "argon," for it is a most astonishingly indifferent body, inasmuch as it is unattacked by elements of very opposite character. . . . It will be interesting to see if fluorine also is without action, but for the present that experiment must be postponed, on account of difficulties of manipulation.

In order to reach a decision regarding the elementary or compound nature of argon, experiments on the velocity of sound in argon were made. This permitted a calculation for the ratio of the specific heats, C_p/C_v or γ, for argon. Using a narrow tube of 2-mm. bore and achieving an exceedingly high pitch by rubbing a projected longitudinal extension rod with an alcohol-wetted rag, the nodes of the waves produced inside the argon-filled tube were registered with lycopodium powder. A calculated value between 1.644 and 1.66 for γ indicated that argon was a monatomic gas, i.e., a gas in which all the kinetic energy imparted to it at constant volume was expended in translational motion. It was well known at this time that diatomic molecules like O_2, N_2, H_2, CO, and HCl had values of γ close to the theoretical value of 1.41, whereas Cl_2, Br_2, and I_2 were somewhat lower. The only case in which a value of 1.66 had been reported by this time was that of mercury at 800° [35]. Since this had been accepted as proof of the monatomic character of the vapor of mercury, the authors felt that the conclusion held equally good for argon. "The only alternative," they remarked, was "to suppose that if argon molecules are di- or polyatomic, the atoms acquire no relative motion, even of rotation—a conclusion improbable in itself and one postulating the sphericity of such complex groups of atoms."

As a consequence of accepting the monatomic nature of argon, which had been found to be 20 times as heavy as H_2, an atomic weight of 40 was determined for argon. This presented obvious difficulties in terms of the position of argon in the series of elements possessing atomic weights near 40, viz., chlorine (35.5), potassium (39.1), calcium (40.0), and scandium (44.0). There was no ready place for A = 40 in Mendeléeff's table:

> If argon be a single element, then there is reason to doubt whether the periodic classification of the elements is complete; whether, in fact, elements may not exist which cannot be fitted among those of which it is composed. On the other hand, if argon be a mixture of two elements, they may find place in the eighth group, one after chlorine and one after bromine. . . .
>
> If it be supposed that argon belongs to the eighth group, then its properties would fit fairly well with what might be anticipated. For the series which contains Si_n(iv), P_4 (iii and v), $S_{8\text{ to }2}$ (ii to vi), and Cl_2 (i to vii), might be expected to end with an element, of monatomic molecules, of no valency, i.e., incapable of forming a compound or if forming one, being an octad; and it would form a possible transition to potassium, with its monovalence, on the other hand. Such conceptions are, however, of a speculative nature; yet they may be perhaps excused, if they in any way lead to experiments which tend to throw more light on the anomalies of this curious element.

The authors remarked that the negative attempts to produce compounds of argon might be likened to the negative attempts to cause combinations between mercury gas and other elements at 800°. As for argon being a gas of atomic weight 40, it was pointed out that "we possess no knowledge why carbon, with its low atomic weight, should be a solid, while nitrogen is a gas, except in so far as we ascribe molecular complexity to the former and comparative molecular simplicity to the latter."

The decision of Rayleigh and Ramsay to accept an experimentally determined γ of 1.66 as evidence for the monatomicity of argon and therefore an atomic weight of 40 was what few scientists were willing to accept.

The discussion following the paper turned almost exclusively on this one issue—the validity of the monatomic character of the gas as deduced from specific heat ratios. Ramsay felt certain about this conclusion. Rayleigh felt that the specific heat ratio was almost conclusive. He said [36] three months later:

> The result is, no doubt, very awkward. Indeed, I have seen some indications that the anomalous properties of argon are brought as a kind of accusation against us. But we had the very best intentions in the matter. The facts were too much for us; and all that we can do now is to apologize for ourselves and for the gas. . . .
>
> It was only after the experiment upon the specific heats that we thought that we had sufficient to go upon in order to make any such suggestion in public [argon as an element]. I will not insist that the observation is absolutely conclusive. It is certainly strong evidence. But the subject is difficult, and one that has given rise to some difference of opinion among physicists. At any rate this property distinguishes argon very sharply from all the ordinary gases.

The presentation of the main paper on argon was followed by the two related papers already mentioned. This work notably helped to characterize the gas beyond what had been accomplished by measurements of density, chemical reactivity, solubility, gaseous diffusion, and specific heat ratio. The examination of the spectrum of argon was entrusted to A. Schuster, professor of physics of Manchester, and William Crookes of London—with the samples of the gas being supplied by the main investigators. Rayleigh was greatly relieved, because of some things that Dewar had claimed, when both spectroscopists reported about the same characteristic spectra for argon.

At the meeting on January 31 Crookes presented the evidence which he had obtained using an accurate spectroscope and spectrograph fitted with complete quartz train. Like nitrogen, the argon revealed "two distinct spectra according to the strength of the induction current employed." A spectrum rich in red rays, with two especially prominent lines (at wave lengths 695.56 and 705.64), was visible at its best at 3 mm. pressure. At lower pressures the discharge changed to a "rich steel-blue," and the spectrum showed "an almost entirely different set of lines." In the "blue glow" produced by the "negative spark" Crookes counted 119 lines, in the "red glow" produced by the "positive spark" he counted 80 lines. Of these, 26 appeared to be common to both spectra.

Unlike the two spectra of nitrogen which were known to be different in character, one showing fluted bands and the other sharp lines, both of the argon spectra were seen to consist of sharp lines. Nevertheless, it turned out to be very difficult, Crookes remarked: "to get argon so free from nitrogen that it will not show the nitrogen flutings superimposed on its own special system of lines." No matter how free Rayleigh's argon was supposed to be from nitrogen, Crookes always detected the nitrogen bands in its spectrum. However, the nitrogen bands invariably disappeared after some time on passing the induction spark through the gas in the ordinary Plücker-type tubes with a capillary section in the middle. "For photographing the higher rays which are cut off by glass," Crookes employed a similar tube with "a quartz window at one end." The results of wave lengths and their intensities were given in tabular form for the red and blue spectra with an accompanying map, which drawn to scale, Crookes remarked, was 40 ft. long with a probable error of line position not greater than 1 mm. The lines of argon were sharper and more brilliant than for nitrogen, with only one or two apparent coincidences between the two. These apparent coincidences with spectra of nitrogen and other elements, Crookes felt, "would probably disappear on using a higher dispersion."

Crookes concluded that his results entirely corroborated "the conclusions arrived at by the discoverers of argon." He mentioned however: "It is not improbable, and I understand that independent observations have already led both the discoverers to the same conclusion, that the gas argon is not a simple body, but is a mixture of at least two elements, one of which glows red and the other blue, each having its distinctive spectrum." Still, Crookes felt that the presence of argon as a single element was supported by the fact that nitrogen also could be made to exhibit two spectra by varying the pressure and the intensity of the spark. Certain analogous phenomena appeared to hold for hydrogen as well.

The study of the properties of argon at low temperatures was relegated to a low-temperature expert—Karol Stanislaw Olszewski, professor of chemistry at the University of Cracow. Using liquid ethylene and liquid oxygen as cooling agents and with a hydrogen thermometer, Olszewski found that the 300 cc. of argon which Rayleigh had sent him condensed readily to a colorless liquid having a density of about 1.5 at its B.P. (−186.9°) under atmospheric pressure. On compressing the gas in the presence of its liquid, the pressure remained sensibly constant until all the gas had condensed to a liquid. At −189.6°* the argon froze to a white solid resembling ice. The critical temperature and pressure were −121° and 50.6 atmospheres. Such sharply defined figures are the criterion for a pure substance, and Olszewski inferred from them that argon as a new substance was therefore hardly to be doubted. He noted that its unexpectedly low critical temperature and B.P. seemed to have some relation to its simple molecular constitution.

After the reading of the three papers by Rayleigh and Ramsay, Crookes, and Olszewski on January 31, a vigorous discussion followed [30]. For example, Henry E. Armstrong of the Guildhall Technical College of London ungraciously challenged the investigators on their "wildly speculative" idea of the monatomicity of the argon based on the value of γ for the gas. Armstrong suggested

* Currently accepted values are: liquid density, 1.3998 at B.P. (−185.88° C.); triple point, −189.37° C.; critical temperature, −122.3° C.; critical pressure, 48.3 atm.; and critical density, 0.536 gm/cm^3. (G. A. Cook, *Argon, Helium and the Rare Gases* [New York, 1961].)

rather that "it was quite likely that the two atoms existed so firmly locked in each other's embrace, that there was no possibility for them to take notice of anything outside, and that they were perfectly content to roll on together without taking up any of the energy that is put into the molecule." The greatest difficulty, Armstrong contended, was to accept the conclusion that the gas was an element of atomic weight 40—because of the "difficulty of placing an element of that kind."

Arthur W. Rücker, professor of physics at the Royal College in London, sprang up "at a white heat," as Mrs. Ramsay later wrote, "and made a most elegant panegyric on the work and discovery," with particular reference to the monatomicity of argon as derived from the determined ratio of the specific heats. On the question of fitting argon into the periodic classification, Rücker said that "whatever the effect might be upon the great chemical generalization of Mendeléeff," that was, after all, "an empirical law based upon no dynamical foundation." If it held its own in this case, that would of course strengthen the belief in it. But, on the other hand, he remarked that the periodic law did not stand on the footing of those great mechanical generalizations which could not be upset "without upsetting the whole of our fundamental notions of science."

Rayleigh remarked that at first sight it seemed rather a strange thing that there should be no rotation in the molecules of the gas. Still, the specific heat measurement required that no margin of energy beyond translatory motion remain to be attributed to intermolecular or interatomic motion.

A letter to Rayleigh from G. F. Fitzgerald, the Irish physicist and professor of natural philosophy at Dublin, stated that the conclusion from the ratio of specific heats for argon might be "not that it is monatomic, but that its atoms are so bound together in its molecule that the molecule behaves as a whole as if it was monatomic."

Rayleigh found it difficult to conceive the possibility of such an eccentrically shaped atom being able to move about without acquiring a considerable energy of rotation. Kelvin remarked that the condition for a ratio of specific heats of exactly 1.66 would not be fulfilled even by an absolutely smooth spherical atom. He suggested that the only kind of atom that could be conceived as giving the ratio 1.66 in the dynamical theory of heat "was the ideal Boscovitch mathematical point endowed with the property of inertia, and with the other property of acting upon neighboring points with a force depending upon distance."

The meeting ended with cheers, congratulations, and hand-shaking that lasted for nearly an hour. A full formal report, with comments probably written by Rücker, was published the following week in *Nature* [37]. The following are excerpts from that report:

> The scene in the theatre of the University of London on Thursday was in all respects unique. It will certainly be historical. . . . It will be long before the same eagerness to obtain a ticket is displayed—long before those who gain admission, will listen to so much worth hearing. . . . All that is known of "Argon" was told to all. . . . As has been

well said, the result is "the triumph of the last place of decimals," that is, of work done so well that the worker knew that he could not be wrong.

On this hypothesis [monatomicity of argon] a very awkward question is no doubt raised. The periodic classification of the elements, cannot, and ought not, to be abandoned at the first challenge, and till further evidence is forthcoming a heavy strain is thrown on the link of the chain of arguments. . . .

Whether in the future other and more convincing evidence will be adduced on the other side, the future alone can show. The courts of science are always open, and every litigant has an unrestricted right of moving for a writ of error.

The published papers brought forth a flurry of theoretical comments and experimental activity. The *Electrical Review* carried on a campaign against the "argon myth" and suggested that the discoverers submit their gas to the "well-known methods of gas analysis" rather than to "spectrum analysis." Kelvin recommended that the writer of that suggestion submit himself to the "silent discharge."

The Irish physicist G. J. Stoney suggested that argon might be a hydrogen compound of infracarbon, one of the six missing elements between H and Li [38].

Bohuslav Brauner, professor of chemistry at the Bohemian University in Prague inferred that the specific heat ratio for argon was a powerful argument for monatomicity but not powerful enough to rule out a polymeric composition of N, like N_3, with its atoms lying close enough to behave as a single atom. "As an orthodox Mendeléefian," he wrote, "I find great difficulty in assuming the existence of a new elementary gas having the atomic weight 20, or 40, or 80, its B. P. being $-187°$." Since it was obvious from the experimentally reported conditions that the N_3 could not be decomposed by heat, Brauner suggested that micro-organisms be used. He indicated how this method had been employed in the preparation of hydroarsenide from arsenic oxide and NH_3 from nitric acid. He also recommended the use of nascent H liberated on contact of H_2 with metallic palladium [39].

On March 14, 1895, the author of the periodic table, Dimitri Mendeléeff, made some comments to the Russian Chemical Society on the relation of argon to his system. The essence of his remarks was that an atomic weight of 40 for argon would be inconsistent with the periodic classification. He did not feel, he said, that the reported experimental evidence would support argon as a compound or as a mixture. He therefore favored the view of an allotropic form of N, "condensed nitrogen," probably N_3—which he believed would be exceptionally stable [40].

Rayleigh could not accept this and said so in his address to the Royal Institution in April 1895:

Most of the chemists with whom I have consulted are of opinion that N_3 would be explosive, or, at any rate, absolutely unstable. That is a question which may be left for the future to decide. We must not attempt to put these matters too positively. The balance of evidence still seems to be against the supposition that argon is N_3, but for my part I do not wish to dogmatize [15].

The celebrated Parisian chemist Marcellin Berthelot announced at the Académie des Sciences in March, 1895, that the sample of argon which Rayleigh had sent him had been induced under the influence of the silent discharge to enter into combination with benzene vapor [32].

Ramsay took great pains to duplicate Berthelot's results, but without success. By that time another discovery had absorbed all of his attention. He had isolated helium from cleveite [41]. Rayleigh, too, had been diverted from argon work to return to physics proper. Finding himself surrounded by carping criticism from chemists, he said "I want to get back again from Chemistry to Physics as soon as I can. The second-rate men seem to know their place so much better" [7].

The discovery of argon, to follow Thomson's reflections of some forty years later, was to claim a new chemical element that "had no chemical properties," that would form "no compound with any other element" and "have nothing to do with the most tempting brides that chemists put before it. . . . As this is the trap on which chemists rely for catching a new element, it is no wonder that argon eluded them" [42].

It might be said that the discovery of argon, although in a sense sensational, had nothing in common with the accidental or favorable surprises which sometimes fall to the scientific investigator. Chance does favor the prepared mind, but the discovery of argon, as Schuster remarked,

> flowed out of a well-designed series of experiments conducted with a definite purpose, and though the most important of the results arrived at was not anticipated, it came as a well-deserved reward for a strictly scientific procedure which concentrated all possible methods of attack upon one object, perfecting these methods until all discrepancies were cleared up.

The discovery of argon admirably illustrates a case of scientific investigation in which the answer to a specific and fairly commonplace question suggested many others. Interest in the discovery itself shifted rapidly to an interest in the results and implications of that discovery—to problems connected with specific heats, kinetic theory, atomicity, spectra, periodic classification, and the internally structured atom. The revolutionary transition from the classical to the non-classical picture in the physical sciences was initiated just before the turn of the century by the discovery of X-rays (1895), radioactivity (1896), and the electron as a discrete particle of negative charge and small mass (1897). The discovery of argon probably gave more impetus to the modern scientific revolution than is generally conceded.

REFERENCES

1. H. Cavendish, *Phil. Trans.* **75**, 372 (1785).
2. L. J. Gay-Lussac and F. H. A. von Humboldt, *J. de Phys.* **60**, 129 (1805).
3. R. W. Bunsen, *Gasometrische Methoden*, p. 77. Braunschweig, 1857.
4. H. V. Regnault, *Ann. Chim. Phys.* **36** (3), 385 (1852).

5. JOLLY, P. VON, *Abh. math.-physik. Classe kgl. bay. Akad. Wiss.* (München) **13,** 53 (1879).
6. *The Scientific Papers of Lord Rayleigh.* Cambridge, 1899–1920.
7. R. J. STRUTT, *John William Strutt, Third Baron Rayleigh.* London, 1924.
8. A. SCHUSTER, *Proc. Roy. Soc., A* **98** (1920).
9. M. W. TRAVERS, *A Life of Sir William Ramsay.* London, 1956.
10. J. J. THOMSON, *Recollections and Reflections,* pp. 239–40. New York, 1937.
11. LORD RAYLEIGH, Address at Southampton, Aug. 24, 1882, *Report of the B.A.A.S.*, 437 (1883).
12. ———, *Proc. Roy. Soc.* **43,** 356 (1888).
13. *Ibid.* **45,** 425 (1889).
14. *Ibid.* **50,** 448 (1892).
15. LORD RAYLEIGH, *Proc. Roy. Inst.* **14,** 524 (1895).
16. ———, *Nature* **46,** 512 (1892).
17. W. RAMSAY, *The Gases of the Atmosphere, the History of Their Discovery.* London, 1896.
18. LORD RAYLEIGH, *Proc. Roy. Soc.* **53,** 134 (1893).
19. *Ibid.* **55,** 340 (1894).
20. *Chemical News* **70,** 87 (Aug. 24, 1894).
21. *Nature* **50,** 410 (1894).
22. J. DEWAR, *Chemical News* **70,** 87 (Aug. 24, 1894).
23. LORD KELVIN, *ibid.* 291 (Dec. 14, 1894).
24. *Chemical News* **70,** 296 (Dec. 21, 1894).
25. *Smithsonian Contributions to Knowledge* **29,** 43 (1896).
26. *Nature* **51,** 337 (Jan. 31, 1895).
27. LORD RAYLEIGH and W. RAMSAY, *Phil. Trans., A* **186** (1), 187 (1895).
28. W. CROOKES, *ibid.* 243.
29. K. OLSZEWSKI, *ibid.* 253.
30. *Nature* **51,** 347 (Feb. 7, 1895).
31. *Proc. Roy. Soc.* **57,** 265 (1895).
32. M. P. E. BERTHELOT, *Ac. Sci. C.R.* **120,** 235, 581 (1895).
33. *Chemical News* **71,** 51 (Feb. 1, 1895).
34. W. RAMSAY, *The Life and Letters of Joseph Black, M.D.* London, 1918.
35. A. KUNDT and E. WARBURG, *Ann. Phys. Chem.* **157,** 353 (1876).
36. LORD RAYLEIGH, *Proc. Roy. Inst.* **14,** 536 (1895).
37. A. RÜCKER, *Nature* **51,** 337 (Feb. 7, 1895).
38. G. J. STONEY, *Chemical News* **71,** 67 (Feb. 8, 1895).
39. B. BRAUNER, *Chemical News* **71,** 79 (Feb. 15, 1895).
40. *Nature* **51,** 543 (April 4, 1895).
41. W. RAMSAY, *Proc. Roy. Soc.* **58,** 65, 81 (1895).
42. J. J. THOMSON, *Recollections and Reflections.* New York, 1937.

A NEW EPOCH IN CHEMISTRY

DON M. YOST
California Institute of Technology

Before the recent discovery at the University of British Columbia [1] and at the Argonne National Laboratory [2, 3] that xenon will combine with PtF_6 and with fluorine itself to form definite, non-transitory compounds in visible and easily weighable amounts, the science of chemistry had reached a stationary state, during which no profoundly fundamental discoveries were reported or even deemed possible. Only artificial radioactivity, nuclear fission and fusion, and the earth and Venus-probing satellites were considered to be the true wonders of our times. This state of affairs has now been changed by two brilliant experimental findings made at the laboratories just cited. So long as man shows any interest whatever in chemistry, the discovery of the xenon and other noble-gas fluorides will not be forgotten.

As is well known, all but one of the noble gases were discovered in the atmosphere by Sir William Ramsay just before the turn of the present century (1894–98). All earlier efforts to make these gases combine with other elements of the periodic system failed, the last such previous effort apparently having been made by Mr. Albert L. Kaye and myself [4] in 1933. To be sure, the spectroscopists had come up with evidence for the existence of transitory noble-gas compounds, but this evidence did not quite satisfy the chemists; for, being skeptical and realistic gentlemen, chemists always want to see something definite in a correctly labeled bottle. Quantum theory—the reigning queen of science—was coy about predicting the existence of stable noble-gas compounds (*varium et mutabile semper femina*). With time it became virtually a *tradition* that such compounds could not exist in non-transitory form. Like many others in science and mathematics, this tradition now finds its resting place in Boot Hill, and we are all glad of it.

My sole excuse for being a contributor to this important book rests in the fact mentioned above, that Albert Kaye (then a graduate student) and I tried without success to bring about reaction between xenon and both chlorine and fluorine; this was thirty years ago at the California Institute of Technology. In some ways conditions were very favorable for carrying out the experiments at that time; there were no redundant administrators, safety officers, or editors to bother us or to suggest delicious sins that we might be tempted to commit. We were on our own throughout. We constructed our own apparatus, blew our own glass, and used cast-off Ford coils as a source of high potentials.

On the other hand, our sole supply of xenon was some 200 cc. at less than one-half atmosphere pressure which had been kindly loaned to us by Dr. Fredrick John Allen of Purdue. Furthermore, we had to construct and operate our own (temperamental) fluorine generator; only visionary scholars ever dreamed then of sometime being able to buy this gas compressed in cylinders. The techniques (or art) of handling fluorine and its generators were in primitive stages of development. In this country and at that time, the list of men actively interested in fluorine chemistry was practically confined to Professors Joel Hildebrand of Berkeley, Joe Simons of Pennsylvania State, Walter Schumb of MIT, and myself at Caltech. Fluorine chemistry was then carried out in the days of wooden ships and iron men, so to speak.

There may, of course, be serpents who say—possibly with some reason—that if in spite of undeveloped techniques we had worked harder and more exhaustively we would have succeeded in preparing one or more xenon fluorides. But the simple fact is that we didn't succeed; the discovery was left for another generation of more daring and more ingenious experimenters whose researches are described in this book. Mr. Kaye and I will have to rest content with the fringe virtue of having said in print that we hadn't proved by our experiments that a xenon fluoride was incapable of existing.

George Santayana is credited with having said that those who do not study history may have to repeat it. Chemists, physicists, and mathematicians are more keenly aware of this truism than are those occupied in other fields of human endeavor; but they are also aware of the fact that when there is little or no history the chances for a capital discovery are much greater. The discovery and studies of the noble-gas compounds described in this book lend vivid emphasis to this point.

But now let us take just a peek into the future of the newly uncovered field of noble-gas chemistry and physics. One can envision a whole sombrero full of studies that can be made on noble-gas compounds and their derivatives, which will enrich our knowledge of nature and her laws. There will also be attempts at applications in various fields, for man never loses hope that he will no longer have to plow with a stick and live like the beasts. And so on. To those who have the high good fortune to be working in this brand new and important field I would like to close by saying *tua res agitur*.

REFERENCES

1. N. Bartlett, *Proc. Chem. Soc. 1962* 218 (1962).
2. H. H. Claassen, H. Selig, and J. G. Malm, *J. Am. Chem. Soc.* **84,** 3593 (1962).
3. C. L. Chernick *et al., Science* **138,** 136 (1962). (See also A. D. Kirshenbaum, L. V. Streng, A. G. Streng, and A. V. Grosse, *J. Am. Chem. Soc.* **85,** 360 [1963].)
4. D. M. Yost and A. L. Kaye, *J. Am. Chem. Soc.* **55,** 3890 (1933).

THE XENON–PLATINUM HEXAFLUORIDE REACTION AND RELATED REACTIONS

NEIL BARTLETT AND N. K. JHA
University of British Columbia

ABSTRACT

Xenon and platinum hexafluoride interact at room temperature to form a red solid of composition $Xe(PtF_6)_x$ where x lies between 1 and 2. That the platinum is present in the +5 oxidation state, no matter what the composition of the adduct, is indicated by the preparation of alkali metal hexafluoroplatinates(V) from material of composition $XePtF_6$ and $Xe(PtF_6)_2$. Material containing more than one mole of platinum hexafluoride per gram atom of xenon combines with more xenon at 130° to approach the composition $XePtF_6$.

Xenon tetrafluoride is liberated in the pyrolysis of platinum hexafluoride–rich adduct, e.g., $Xe(PtF_6)_{1.8}$. The residual, brick-red, diamagnetic, xenon-containing platinum compound has a composition close to $XePt_2F_{10}$.

Rhodium hexafluoride forms a deep-red adduct with xenon, the composition of which is close to $XeRhF_6$. Krypton does not react with either platinum hexafluoride or with rhodium hexafluoride at temperatures below 50°.

INTRODUCTION

In the preliminary communication [1] reporting the interaction of xenon with platinum hexafluoride, the adduct was given the empirical formula $XePtF_6$ and was assumed to be a derivative of pentapositive platinum. Subsequent work has shown that although the gases may interact in a 1:1 ratio, the product usually contains more platinum and can have the composition $Xe(PtF_6)_2$. That the platinum in the xenon–platinum hexafluoride adduct $Xe(PtF_6)_x$ is in the +5 oxidation state has now been established.

EXPERIMENTAL

Reagents.—Platinum hexafluoride and rhodium hexafluoride were prepared from 0.03-in. diameter pure metal wire supplied by Johnson, Matthey, and Mallory Ltd., Toronto. The reaction with fluorine was initiated by electrical heating of the wire, the metal reactor used being similar to that described by Weinstock, Malm, and Weaver [2]. Airco "Reagent Grade" xenon was used.

The vacuum system.—The hexafluorides and xenon were manipulated in an

Research reported in this chapter was supported by the Research Corporation, the President's Research Fund of the University of British Columbia, and the National Research Council.

all metal vacuum system which consisted of a brass main line connected by a Hoke A431 valve to a glass, liquid air–cooled trap. A vacuum of better than 10^{-6} mm. of mercury was maintained by a Balzers DU05 rotary pump and metal oil-diffusion pump.

The noble-metal fluorides were stored in welded nickel bottles joined to the main line through Hoke A431 valves. Xenon was fed from a glass storage bulb into the metal system through a similar valve.

Gas pressure in the metal vacuum system was measured using a nickel bellows gauge. This consisted of a small nickel bellows which terminated one arm of a nickel U-trap, the bellows being silver-soldered to the trap. The other arm of the trap was provided with a Hoke A431 valve. In order that the gauge might be easily thermostated, it was suspended below the main line and joined to it by a nickel tube 15 in. long. An optical lever was used to magnify the bellows movements, and the gauge was used as a null instrument, the pressure developed being balanced by adjustment of the pressure in the bellows housing

TABLE 1

PtF_6(G) AND Xe(G) IN GLASS BULB

	1	2
PtF_6 pressure, mm.	95.0	70.0
Initial Xe pressure, mm.	156.6	72.0
Residual Xe pressure, mm.	82.0	23.0
Combining ratio Xe/PtF_6	1:1.27	1:1.42

and this pressure being registered by a mercury manometer. An accuracy of better than 0.5 mm. of mercury was attainable with this arrangement.

In most of the experiments the gases were measured out in the volume of the gauge and main line, which amounted to 53.98 ml. (22°). The various reactions were carried out in apparatus attached by means of Swaglock compression fittings to a Hoke A431 valve, which was part of the main-line assembly. Teflon front-ferrules and nylon back-ferrules were used in the compression fittings.

Xenon–platinum hexafluoride in glass apparatus.—A Pyrex glass break-seal bulb, the seal of which was bypassed by a small bore tube, was evacuated by way of a glass side arm to avoid introducing moisture to the metal system. The glass was flamed out under vacuum, and the side arm drawn off. Platinum hexafluoride of known volume, temperature, and pressure was condensed from the metal system into the glass bulb. The break-seal bypass was sealed to isolate the fluoride, which was allowed to vaporize. Xenon, also measured out in the metal system, was admitted by breaking the glass seal. To minimize photolysis of the hexafluoride, the experiments were conducted in near darkness.

The results for the reactions of PtF_6 (gas) and Xe (gas) in a glass bulb are summarized in Table 1.

Xenon–platinum hexafluoride in silica apparatus.—In other experiments the reactants were condensed in succession in a silica bulb. On warming, the platinum hexafluoride reacted both as a solid and as a vapor with the gaseous xenon.

The results for the reactions of PtF_6 (solid + gas) and Xe (gas) in a silica bulb are summarized in Table 2.

Xenon–platinum hexafluoride in nickel apparatus.—A series of reactions were carried out using a nickel weighing-can with a capacity of approximately 113 ml., weight 174.5 gm. This bottle was fitted with a Hoke A431 valve and was joined to the vacuum system by a compression fitting. The can was conditioned by exposure to platinum hexafluoride gas for several hours. In each experiment tensimetrically measured platinum hexafluoride was condensed into the can, which was then weighed. Next a tensimetered sample of xenon was condensed in the can, which was then warmed to room temperature. Gaseous residues were transferred to the gauge for measurement, and the evacuated can reweighed.

TABLE 2

PtF_6(G+S) AND XE(G) IN SILICA BULB

	1	2	3
Initial Pt_6 pressure, mm.	93.0	117.5	56.0
Residual PtF_6 pressure, mm.	0	0	12.0
Initial Xe pressure, mm.	108.0	117.5	27.5
Residual Xe pressure, mm.	17.0	59.0	0
Combining ratio Xe/PtF_6	1:1.02	1:2.0	1:1.6

Sufficient xenon (pressure 311 mm.) was added to the weighing-can containing 0.6464 gm. of adduct of composition $Xe(PtF_6)_{1.89}$ to bring the composition to $XePtF_6$, and the can was heated to 130° for one hour. Most of the xenon was consumed (residual pressure 128 mm.) and amounted to 0.0621 gm., this representing a change in composition to $Xe(PtF_6)_{1.24}$. Prolonged heating with excess xenon did not change this composition significantly.

The results for a sequence of PtF_6 (solid + gas) and Xe (gas) reactions in a nickel can are summarized in Table 3.

The results for a sequence of PtF_6 (gas) and Xe (gas) reactions in a nickel bulb are summarized in Table 4.

Spectra of Xe-PtF_6 adducts.—The infrared spectrum of material deposited on silver chloride windows in a nickel-bodied gas cell was recorded. The composition of the adduct was $Xe(PtF_6)_{1.72}$. Only two peaks in the region 400–4000 cm.$^{-1}$ were assignable to the adduct:652 *vs*, 550 *s*. cm.$^{-1}$. The visible and ultraviolet spectrum of material deposited on the windows of a silica gas cell was recorded. A single peak at 3825 A. was observed. The material absorbed strongly beyond 4000 A. No differences in the absorption spectra were noted for several separate adduct samples.

Physical properties of the Xe-PtF_6 adduct.—The adduct is yellow when deposited in thin films but in bulk is deep red. The solid becomes glassy in appearance when heated to 115°, but does not melt below 165°, when it decomposes to produce xenon tetrafluoride. X-ray powder photographs of the adduct of composition $XePtF_6$ show no diffraction pattern. Complex patterns are observed with samples of material richer in platinum. The plasticity of the material made the preparation of good powder samples difficult. Even well-cooled samples did not grind well.

TABLE 3

PtF_6(S+G) AND XE(G) IN NICKEL CAN

	1	2	3	4	5	6
Pressure of PtF_6, mm.	....	97.0	73.0	90.0	96.0	95.0
Weight of PtF_6, gm.	0.1117	0.0829	0.0714	0.0828	0.0897	0.0893
Initial pressure of Xe, mm.	....	58.0	46.0	53.0	59.0	74.0
Residual pressure of Xe, mm.	....	6.0	7.0	3.0	5.0	23.0
Weight of Xe, gm.	0.0175	0.0212	0.0144	0.0188	0.0255	0.0212
Temperature	20.8	22.2	23.6	23.6	23.6	23.6
Combining ratio:						
Tensimetric	....	1:1.87	1:1.87	1:1.80	1:1.78	1:1.86
Gravimetric	1:2.3	1:1.66	1:2.1	1:1.87	1:1.49	1:1.78

The over-all combining ratio:
Tensimetric 1:1.83
Gravimetric 1:1.89

TABLE 4

PtF_6(G) AND XE(G) IN NICKEL BULB

	1	2	3	4
PtF_6 pressure, mm.	86.0	76.5	88.5	88.5
Initial Xe pressure, mm.	112.0	94.5	154	146.0
Residual Xe pressure, mm.	12.0	53.0	114	95.5
Combining ratio Xe/PtF_6	1:0.86	1:1.84	1:2.2	1:1.75

The thermal decomposition of Xe-PtF_6 adduct.—A sample of material of composition $Xe(PtF_6)_{1.8}$ contained in a silica bottle was heated under an atmosphere of nitrogen. No change was observed up to 115°, when the red solid deepened in color and became glassy. There was no further change up to 165°, when the solid fell to a brick-red powder, and a deposit of colorless crystals collected on the cooler silica. The temperature was maintained at or close to 165° for one hour, when decomposition appeared to be complete.

An infrared spectrum of the volatile white solid over the range 400–4000 cm.$^{-1}$ showed peaks at 595 *vs* and 580 *vs* cm.$^{-1}$, values in close agreement with those given by Claassen, Chernick, and Malm [3]—591, 581 cm.$^{-1}$—for the strong doublet of xenon tetrafluoride.

The residual platinum compound was analyzed for xenon by the conventional Dumas method for nitrogen; and, for platinum and fluorine by the pyrohydrolytic technique previously described [4]. Found: F, 27.0 per cent; 26.4 per cent; Pt, 54.5 per cent; and Xe, 14.7 per cent; 14.7 per cent. $XePt_2F_{10}$ requires F, 26.7 per cent; Pt, 54.8 per cent; and Xe, 18.5 per cent.

The preparation of $RbPtF_6$ from $XePtF_6$ in iodine pentafluoride.—Platinum hexafluoride (total pressure in the known volume at 21.3° of 186 mm.) was condensed, followed by xenon (pressure 216 mm.), on rubidium fluoride (0.10 gm.) in a silica bulb. The mixture was warmed slowly until the platinum hexafluoride was seen to react with the xenon. The reaction was moderated by judicious cooling. A residual pressure of xenon of 34 mm. indicated a composition $Xe(PtF_6)_{1.02}$ for the adduct. Iodine pentafluoride was distilled onto the rubidium fluoride–xenon fluoroplatinate mixture and the solids dissolved; some gas (probably xenon) evolved; and an orange-yellow solution formed. Removal of the iodine fluoride under vacuum left an orange-yellow solid. X-ray powder photographs of this material revealed a strong pattern of lines which were indexed on the basis of a rhombohedral unit cell a, 5.08 A.; α, 96°58′ indicative of $RbPtF_6$ [5].

The preparation of $CsPtF_6$ from $Xe(PtF_6)_{2.0}$ in iodine pentafluoride.—An experiment similar to the above, involving interaction between material of composition $Xe(PtF_6)_2$ and cesium fluoride in iodine pentafluoride, produced an orange-yellow solid, X-ray powder photographs of which identified it as cesium hexafluoroplatinate(V). The rhombohedral unit cell a, 5.27 A., α, 96°25′ being very similar in dimensions to those reported for cesium hexafluoroiridate(V) and osmate(V) [5].

Xenon tetrafluoride with platinum tetrafluoride in iodine pentafluoride solution.—Although xenon tetrafluoride dissolved in iodine pentafluoride without reaction, no adduct could be isolated. Furthermore, this solution failed to react with platinum tetrafluoride even on prolonged reflux at $\sim$100°.

Reaction of xenon with rhodium hexafluoride.—The hexafluoride was purified by vacuum sublimation from the reactor at room temperature to the storage bottle at −75° and was maintained at this temperature for several hours with the bottle open to the high vacuum line.

Since the rhodium hexafluoride was reduced in the gauge and main line so rapidly that tensimetric measurement of the quantity of hexafluoride was unreliable, the stoichiometry of the reaction with xenon was determined from the weight of adduct formed in the consumption of a tensimetrically measured quantity of xenon. Rhodium hexafluoride was roughly measured out tensimetrically and quickly condensed in a nickel weighing-can at −196°. Xenon, in roughly twofold molar excess of the rhodium fluoride, was accurately measured tensimetrically and also condensed in the can, which was subsequently warmed to room temperature. Residual xenon was transferred from the can to the gauge by cooling the well of the gauge to −196°. The weight of xenon consumed was computed from the tensimetric measurements, the volume of the line and gauge being accurately known.

1. *Results:* (1) Wt. of xenon consumed, 0.0050 gm.; wt. of adduct, 0.0139 gm.; combining ratio, $XeRhF_6$ = 1:1.05. (2) Wt. of xenon consumed, 0.0120 gm.; wt. of adduct, 0.0337 gm.; combining ratio, $XeRhF_6$ = 1:1.10.

2. *Properties of the adduct:* Samples prepared in silica and of uncertain composition were deep red. X-ray powder photographs of this material showed only a faint, sharp pattern, perhaps belonging to a minor phase. A similar faint pattern appeared on photographs of some xenon–platinum hexafluoride specimens.

Attempts to oxidize krypton with noble-metal hexafluorides.—Platinum hexafluoride and a large krypton excess were separately condensed in a silica bulb. The red vapor of platinum hexafluoride was observed on warm up and was still observable at 50°.

A similar experiment with rhodium hexafluoride established that krypton does not react with that fluoride below 50°.

DISCUSSION

With the transition-metal hexafluorides, the oxidizing power increases and the thermal stability decreases with molecular weight in each transition series. Thus, platinum hexafluoride [2] and rhodium hexafluoride [6] are the least stable and most powerfully oxidizing members of their series. This situation is consistent with the electronic configurations of the transition-metal atoms in these molecules. For the transition-metal hexafluorides, other than molybdenum and tungsten, the central atom possesses *d* electrons which are not needed in bonding and which are located in t_{2g} orbitals. Since these orbitals do not screen the ligands from the charge on the central atom, the effective nuclear charge $(Z\text{-}S)$ increases with atomic number (Z). The fluorine ligands about the platinum atom in the hexafluoride are thus highly polarized and, as the observed dissociation [2] of the molecule into fluorine and lower fluorides shows, can lose electrons completely to the platinum. Clearly, the capture of an electron from another source will stabilize the hexafluoride unit. That this electron capture is a highly exothermic process is demonstrated by the formation of dioxygenyl hexafluoroplatinate(V). Since the thermal stability of rhodium hexafluoride [6] is lower than that of the platinum compound, there was the possibility that this molecule could possess a sufficiently high electron affinity to oxidize krypton. Neither platinum hexafluoride nor rhodium hexafluoride will do this, however.

Unfortunately, the absence of X-ray structural information and magnetic evidence has prevented the establishment of the formula $Xe^+[PtF_6]^-$ for the 1:1 adduct, although the derivation of rubidium hexafluoroplatinate(V) from this compound in the oxidatively and reductively neutral solvent iodine pentafluoride indicates that it is a hexafluoroplatinate(V). Some support for the

$Xe^+[PtF_6]^-$ formulation is also given by the similarity of the spectra with those of potassium and dioxygenyl hexafluoroplatinates(V). The data are compared in Table 5. That the material containing more platinum than $XePtF_6$ contains platinum pentafluoride, arising by the reaction

$$Xe + 2PtF_6 \rightarrow XeF_2 + 2PtF_5 ,$$

is consistent with the plastic consistency of such products and with their glassy appearance, but there is no evidence to support the presence of free xenon difluoride. It is possible, since such material contains pentapositive platinum, that the material of composition $Xe(PtF_6)_2$ is xenon(II) hexafluoroplatinate(V).

The diamagnetic brick-red solid of composition $XePt_2F_{10}$, formed along with xenon tetrafluoride in the pyrolysis of $Xe(PtF_6)_{1.8}$, contains the platinum in the +4 oxidation state, since interaction of the material with cesium fluoride in iodine pentafluoride gives a yellow product, X-ray powder photographs of which

TABLE 5

ABSORPTION BANDS

	COMPOUND		
	$Xe(PtF_6)_{1.7}$	$KPtF_6$	O_2PtF_6
I.R. ($cm.^{-1}$)................	652(*vs*); 550(*s*)	640(*vs*); 590(*s*)	631(*vs*); 545(*s*)
Visible and U.V. (A.).......	3825	....	3500

indicate the presence of cesium hexafluoroplatinate(IV). The powder photographs show the pentapositive salt, $CsPtF_6$, to be absent. Although the X-ray powder pattern of $XePt_2F_{10}$ is complex, it does not contain lines attributable to platinum tetrafluoride. Formulation of this material as the mixture $PtF_4 + XePtF_6$ is therefore inadmissible.

The departure from the 1:1 reaction stoichiometry in the xenon–rhodium hexafluoride system is less than for the platinum system. This is surprising in view of the greater instability and chemical reactivity of the rhodium fluoride. Ruthenium hexafluoride, which is less reactive than rhodium hexafluoride, has been reported [7] to react non-stoichiometrically with xenon. Perhaps the use of small quantities of rhodium fluoride favored the 1:1 addition. There is as yet no evidence for the oxidation state of rhodium in the adduct, although the formulation $Xe + [RhF_6]^-$ would, as in the corresponding platinum case, appear to be energetically more favorable than $Xe^{2+}[RhF_6]^{2-}$.

Acknowledgment.—Cesium and rubidium hexafluoroplatinate(V) were prepared by Michael Booth.

REFERENCES

1. N. Bartlett, *Proc. Chem. Soc.* 218 (1962).
2. B. Weinstock, J. G. Malm, and E. E. Weaver, *J. Am. Chem. Soc.* **83**, 4310 (1961).
3. H. H. Claassen, H. Selig, and J. G. Malm, *ibid.* **84**, 4164 (1962).
4. N. Bartlett and D. H. Lohmann, *J. Chem. Soc.* 5253 (1962).
5. M. A. Hepworth, W. M. Jack, and G. J. Westland, *J. Inorg. Nuclear Chem.* **2**, 79 (1956).
6. C. L. Chernick, H. H. Claassen, and B. Weinstock, *J. Am. Chem. Soc.* **83**, 3165 (1961).
7. C. L. Chernick *et al.*, *Science* **138**, 136 (1962).

REMARKS RELATIVE TO THE FIRST XENON TETRAFLUORIDE PREPARATION

H. H. Claassen, John G. Malm, and H. Selig
Argonne National Laboratory

The reaction between xenon and PtF_6 first reported by Bartlett [see this volume, p. 23] had been repeated here and extended to RuF_6. It had been found that the simple one-to-one ratio of hexafluoride to xenon did not generally hold. For RuF_6 the ratio was nearly three to one. Since ruthenium has a stable pentafluoride, this suggested that the primary role of the unstable PtF_6 or RuF_6 might be that of a fluorinating agent. If this were correct, it might be possible under some conditions to fluorinate xenon with elemental fluorine. Hence, we decided to heat a mixture of xenon and fluorine.

The first experiment was conducted on August 2, 1962, and was essentially qualitative. Xenon (1.06 atm.) and fluorine (9.6 atm.) at room temperature were introduced into a 90-ml. nickel can. This was heated for an hour at 400° C., cooled to −195° C., and the fluorine was removed by pumping to a good vacuum. This would leave the xenon, even if not reacted, because xenon is solid and non-volatile at −195° C. The can was now warmed to −100° C. to transfer the xenon to a U-tube kept at −195° C. When this U-tube on the gas-handling line was warmed to room temperature the 0 to 1000 mm. bourdon pressure gauge should have gone to 660 mm., the original xenon pressure as observed in the line. In fact, the gauge stayed right at zero. Although the specific intention of this experiment was to react xenon with fluorine directly, we were nevertheless surprised by the results. In our notebook that afternoon we wrote, "Go home somewhat puzzled. Where is the Xe?" The possibilities were either that we had made a mistake in turning valves or that all the xenon had been formed into a compound that was non-volatile at −100° C.

The experiment was next done quantitatively, using a thin-walled nickel can that could be weighed. From weights combined we arrived at the formula XeF_4 and also obtained evidence for existence of a lower fluoride. Visual evidence was obtained by subliming some of the material into glass. The crystals of XeF_4 (Plate I) in a silica tube soon became the standard exhibit used to convince local and visiting chemists that xenon compounds are real.

The formula was verified by chemical analysis and the communication prepared and sent to the *Journal of the American Chemical Society* [1]. Less than one month elapsed between mailing the manuscript and seeing it in print in our library.

Acknowledgments.—In this work we had help from several members of the Chemistry Division. We want to thank W. M. Manning, director of the Division, for encouragement and support. J. J. Katz suggested reduction with hydrogen as a method of analysis. Fluoride analyses by I. M. Fox and xenon-purity measurements by C. Plucinski were of essential value. Important help in the laboratory was given by R. Lindholm and R. Springborn.

REFERENCE

1. H. H. Claassen, H. Selig, and J. G. Malm, *J. Am. Chem. Soc.* **84**, 3593 (1962).

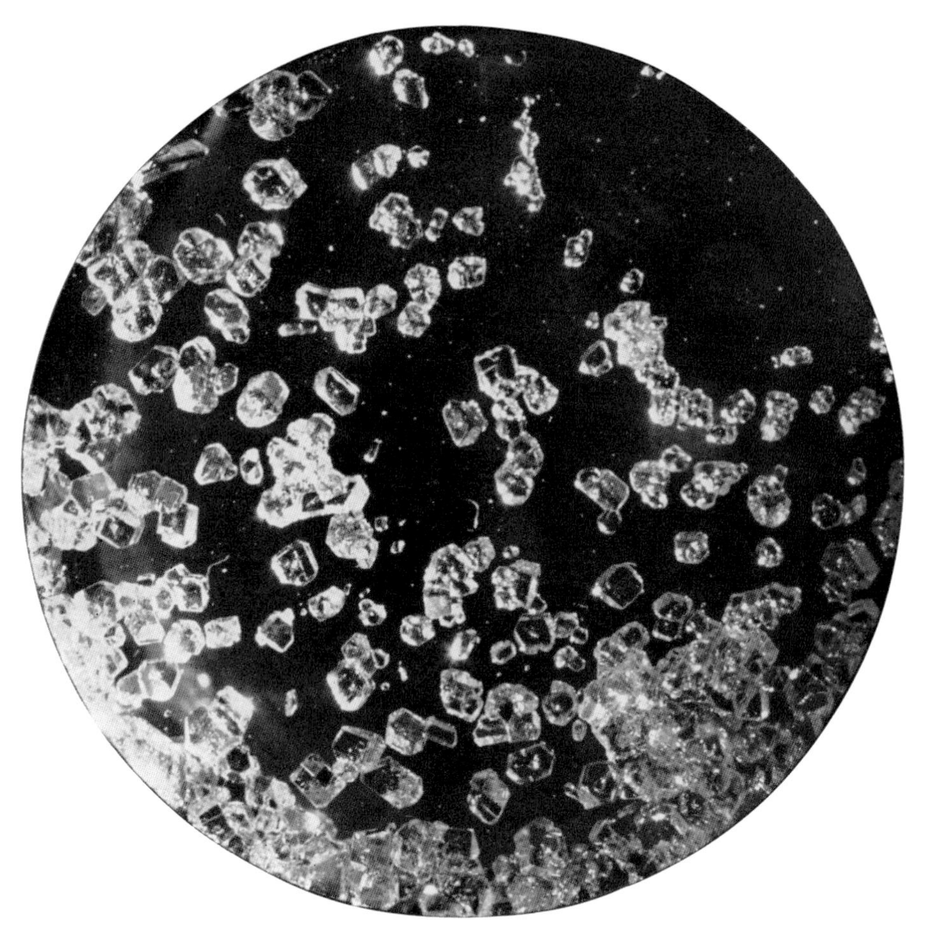

PLATE I.—Crystals of XeF_4 in 22 mm. diameter silica spectrophotometer cell. (This photograph appears on the cover of *Science* [Vol. 138, No. 3537, October 12, 1962].)

PART 2 PREPARATION AND SOME PROPERTIES OF NOBLE-GAS FLUORIDES

The essential chemical manipulations used in the production and purification of the noble-gas fluorides and in the determination of some of their properties are reported in Part 2. The papers vary widely in scope. The general methods of preparation and the simple physical properties are reviewed by Chernick. Smith and Studier and Sloth summarize in this context the uses of the infrared spectrophotometer and the mass spectrometer, which are major analytical tools in the synthesis and purification of the noble-gas fluorides.

Weinstock, Weaver, and Knop emphasize the problems of preparing purer species and discuss, in particular, the equilibrium between xenon-fluorine compounds of different valence. We may note that the authors, on the basis of experimental studies immediately following the conference, now question the tentative explanation they offered for their observations. This preliminary explanation involved a more stable pentavalent xenon species than seemed reasonable to most conferees.

In this section, the reader will also find discussions of the fluorination of krypton and radon as well as xenon. The scarcity of data reflects the difficulties encountered in these studies. Krypton yields compounds, but much more reluctantly than xenon, while radon is available only as an intensely radioactive species, posing formidable experimental difficulties.

Perhaps it is appropriate at this point to comment on the European contribution to the work on noble-gas compounds. The research institutes in Yugoslavia have brought an unexpected abundance of manpower and resources to bear on problems of noble-gas chemistry. This effort began shortly after the first Argonne publication appeared, and by the time the conference was held a number of papers had been published. The Yugoslavian contributions to this new area of knowledge, the conference, and this volume are comparable to those of an American national laboratory with a similar interested group.

Professor Peacock also initiated his research shortly after the announcement of the preparation of xenon tetrafluoride. The contribution from the University of Birmingham in England is all that might be anticipated from a first-class university laboratory with experience in manipulating fluorine compounds.

In correspondence with the editor, the senior author of the German contribution, Dr. Hoppe, emphasized the long-continuing interest of workers in his laboratory in synthesizing unusual compounds of fluorine—unusual in the sense that the combining element or elements are not in their expected valence state. This group, according to Dr. Hoppe, has long speculated on the existence of noble-gas fluorides along with a variety of other unusual compounds. Unfortunately, as is pointed out in the paper of Hoppe, Mattauch, Rödder, and Dähne, the resources available to their laboratory have been so limited that they have not been able to establish which of these potentially interesting reactions are in fact experimentally realizable, and they have not always been able to verify their interpretations adequately before publication. On learning of the Argonne discovery of xenon tetrafluoride, the München group proceeded with their own approach and published some inadequately established preliminary observations. Their contribution to this volume is a revised and somewhat enlarged version of this earlier communication.

FLUORIDES OF XENON, PREPARATION AND PROPERTIES AN INTRODUCTION AND REVIEW

Cedric L. Chernick
Argonne National Laboratory

The first preparation of a binary fluoride of xenon was reported by Claassen, Selig, and Malm [1] shortly after Bartlett's announcement of the reaction between xenon and platinum hexafluoride [2]. They identified their compound as xenon tetrafluoride. This work led to the discovery of two other binary fluorides, a difluoride and a hexafluoride. This chapter will summarize the methods used for preparing the three fluorides, giving in more detail the preparative methods used at Argonne National Laboratory. Table 1 lists the methods used for preparing these fluorides from elemental xenon and elemental fluorine. In Table 2 some methods are given which involve the use of elemental xenon and a fluorine-containing compound. In this case the first step is no doubt the dissociation of the fluoride to give fluorine atoms. Some of the properties of XeF_2, XeF_4, and XeF_6 will be compared.

XENON DIFLUORIDE

Preparation of xenon difluoride by irradiating mixtures of xenon and fluorine with ultraviolet light has been reported by Weeks, Chernick, and Matheson [13]. [See also this volume, p. 89.] The gases were contained in a nickel cell with two sapphire windows through which was passed the beam from a high-pressure mercury arc. A schematic drawing of the reaction cell is shown on page 91. A known pressure of xenon gas was introduced into the cell and condensed by placing a liquid-nitrogen bath around the bottom of the U-bend. An excess pressure of fluorine was measured out in a vacuum manifold similar to the one shown in Figure 1, and an equimolar amount of fluorine was bled into the cell, with the liquid-nitrogen bath still on the U-bend. The valve to the filling system was closed and the xenon-fluorine mixture warmed to room temperature. Irradiations were then carried out, as described on pages 89–90. After completion of the reaction, any unreacted xenon or fluorine was removed by pumping on the cell with the U-bend at $-78°$ C. The cold bath was then removed from the U-bend, and the XeF_2 purified by distilling, while pumping, into a further U-tube that had been cooled to $-78°$ C. Yields of xenon difluoride of greater than 99 per cent, based on the amount of starting material, have been obtained in quantities up to 8 gm. per preparation. The purity, checked by infrared analysis, showed neither the bands associated with XeF_4 nor those associated with XeF_6. (It is estimated that about 1 per cent of either of these materials could be detected.) The xenon difluoride was analyzed by reduction with hydrogen, which proceeded quantitatively according to the reaction,

$$XeF_2 + H_2 \rightarrow Xe + 2HF .$$

TABLE 1

PREPARATION OF XENON FLUORIDES BY REACTION OF XENON AND FLUORINE: REPORTS OF METHODS

METHOD	REFERENCES		
	XeF_2	XeF_4	XeF_6
Heating in a closed system	....	[1], [3]	[4], [5], [6], [7]
Heating in a flow system	[8]	[9], [10], [11]	....
Electric discharge	....	[12]	....
Ionizing radiation	*	*	....
Ultraviolet irradiation	[13]	....	....

* MacKenzie and Wiswall, this volume, p. 81.

TABLE 2

PREPARATION OF XENON FLUORIDES BY OTHER METHODS

Method	Fluoride Produced
Heating xenon with CF_3OF or SO_3F_2	XeF_2*
Electric discharge through xenon + CF_4	XeF_2†

* See Gard, Dudley, and Cady, this volume, p. 109.
† See reference 14.

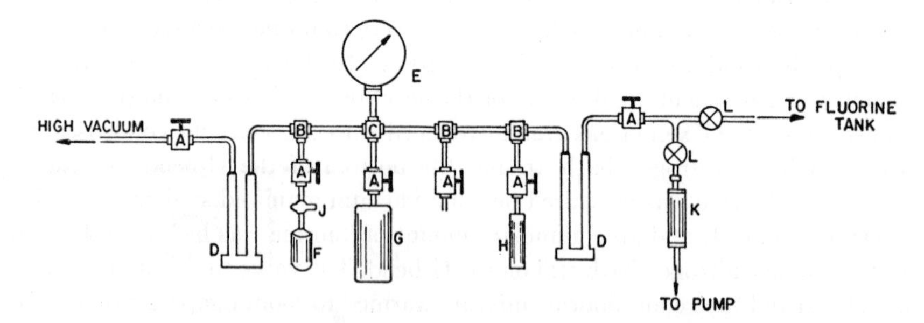

FIG. 1.—Preparation and purification system for reactive fluorides: (*A*) 30,000-p.s.i. valve, Monel body, Teflon packing; (*B*) 30,000-p.s.i. Monel Tee; (*C*) 30,000-p.s.i. Monel Cross; (*D*) welded nickel U-tube; (*E*) welded Monel bourdon gauge (0–1000 mm.); (*F*) 130-cc. welded nickel reactor can; (*G*) 1500-cc. welded nickel storage and measuring can; (*H*) 85-cc. welded nickel storage and measuring can; (*J*) brass valve: Hoke type A431; (*K*) soda-lime trap; (*L*) Monel valve: Hoke type 413.

The xenon and hydrogen fluoride were separated and weighed. The hydrogen fluoride was also dissolved in water, and the aqueous solution analyzed for both acid and fluoride-ion content.

XENON TETRAFLUORIDE

Xenon tetrafluoride was first prepared in an apparatus similar to the one shown in Figure 1. One part xenon and five parts fluorine, by pressure, were condensed in the can, shown at *F*, which was then heated to 400° C. for one hour. The pressure in the can at 400° C. was about 6 atms. The can was then cooled and immersed in a −78° C. bath, and the excess fluorine pumped off through the trap (*D*) immersed in a −195° C. bath. Any unreacted xenon would have been trapped at −195° C., but none was found on warming this trap. The xenon tetrafluoride was purified and analyzed. Its purity was checked in the same manner as already described for the difluoride. Using a 1600-cc. nickel can as a reactor, up to 10 gm. of xenon tetrafluoride have been prepared in one loading, with yields of better than 99 per cent, based on the initial amount of xenon used.

XENON HEXAFLUORIDE

Xenon hexafluoride was independently prepared at several laboratories at the same time by heating xenon with a large excess of fluorine [4–7]. One part xenon and twenty parts fluorine when heated in a nickel vessel to 300° C. for sixteen hours yield xenon hexafluoride. In a typical experiment the pressure at reaction temperature was of the order of 60 atms. The formula has been checked by hydrogen reduction and by studying the combination of xenon- and fluorine-containing Xe^{133} and F^{18} tracers. The hexafluoride may be purified by flash distillation, the less volatile fractions being discarded as possibly containing lower fluorides. Spectroscopic checks on purity can be used to show the absence of XeF_2, XeF_4, and $XeOF_4$. A peak is observed at 520 cm.$^{-1}$, which may be due to an impurity but is difficult to eliminate. Amounts up to 10 gm. have been prepared at one time, with yields of over 95 per cent, based on the amount of xenon taken.

PROPERTIES

The three fluorides are all colorless, crystalline solids at room temperature. The vapors of XeF_2 and XeF_4 are colorless, but that of XeF_6 is pale yellow-green in color. The solubilities of the fluorides in anhydrous HF [this volume, p. 275] and their reactions in aqueous media [this volume, p. 153] are the subject of other chapters. As already indicated in the analytical procedures, all three fluorides are quantitatively reduced by hydrogen to give elemental xenon and hydrogen fluoride, and by mercury to give xenon and either Hg_2F_2 or HgF_2. [Cf. reference 12.] They are thermodynamically stable at room temperature with respect to dissociation into fluorine and lower fluorides or xenon. All three fluorides can be stored at room temperature in nickel or Kel-F containers, and the two lower fluorides can also be stored indefinitely in thoroughly dried quartz vessels. The hexafluoride reacts with quartz to produce $XeOF_4$ [see this volume, p. 106]. The difluoride and tetrafluoride have very similar vapor pressures (about 3 mm.) at room temperature, which makes separating them by simple sublimation techniques very difficult. The hexafluoride is about ten times more volatile. Some of the physical properties of XeF_2, XeF_4, and XeF_6 are summarized in Table 3.

TABLE 3

PROPERTIES OF XENON FLUORIDES

	XeF_2	XeF_4	XeF_6
Color of solid	Colorless	Colorless	Colorless
Color of gas	Colorless	Colorless	Yellow-Green
M. P., ° C.	140*	~114	46
Vapor pressure, mm. at 25° C.	3.8*	3	29
Heat of vaporization, kcal/mole	12.3†	15.3†	9.0‡

* See reference 10.
† See reference 15.
‡ Calculated from vapor pressure data.

REFERENCES

1. H. H. CLAASSEN, H. SELIG, and J. G. MALM, *J. Am. Chem. Soc.* **84,** 3593 (1962).
2. N. BARTLETT, *Proc. Chem. Soc.* 218 (1962).
3. J. SLIVNIK *et al.*, *Croat. Chem. Acta* **34,** 187 (1962).
4. J. G. MALM, I. SHEFT, and C. L. CHERNICK, *J. Am. Chem. Soc.* **85,** 110 (1963).
5. E. E. WEAVER, B. WEINSTOCK, and C. P. KNOP, *ibid.* 111 (1963).
6. F. B. DUDLEY, G. GARD, and G. H. CADY, *Inorg. Chem.* **2,** 228 (1963).
7. J. SLIVNIK *et al.*, *Croat. Chem. Acta* **34,** 253 (1962).
8. D. F. SMITH, *J. Chem. Phys.* **38,** 270 (1963). See also this volume, p. 39.
9. J. H. HOLLOWAY and R. D. PEACOCK, *Proc. Chem. Soc.* 389 (1962).
10. P. A. AGRON *et al.*, *Science* **139,** 842 (1963).
11. D. H. TEMPLETON *et al.*, *J. Am. Chem. Soc.* **85,** 242 (1963).
12. A. D. KIRSHENBAUM *et al.*, *ibid.* **85,** 360 (1963).
13. J. L. WEEKS, C. L. CHERNICK, and M. S. MATHESON, *ibid.* **84,** 4612 (1962).
14. D. E. MILLIGAN and D. SEARS, *ibid.* **85,** 823 (1963).
15. J. JORTNER, E. G. WILSON, and S. A. RICE, *ibid.* **85,** 814 (1963).

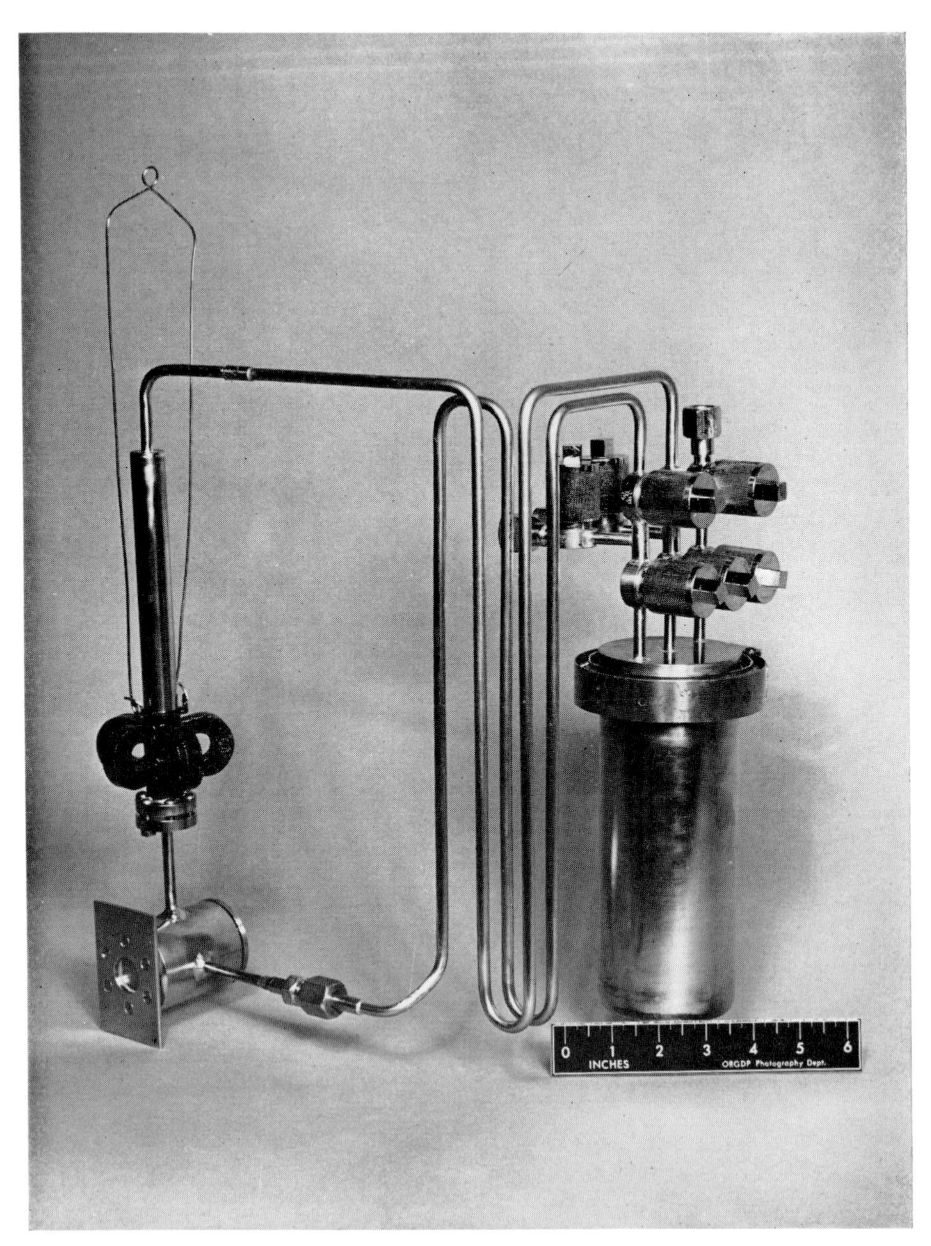

PLATE I.—Photograph of circulating loop for studying systems containing corrosive fluorinating gases.

THE USE OF INFRARED SPECTROSCOPY IN THE PREPARATION AND STUDY OF XENON COMPOUNDS

D. F. Smith
Oak Ridge Gaseous Diffusion Plant

The basic device used in this work is a circulating loop containing an infrared cell located in a spectrophotometer beam. This was designed for studying by infrared absorption the corrosive fluorine-containing gases and their volatile reaction products. The loop has been used to study the Xe-F_2 and the XeF_2-F_2 reactions and to prepare small quantities of fairly pure XeF_2. It also has been used to test the purity and to improve the purity of XeF_2, XeF_4, and XeF_6 preparations. Small quantities of $XeOF_4$ have been prepared by the hydrolysis of XeF_6, and the vapor pressure as a function of temperature has been determined for $XeOF_4$ and the xenon fluorides.

A schematic diagram of the loop used is shown in Figure 1; and a photograph of a slightly different loop, in Plate I. The essential parts of the loop include a piston pump actuated by external magnets, a ballast volume, an infrared cell located in the beam of a spectrophotometer, two U-tubes of $\frac{1}{4}$-in. nickel tubing, and valves to permit passing the gas through the ballast volume or bypassing the isolated ballast volume. The loop is connected with $\frac{1}{4}$-in. copper tubing to a supply-and-evacuation manifold, which is equipped with several kinds of pressure gauges.

The piston pump diagrammed in Figure 2 is a modification of the double-action pump described by Rosen [1]. The simplification results in a more compact, single-action pump with a smaller dead volume. Normally the pump is operated at about 100 strokes per minute, but the speed can be reduced to less than 25, or increased to more than 150, strokes per minute. Since the displacement is about 25 cc. per stroke, the gases in the loop are well mixed.

The infrared cell (Fig. 3) has a volume less than 30 cc. A tapered hole is cut through a Monel bar 2 in. in diameter to give a cell 6 cm. long. Silver chloride windows 2 mm. thick are used. The larger window, 35 mm. in diameter, is sealed by a circular flange screwed to the cell body. The smaller window, 25 mm. in diameter, is sealed with a rectangular flange which fits in the spectrophotometer cell holder. Usually, Teflon gaskets 1 or 2 mils thick are used to isolate the AgCl

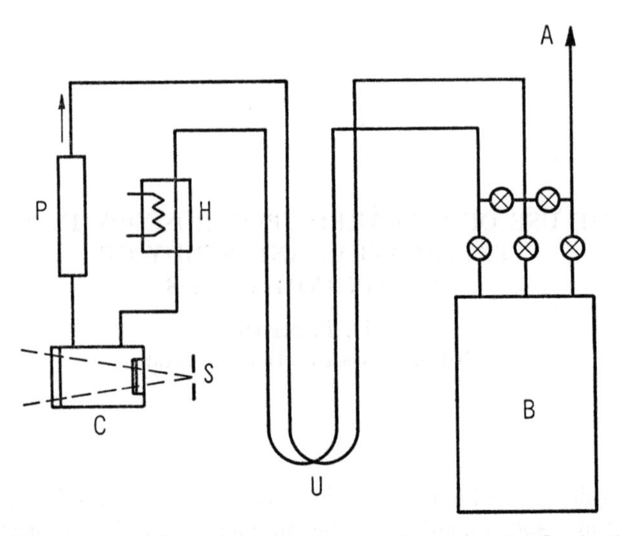

Fig. 1.—Circulating loop for studying systems containing corrosive fluorinating gases

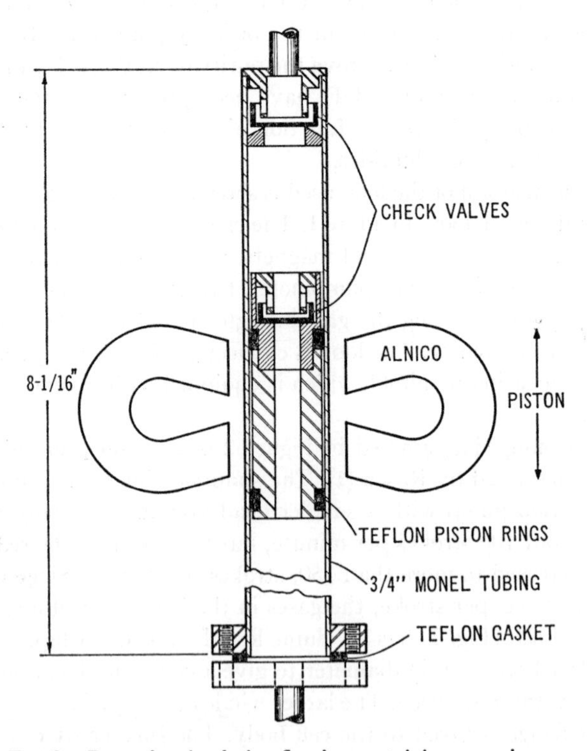

Fig. 2.—Pump for circulating fluorine-containing corrosive gases

from the Monel to avoid corrosion, but these may be eliminated. When working with corrosive materials, replacement of cell windows must be done frequently. With this design the windows can be replaced inexpensively and quickly.

As can be seen in the photograph (Plate I), the U-tubes are mounted side by side so that they may be accommodated in baths either in the same or in two different Dewar flasks. The volume of the primary loop, exclusive of the 687-cc. ballast volume, is 111 cc. This means that a condensable material present in the loop and ballast volume can be condensed in the U-tubes, and the ballast volume can then be isolated. On warming the U-tubes, the partial pressure of the condensable material is increased by a factor of 7.19. In many cases the pumping is more effective if a "non-condensable" gas is present to function as a carrier. For most applications of this loop, fluorine serves this purpose well. In trapping a condensable material, the Dewar with the cold bath is placed on the U-tube

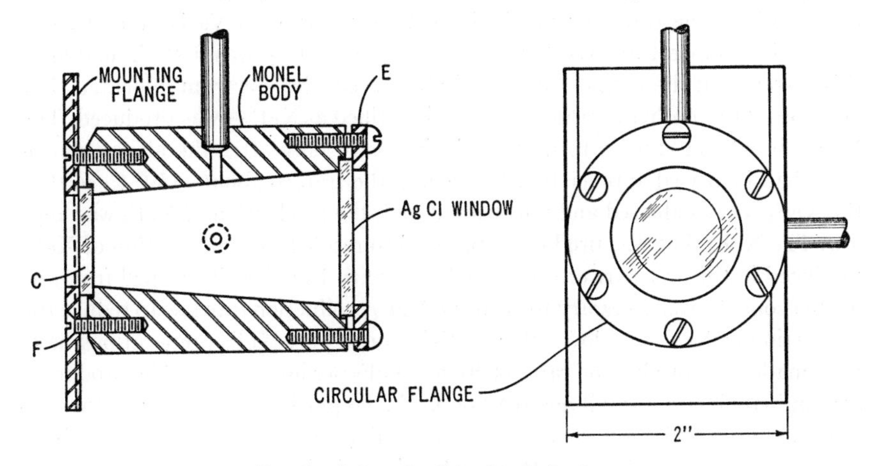

Fig. 3.—Infrared cell in circulating loop

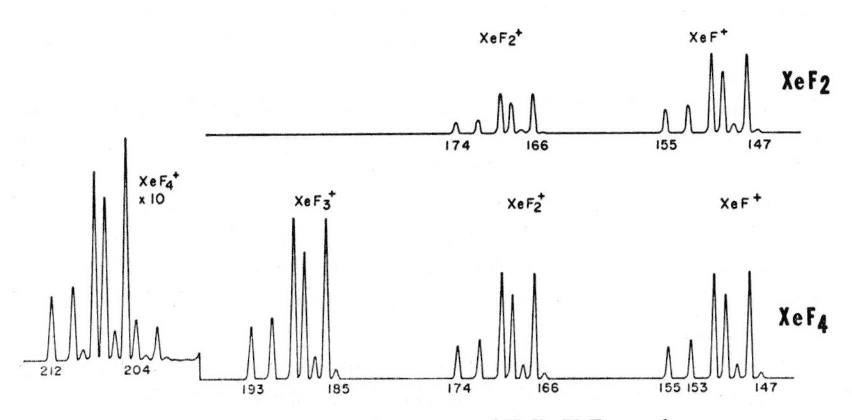

Fig. 4.—Mass-spectrometer scans of XeF_2, XeF_4 samples

on the exit side of the pump, and the spectrophotometer wave length is adjusted to a strong infrared band. The flow and trapping is continued until the infrared absorption of the condensable material is negligible.

The rubbing of the piston on the cylinder wall scrapes the nickel fluoride off the wall, exposing a fresh nickel surface, which then reacts with fluorine. The nickel fluoride dust so formed accumulates in the infrared cell and impairs the loop efficiency. For this reason the loop is taken apart and cleaned with warm water after every few hundred hours of operation.

For studies of the xenon-fluorine reaction, a section of the nickel tubing ahead of the infrared cell could be heated to desired temperatures. This was accomplished by binding a small electrical heater and a thermocouple junction to the nickel tubing with Fiberglas tape. Additional insulation was provided by overwrapping with alternate layers of aluminum foil and Fiberglas tape. Care was taken to keep the transition to the heated zone abrupt. A Variac was used to adjust the temperature. When a xenon-fluorine mixture was first circulated in this system and the temperature slowly increased, the spectrum indicated that not XeF_4, but another compound—later identified as XeF_2—was produced. The XeF_4 absorption started to increase in intensity only after the absorption of XeF_2 had increased appreciably. By cooling the trap to about $-50°$ C. so that the XeF_2 was condensed and not recirculated, the production of XeF_4 was kept low. The XeF_4 that was produced appeared to result from the reaction of XeF_2 produced in the first part of the heated zone with F_2 before it escaped from the heated zone. It was necessary to keep the XeF_2 production rate low to reduce the proportion of XeF_4 to a low value. Satisfactorily pure XeF_2 preparations have been made with production rates of 50 mg. XeF_2 per hour or less. The product is contaminated with a small quantity of XeF_4. To purify it, the vapor in the system (along with fluorine as a carrier) is equilibrated with the solid at about 0° C. The ballast volume is then isolated, and the vapors trapped therein pumped away. The gas mixture is again expanded into the ballast volume, and the process repeated until the XeF_4 impurity has been exhausted, and only XeF_2 remains. A mass-spectrometer scan of this product is shown in Figure 4 with a scan of XeF_4.

Attention should be called to this specific sequence of techniques as a general procedure for isolating and identifying unknown compounds. First, a characteristic infrared band not previously identified is noted in the vapor. The sample is then purified on the basis of this spectral feature. There is left a pure compound with the infrared spectrum giving some indication of the kinds of bonds present in the compound, but usually no identification. The mass spectrum of this pure compound almost always presents enough information to identify the compound unambiguously.

Some details of the Xe-F_2 reaction were readily demonstrated with the loop. By simply monitoring the Xe-F_2 absorption, it was observed that increasing the heated zone temperature increased the reaction rate, while decreasing the temperature decreased the reaction rate. Somewhere above 600° C. the rate of reaction of XeF_2 with nickel becomes rapid enough to compete with the Xe-F_2 reaction, and the production rate appears to decrease with temperature. It was demonstrated that the reaction rate increases with increasing partial pressure of F_2 simply by monitoring the XeF_2 absorption after adding F_2 to the loop. By concentrating the xenon in the primary loop, thereby increasing the partial pressure of xenon by the factor of 7.19 noted previously, the large XeF_2 production-rate increase could be noted.

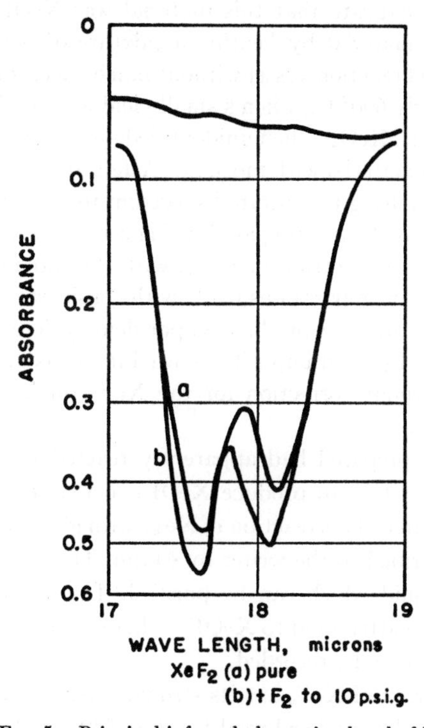

FIG. 5.—Principal infrared absorption band of XeF_2

By studying mixtures of XeF_2 and F_2 with no xenon present, it could be shown that the reaction rate to produce XeF_4 increased with increasing XeF_2 or F_2 partial pressure or increased with increasing reaction-zone temperature. Efforts were made to obtain evidence for an equilibrium with null results. When the xenon fluorides were circulated through the heated zone in the absence of xenon or fluorine, the decrease in pressure was noted only at high temperatures where such a decrease could be accounted for by reaction with nickel. It seems clear that the reactions,

$$Xe + F_2 \rightarrow XeF_2 \qquad (1)$$

and

$$XeF_2 + F_2 \rightarrow XeF_4\,, \qquad (2)$$

proceed with the equilibrium strongly to the right and with the production of the lower xenon fluorides determined essentially by the reaction rates. We have, as yet, made no effort to modify these rates with catalysts.

The loop has also been used to examine XeF_4 preparations for purity. Occasionally XeF_2 has been found. When the concentration of XeF_2 has been low, it was worthwhile to convert the XeF_2 to XeF_4 by reaction with F_2 in the loop. In one of the early preparations of XeF_4 an absorption at 610 cm.$^{-1}$ was noted, and the material responsible was present in sufficient amount to purify it. A mass scan seemed to indicate that this material was $XeOF_4$. Attempts were made to prepare the material by heating a mixture of Xe, F_2, and O_2 in a 1:2:10 ratio in a nickel reaction vessel without much success until the temperature was raised to nearly 600° C., when a small yield was obtained. Before much had been done with this, XeO_3—then unidentified—was inadvertently prepared by the reaction of this material with moist air. Attention was then turned to the identification of XeO_3, using the infrared spectrum to characterize the starting material. Later, when XeF_6 was prepared, using a method described by Malm, this was found to have a spectrum identical with the starting material for the XeO_3 preparation. Other preparations made with considerable care to eliminate sources of oxygen gave the same product. Vapor-density determinations of preparations containing XeF_4 as an impurity showed it to have a molecular weight of more than 238 without correction for the XeF_4 (mol. wt. $XeOF_4$ = 223; mol. wt. XeF_6 = 245).

The XeF_6 initially prepared had apparently reacted with the glass in the mass-spectrometer manifold to produce $XeOF_4$ so that no XeF_6 reached the mass-spectrometer source. This reaction of XeF_6 with glass or quartz to produce $XeOF_4$ had been described to the writer by Malm. This reaction of XeF_6 with quartz produces a liquid which can be purified. The vapors yielded a mass spectrum that could be attributed to $XeOF_4$; whereas with the XeF_6 vapor, ion peaks of XeF_5^+ could now be recorded.

The infrared spectrum of XeF_6 has its strongest band, which is abnormally broad, at 610 cm.$^{-1}$ and a weaker broad band at 520 cm.$^{-1}$. $XeOF_4$ has, as its strongest band, a band of normal breadth at 608 cm.$^{-1}$, so the strongest bands of XeF_6 and $XeOF_4$ overlap. $XeOF_4$ has negligible absorption at 520 cm.$^{-1}$ and a characteristic band with a strong *Q*-branch at 927 cm.$^{-1}$.

While the reaction of XeF_6 with quartz produced a small but sufficient quantity of $XeOF_4$ for identification, the quartz became almost inert to XeF_6 as the reaction proceeded, and the rate of production of $XeOF_4$ slowed down. With CaO, $XeOF_4$ was at first produced, but it reacted with the CaO. A vapor-phase hydrolysis was then used for preparation. XeF_6 at nearly saturated vapor pressure (about 20 mm. Hg) was admitted to the loop; then air that had been bubbled through water was slowly admitted to the loop. The 520 cm.$^{-1}$ absorption of XeF_6 decreased in intensity as the XeF_6 reacted with the water, and the addition of wet air was discontinued when the 520 cm.$^{-1}$ band had been nearly eliminated. Reasonably high yields of small amounts of $XeOF_4$ were obtained with this hydrolysis reaction. $XeOF_4$ is a clear, colorless liquid, freezing at $-41°$ C.

The loop is easily used to measure the temperature variation of the saturated vapor pressure on a relative pressure scale. If the saturated vapor pressure of the pure compound is measured at a single temperature on an absolute scale, the relative pressure scale can be converted to an absolute scale. The results obtained for XeF_2 and XeF_4 are in satisfactory agreement with those of Jortner, Wilson, and Rice [2]. The results obtained for $XeOF_4$ and XeF_6 are given in Figure 6. The heat of sublimation of XeF_6 is about the same as the heat of

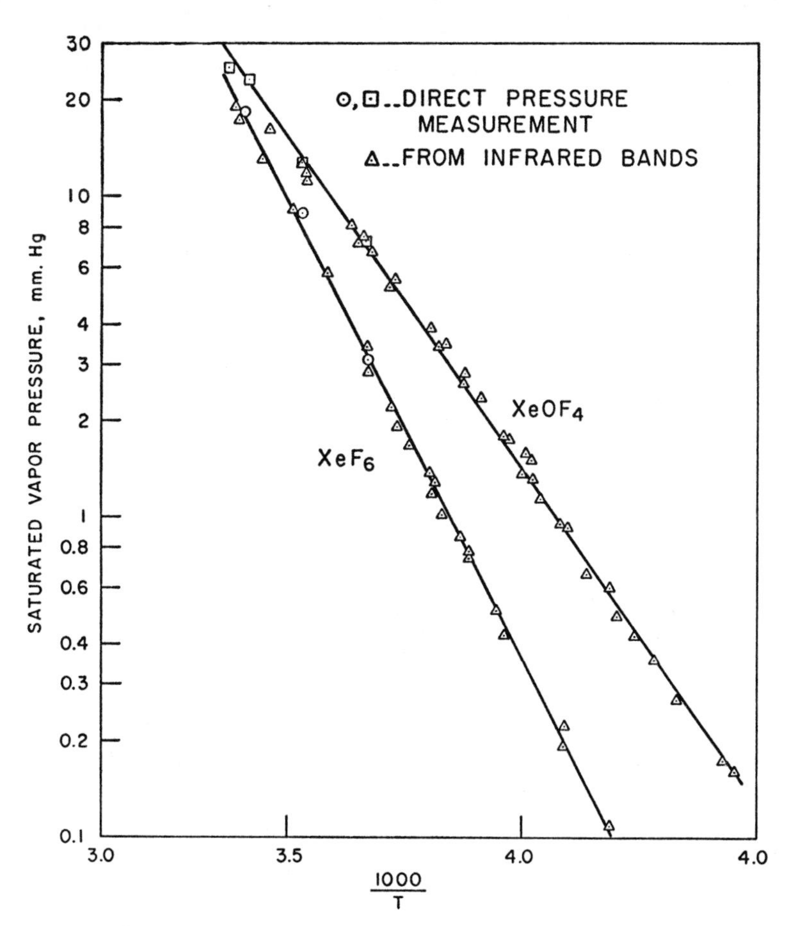

FIG. 6.—Infrared absorption spectrum of $XeOF_4$ and XeF_6

sublimation of XeF_2 and XeF_4, so the xenon fluorides are not readily separable. $XeOF_4$ has a smaller heat of vaporization and can be separated from the xenon fluorides by vaporization at lower temperatures.

REFERENCES

1. F. D. Rosen, *Rev. Sci. Inst.* **24,** 1061 (1953).
2. J. Jortner, E. Wilson, and S. A. Rice, *J. Am. Chem. Soc.* **85,** 814 (1963).

MASS-SPECTROMETRIC STUDIES OF NOBLE-GAS COMPOUNDS

MARTIN H. STUDIER AND ERIC N. SLOTH
Argonne National Laboratory

Observations on the vapor phase of xenon compounds in a modified Bendix Time-of-Flight Mass Spectrometer [1, 2] have demonstrated the existence of independent xenon oxyfluorides and have verified the existence of several stable xenon fluoride species (Table 1).

Numerous samples prepared by others [3–7 and Chernick *et al.*, p. 35, this volume] were examined for xenon compounds and impurities. Each sample was cooled in liquid nitrogen and evacuated with an auxiliary vacuum system. Small portions of gaseous species were introduced into the spectrometer as the sample was warmed, and the vapors were recondensed in a cooled U-tube. This technique was used to identify independent species from mixtures as various fractions of the samples were sublimed.

Ions produced by electron bombardment of sample vapors were identified by their masses and the characteristic xenon isotopic abundance pattern (Plate I). Since fluorine is monoisotopic, this pattern is preserved in fluorides of xenon. The isotopes of xenon and mercury as well as hydrocarbon peaks (which occur at every mass from impurities in the spectrometer) were used to determine precise masses. The observed fragmentation patterns of the xenon fluorides were typical of those of other fluorides. Doubly charged species were also observed.

Plates II, III, and IV are the mass spectra obtained from samples of xenon difluoride, xenon tetrafluoride, and xenon hexafluoride. As in the case of uranium hexafluoride, the ion of the parent molecule, XeF_6, is difficult to observe. However, in Plate V, where the peaks of the fragments are so large that their images are distorted, the ion XeF_6^+ is barely visible.

The formula of the oxyfluoride $XeOF_4$ was deduced from its mass and the complexity of the fragmentation pattern caused by loss of either oxygen or fluorine atoms (Plates VI, VII, and VIII). The compound $XeOF_3$ from a xenon tetrafluoride sample was first thought to be an independent species [4, 6] because of some apparent change in the $XeOF_3^+/XeOF_4^+$ ratio and the absence of a detectable XeF_4^+ mass spectrum, even though XeF_3^+, XeF_2^+, and XeF^+ as well as $XeOF_2^+$, $XeOF^+$, and XeO^+ were observed. More careful studies on samples of $XeOF_4$ prepared from xenon hexafluoride [see Chernick *et al.*, p. 106, this volume] show about 2 per cent XeF_4^+ present and a constant $XeOF_3^+/XeOF_4^+$ ratio of about 100:1. In addition, negative-ion spectra of the same sample (Plate IX) show XeF_4^- to be of greater intensity than $XeOF_3^-$, indicating that a parent with four fluorine atoms exists. It is probable, therefore, that in samples of $XeOF_4$ prepared from XeF_6 little or no independent $XeOF_3$ exists.

Plate X illustrates the presence of XeO_2F_2 and its fragmentation products. This compound was prepared by distilling xenon hexafluoride into a nickel weighing-can containing water. Subsequent evacuation, heating, and distillation of the products of the reaction into the spectrometer revealed XeO_2F_2 to have an independent existence. It appeared and persisted after warming above room

TABLE 1

REPORTS OF PREPARATION OF INDEPENDENT GASEOUS XENON COMPOUNDS

Species	Reference
XeF_2	[3–6]
XeF_4	[3], [4], [6]
XeF_6	[6], [7]
$XeOF_4$	[4], [6]
XeO_2F_2	This work

TABLE 2

OBSERVED DINUCLEAR SPECIES FROM SAMPLES OF XeF_2 AND XeF_4

XeF_2	XeF_4
$Xe_2F_3^+$	$Xe_2F_7^+$
$Xe_2F_2^+$	$Xe_2F_5^+$
Xe_2F^+	$Xe_2F_3^+$
Xe_2^+	

temperature. No spectrum attributable to xenon trioxide was observed, although copious evolution of xenon and oxygen suggested the presence of an oxide. Also, an explosion which ruptured the nickel can implied its presence [8–10]. These observations suggest as a possible mode of formation the following reactions:

$$XeF_6 + H_2O \rightarrow XeOF_4 + 2HF\,, \tag{1}$$

$$XeOF_4 + H_2O \rightarrow XeO_2F_2 + 2HF\,, \tag{2}$$

$$XeO_2F_2 + H_2O \rightarrow XeO_3 + 2HF\,. \tag{3}$$

A number of charged species containing two atoms of xenon were observed at pressures as low as 5×10^{-6} mm. Hg (Table 2). Variations of intensity with pressure demonstrated that these are formed in the spectrometer by ion-molecule association, but the independent existence of one or more of these was not ruled out by the experiment. At least four additional dinuclear species were ob-

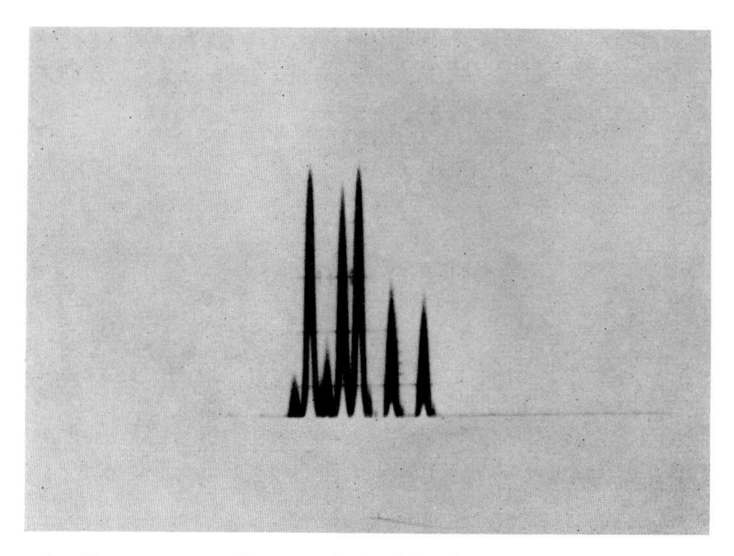

Plate I.—Mass spectrum of xenon, obtained by photographing the mass pattern displayed on an oscilloscope. (From *J. Phys. Chem.*)

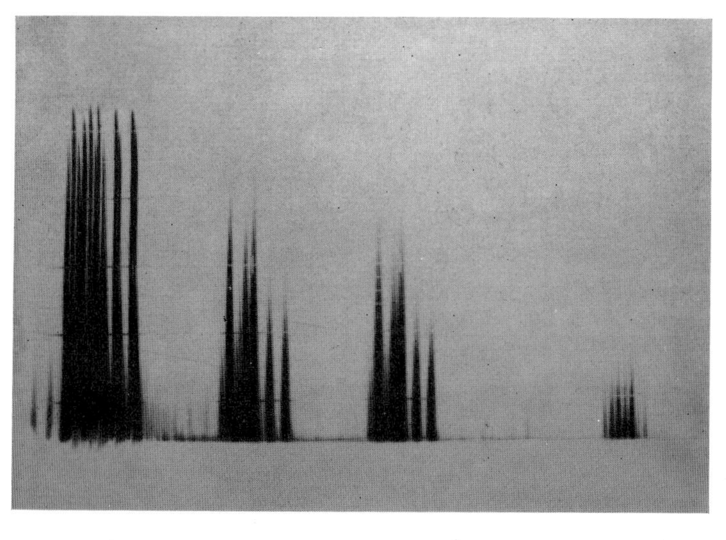

Plate II.—Mass spectrum of XeF_2

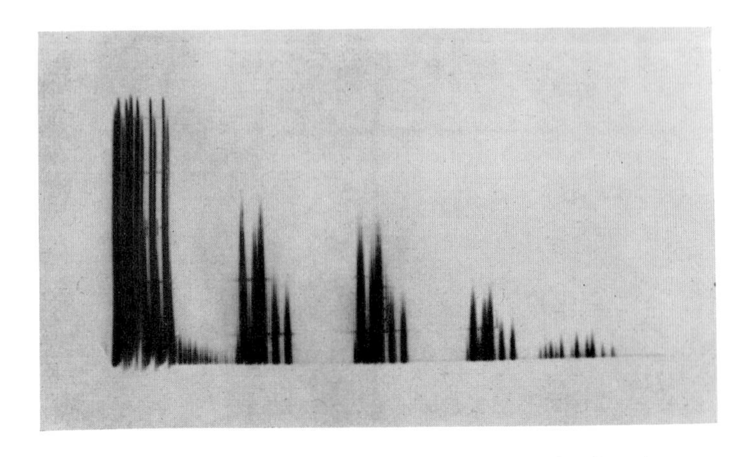

PLATE III.—Mass spectrum of XeF_4 (From *J. Phys. Chem.*)

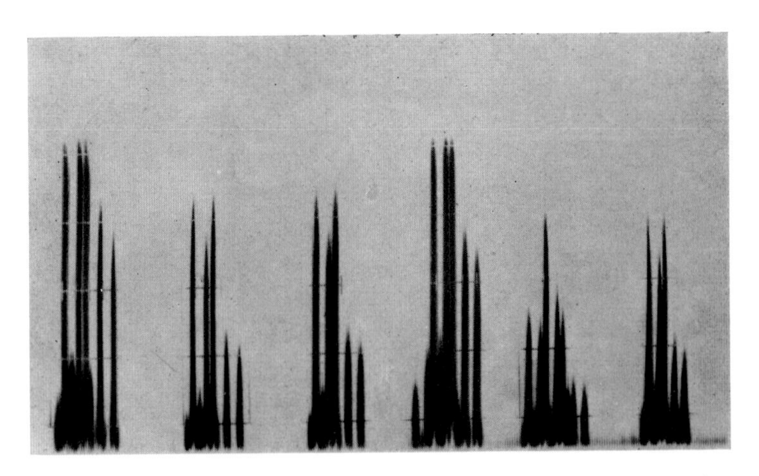

PLATE IV.—Fragmentation pattern from XeF_6

XeF_6^+

PLATE V.—Mass spectrum of XeF_6

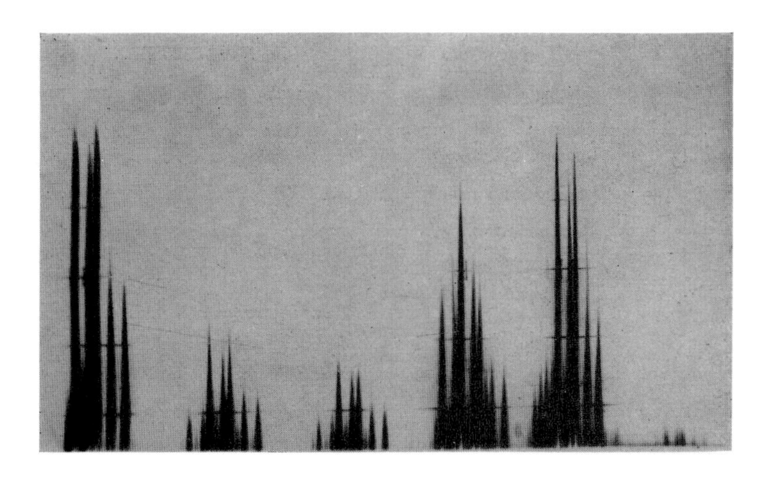

$XeOF_4$

PLATE VI.—Mass spectrum of $XeOF_4$

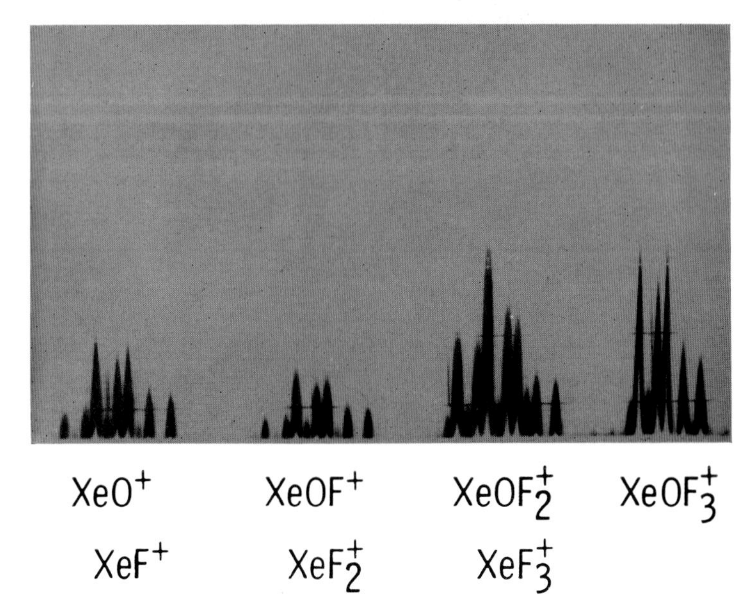

PLATE VII.—Fragmentation pattern from $XeOF_4$

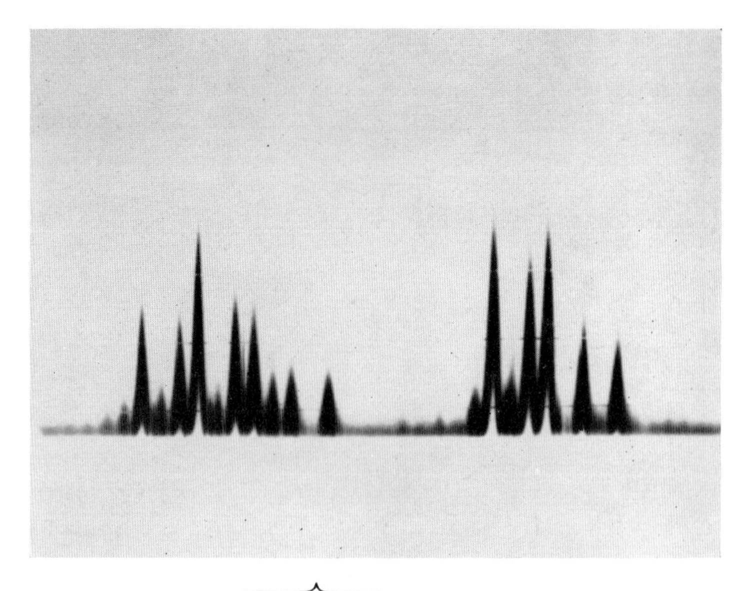

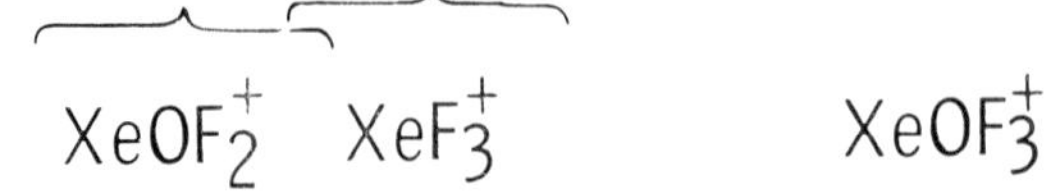

PLATE VIII.—Fragmentation pattern from $XeOF_4$ at XeF_3 position

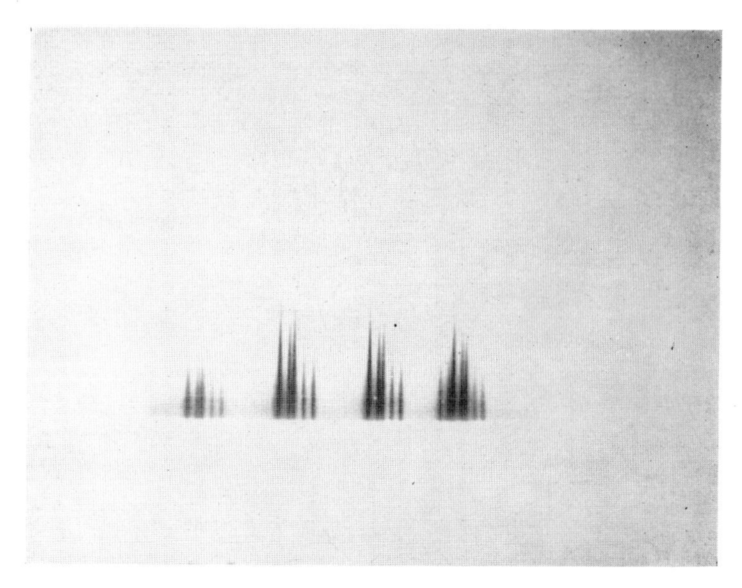

PLATE IX.—Fragmentation pattern from $XeOF_4$; negative ions

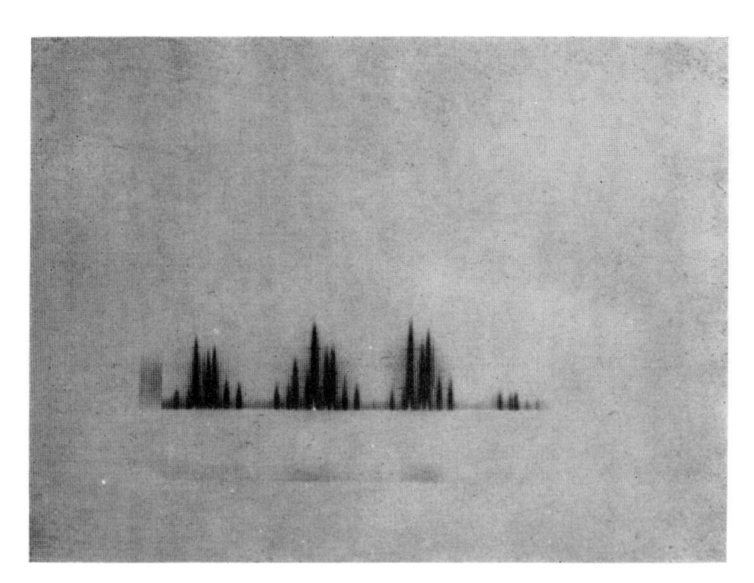

PLATE X.—Mass spectrum of XeO_2F_2

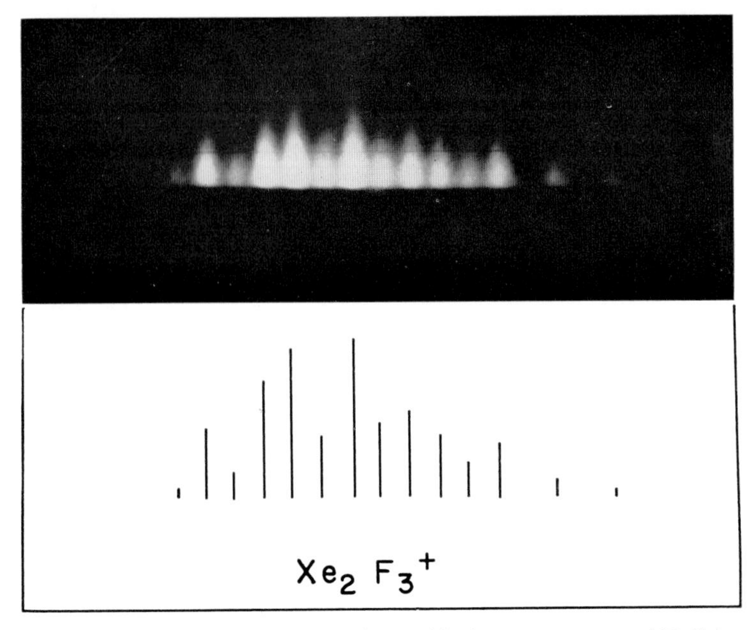

PLATE XI.—Calculated and observed intensities in mass spectrum of $Xe_2F_3^+$

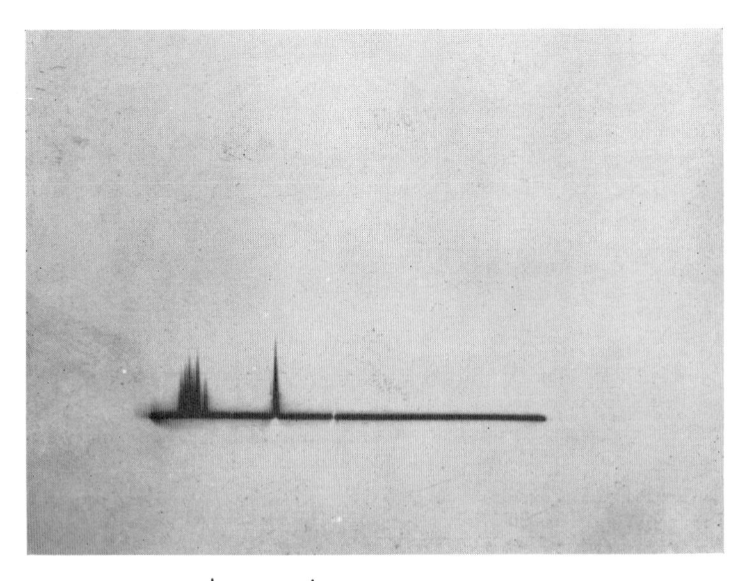

PLATE XII.—Mass spectrum of radon (222)

served in samples containing $XeOF_4$ and XeF_6, but the exact formulas for these have not been determined. Plate XI shows a spectrum of $Xe_2F_3^+$ compared with the calculated isotopic intensities for a molecule with two xenon atoms.

In preparation for the mass-spectrometric analysis of radon compounds [4, 11], 1 mc. of Rn(222) was analyzed in the mass spectrometer (Plate XII) and found to give a mass peak roughly 10,000 times the limit of detection of the instrument.

Acknowledgment.—The authors thank Leon P. Moore for technical assistance.

Comment.—

In our paper on XeF_6 we reported our inability to observe the XeF_6^+ ion in the mass spectrum of XeF_6. We likened the behavior of XeF_6 to that reported for WF_6. It is gratifying to us to hear that the Argonne study of the mass spectrum of XeF_6 is in agreement with our experience and that the XeF_6^+ ion is found only as a very minor species in the XeF_6 mass spectrum.—*B. Weinstock.*

REFERENCES

1. D. B. Harrington, *Encyclopedia of Spectroscopy*, p. 628. New York: Reinhold Publ. Corp., 1960.
2. M. H. Studier, Eleventh Annual Conference of the A.S.T.M. Committee E-14 on Mass Spectrometry, May, 1963.
3. H. H. Claassen, H. Selig, and J. G. Malm, *J. Am. Chem. Soc.* **84**, 3593 (1962).
4. C. L. Chernick *et al.*, *Science* **138**, 136 (1962).
5. J. L. Weeks, C. L. Chernick, and M. S. Matheson, *J. Am. Chem. Soc.* **84**, 4612 (1962).
6. M. H. Studier and E. N. Sloth, *J. Phys. Chem.* **67**, 925 (1963).
7. J. G. Malm, I. Sheft, and C. L. Chernick, *J. Am. Chem. Soc.* **85**, 110 (1963).
8. D. F. Smith, *ibid.* 816 (1963).
9. D. H. Templeton, A. Zalkin, J. D. Forrester, and S. M. Williamson, *ibid.* 817 (1963).
10. N. Bartlett and R. R. Rao, *Science* **139**, 506 (1963).
11. P. R. Fields, L. Stein, and M. Zirin, *J. Am. Chem. Soc.* **84**, 4164 (1962).

THE XENON-FLUORINE SYSTEM

BERNARD WEINSTOCK, E. EUGENE WEAVER, AND CHARLES P. KNOP
Ford Motor Company

The initial experiments on the fluorination of xenon led to the discovery of XeF_4 [1] and XeF_2 [2, 3]. No indication of possible contamination of these compounds by other binary fluorides was reported. We therefore concluded that if XeF_6 existed, it would be an unstable compound and require a non-equilibrium process for its synthesis. Consequently, in the method of preparation for XeF_6 that we used, xenon and fluorine were reacted at a high temperature and the products formed were rapidly cooled to a lower temperature [4]. In our initial preparations the ratio of fluorine to xenon was varied from 6 to 40 and their combined starting pressure was 1000 p.s.i. Reaction temperatures in the range, 350°–450° C., were provided by an electrically heated nickel-gauze filament and the reactor wall kept either at $-115°$ C. or at 80° C. to provide the quenching. XeF_6 was produced in all of the experiments as well as XeF_2, XeF_4, and another xenon fluoride that was not immediately characterized. The presence of these several compounds was qualitatively monitored by infrared spectroscopy. XeF_6 was identified with a strong absorption at 610 $cm.^{-1}$, XeF_2 with bands at 551 $cm.^{-1}$ and 567 $cm.^{-1}$ [3], and XeF_4 with bands at 582 $cm.^{-1}$ and 592 $cm.^{-1}$ [1, 2]. The spectrum of the unidentified compound showed a broad absorption system with a characteristic maximum at 520 $cm.^{-1}$. The infrared spectrum of a mixture of these compounds is given in Figure 1.

One of the important features of this report will be the identification of the compound associated with the absorption band at 520 $cm.^{-1}$. At the present time our evidence strongly suggests that this compound is XeF_5, referred to as the pentafluoride. However, it should be pointed out that while our evidence is strongly suggestive of this formula, it is still incomplete.

The use of rapid quenching for the preparation of XeF_6 was actually unnecessary, since this compound appears to be relatively stable. In fact, three other independent syntheses of XeF_6 were also reported at about the same time [5, 6, 7], in which xenon and fluorine were heated together at a high pressure of fluorine (60–500 atm.) and a variety of temperatures (227°–700° C.) without provision for quenching. None of these reports, however, mention the presence of "XeF_5," although it should have been an important constituent of their preparations.

Prior to the chemical analysis of XeF_6 reported in our earlier paper [4], the mixtures of xenon fluorides produced were purified by bulb-to-bulb fractional distillation. It was observed that XeF_2 and XeF_4 were substantially less volatile than XeF_6 and were readily separated. XeF_5 was more difficult to separate from

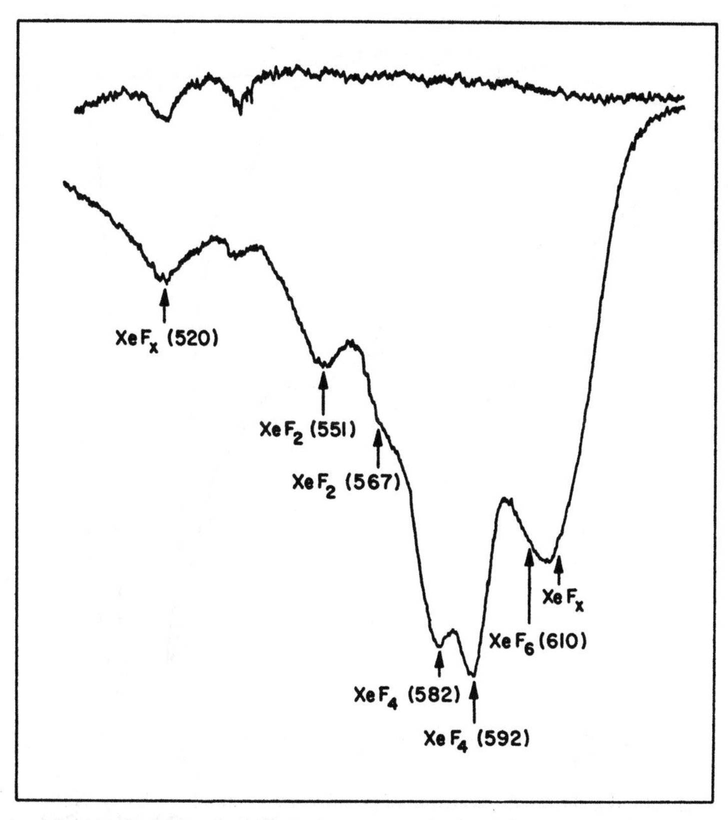

FIG. 1.—The infrared spectrum of the reaction products obtained by heating a xenon-fluorine mixture at 205° C. for 330 hours. The final fluorine pressure was 3.7 atm. The numbers in parentheses correspond to the wave number ($cm.^{-1}$) indicated by the arrows.

XeF_6 and, in fact, proved to be very difficult to distinguish from XeF_6. Mixtures that have substantial XeF_5 and XeF_6 concentrations exhibit infrared spectra that are similar to curve *A* of Figure 2. By comparison with Figure 1 it is clear that XeF_2 and XeF_4 are not present in significant concentration. After removal of XeF_5, infrared spectra similar to curve *B* of Figure 2 are obtained; these spectra are representative of relatively pure XeF_6.

The separation of XeF_5 from XeF_6 was effected in a variety of ways. In one method, a reservoir of the mixture was maintained at $-25°$ C. and the volatile head collected. This procedure was quite tedious and required four hours to produce 192 mg. of volatile material; chemical analysis of this material showed a F/Xe ratio of 5.9 [4]. In a second method of separation, a mixture of XeF_5 and XeF_6 was exposed to glass that had been flamed under high vacuum conditions. When the xenon fluoride sample was first condensed in the glass, a mixture of a yellow-green and a white solid was observed. The green color soon faded, and

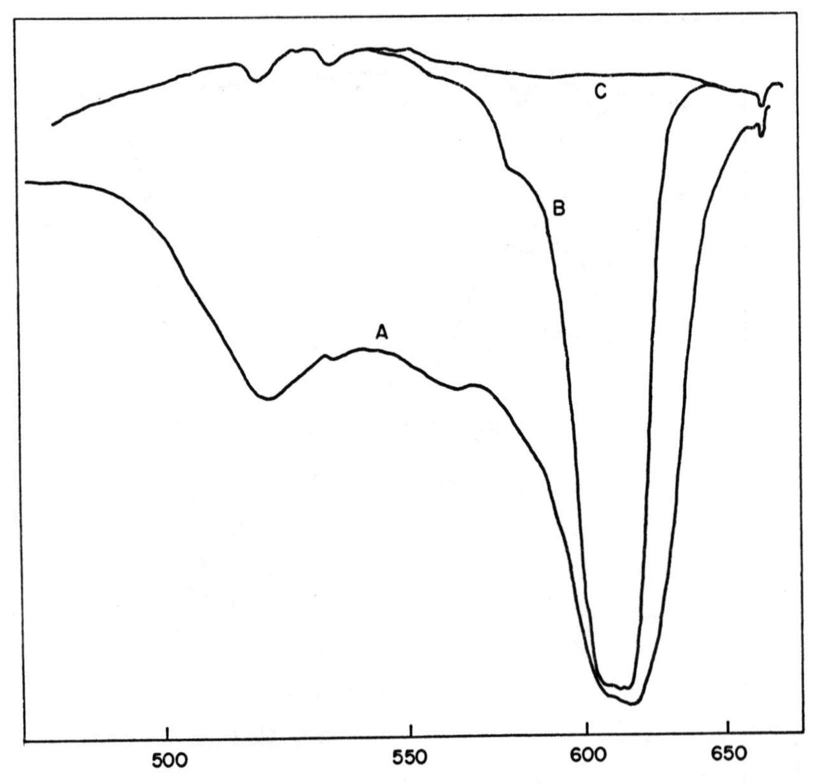

FIG. 2.—*A*—the infrared spectrum of a mixture containing chiefly XeF_5 and XeF_6. *B*—the infrared spectrum of XeF_6 after removal of XeF_5 by adsorption. *C*—background. The spectral range covered is 480–650 cm.$^{-1}$.

the remaining material was colorless. After a short exposure in the glass system, the infrared spectrum of the volatile material no longer showed the broad absorption at 520 cm.$^{-1}$. Presumably XeF_5 had been removed by reaction with the glass.

Another example of the removal of the 520 cm.$^{-1}$ band by chemical reaction is found when infrared cells (using AgCl windows) are seasoned with similar xenon fluoride mixtures. In a cell that had been previously seasoned only with fluorine the band at 520 cm.$^{-1}$ is at first observed to diminish in intensity rapidly. Eventually, when the seasoning has been completed, the intensity of this band will remain unchanged with time and be reproducible for fillings of different portions of the same sample.

The study and identification of XeF_5 has been greatly hampered by its reactivity. A particular difficulty appears to be its rapid hydrolysis by traces of moisture. Thus, great care must be exercised in the transfer of mixtures containing XeF_5 and XeF_6 to avoid exposure of even minor parts of the system to the atmosphere and subsequent erroneous assessment of the initial composition of the mixture.

The particularly pure sample of XeF_6 that gave trace *B* in Figure 2 was obtained in an unusual way. A small sample that had the usual spectrum of a XeF_5-XeF_6 mixture, curve *A*, was stored in a carefully seasoned nickel tube for three hours. Upon returning this sample to the infrared cell, curve *B* was observed. Apparently all of the XeF_5 had been removed by absorption on the nickel surface. This separation was studied in more detail with a 252-mg. sample of a XeF_5-XeF_6 mixture. The results were complex and will not be reported in detail here. A substantial fractionation of XeF_5 was observed, which confirmed that the 520 band and the 610 band did not belong to the same compound. Not all of the material could be removed by prolonged pumping under high vacuum conditions even at room temperature. One of the volatilized fractions showed a spectrum similar to *B*, but still contained a small amount of XeF_5. The final 40 mg. of sample was recovered by pumping for one hour at 150° C. The spectrum of this material was similar to that of the starting material but showed increased proportions of XeF_5, XeF_4, and XeF_2. Our tentative conclusion from this work is that XeF_5 is strongly adsorbed by nickel, Monel, and oxidized or fluorinated surfaces of these materials. Quite probably XeF_5 is polymerized on these surfaces. Removal of adsorbed XeF_5 by heating and pumping may result in disproportionation into XeF_6 and lower xenon fluorides.

In spite of the care that we took to purify XeF_6 prior to its study, we were nevertheless misled into assigning two weak absorptions at 1154 $cm.^{-1}$ and 1189 $cm.^{-1}$ to this compound [4]. These bands are associated with a xenon fluoride containing oxygen identified by others as $XeOF_4$ [this volume, pp. 290 and 302]. We failed to notice that these bands always had intensities that corresponded to a sharp and stronger infrared absorption at 928 $cm.^{-1}$ and only coincidentally appeared when higher pressures of XeF_6 were studied. To complicate matters further, $XeOF_4$ also has a sharp absorption in the neighborhood of 610 $cm.^{-1}$.

We had observed in our fractionation of XeF_5-XeF_6 mixtures that XeF_5 also absorbs near 610 $cm.^{-1}$ to increase the complexity of the problem. It then appeared necessary to demonstrate that $XeOF_4$ was not an important constituent of our preparations. We therefore reacted a Xe-F_2 mixture containing a large excess of oxygen under conditions identical with an experiment in which no oxygen had been added. There was no indication that the concentration of $XeOF_4$ had been increased by the addition of oxygen to the system. The F/Xe ratio in the product was consistent with the value obtained without the addition of oxygen. The infrared spectra still showed the 928 band as a minor absorption. This result also removed a related uncertainty about the effect of the purity of our fluorine, which contained 1–2 per cent O_2. It is a surprising fact that the above experiment did not produce $XeOF_4$ in increased amount.

Since the sum bands at 1154 cm.$^{-1}$ and 1189 cm.$^{-1}$ have now been assigned to another compound, our previous conclusion about the stability of XeF_6 based on a value of ν_1 derived from one of these bands must also be withdrawn [4]. XeF_6 does not appear to be an unusually unstable hexafluoride.

From these observations it was apparent that the Xe-F_2 system was rather complex. It therefore seemed worthwhile to study this system in a thorough manner in order to determine the relative stability of the various compounds as well as to learn how to prepare them directly in a very pure state.

Our first objective was to demonstrate the existence of XeF_5 by the classical methods of physical chemistry. Consider the equilibria:

$$Xe + F_2 = XeF_2\,, \quad (1)$$

$$Xe + 2F_2 = XeF_4\,, \quad (2)$$

$$Xe + 2.5F_2 = XeF_5\,, \quad (3)$$

and

$$Xe + 3F_2 = XeF_6\,. \quad (4)$$

These can be expressed in terms of equilibrium constants:

$$K_i = (XeF_i)/(Xe)(F_2)^{i/2} = (x_i)/(x_o)P^{i/2}\,. \quad (5)$$

In these relations x_i and x_o are the pressures of the appropriate xenon fluoride and xenon respectively, and P is the pressure of fluorine. No correction for non-ideality has been made.

The mass balance equations are:

$$n_o = x_o + x_2 + x_4 + x_5 + x_6\,, \quad (6)$$

and

$$rn_o = 2x_2 + 4x_4 + 5x_5 + 6x_6\,; \quad (7)$$

where n_o is the initial pressure of xenon and r is the ratio of total number of fluorine atoms to total number of xenon atoms in the equilibrium mixture.

Combination of equations 5, 6, and 7 yields:

$$r + (r-2)K_2P + (r-4)K_4P^2 + (r-5)K_5P^{5/2} + (r-6)K_6P^3 = 0\,. \quad (8)$$

Under high pressure conditions the last two equilibria will be determining, and the relation

$$(6-r)P^{1/2}/(r-5) = K_5/K_6 \quad (9)$$

should approach a constant value if XeF_5 is an important species.

To test this relation we undertook to evaluate r over a wide range of F_2 pressures. The initial quantity of Xe was measured by weight and of F_2 by PVT determination. The mixtures were reacted at 250° C. for periods of time varying from 65 to 184 hours. At the end of each experiment the Monel reaction vessels were cooled rapidly. The reactor was then immersed in liquid oxygen, and the unreacted fluorine determined by PVT measurement. The xenon fluoride product was purified by distillation and weighed. The fluorine content of the product was determined from the increase in weight of the xenon fluorides over the weight of the initial xenon. The value of r was calculated from these weighings. The results of the equilibrium measurements are summarized in Table 1.

Mass balances for fluorine always showed an additional loss of fluorine that was attributed to corrosion of the reactor. Chemical analysis of the F/Xe ratio in the product was made in two instances, and the analyses were not in exact agreement with the results of the weighings. However, for the purposes of this preliminary report, this discrepancy is not taken into account.

TABLE 1

THE VARIATION OF r WITH FLUORINE PRESSURE AT 250° C.

r	F_2 pressure (atm.)	$(6-r)P^{1/2}/(r-5)$ (atm.$^{1/2}$)
5.367	4.102	3.50
5.632	6.628	1.50
5.745	8.817	1.02
5.775	12.26	1.02
5.818	16.18	.89
5.842	18.69	.81
5.800	41.31	1.60
5.916	107.4	.95
5.860	166.9	2.10
5.818	171.9	2.92

Prior to chemical analysis most of the samples were vaporized completely to avoid fractionation and a small portion taken for infrared study. The XeF_4 absorption bands were a prominent component in the spectrum of the 4-atm. experiment but were appreciably reduced in intensity in the spectrum of the 12-atm. experiment. These observations are in accord with the equilibria involved; the relative concentration of XeF_4 to XeF_6 should vary inversely with the pressure of F_2.

Qualitatively, r is seen to approach the value of 6 in the limit of highest pressure. This suggests that under equilibrium conditions at 250° C., XeF_6 is the highest fluoride of importance. Quantitatively, the data are seen to be somewhat inconsistent, and, therefore, we have not made a thorough mathematical treatment of these results. An important source of difficulty in these measurements has been the unusual reactivity of XeF_6 and its adsorption on metal surfaces. Methods for overcoming these difficulties have been developed, and the studies are continuing.

The value of the quantity given by equation 9 is tabulated in the third column. In the limit of high pressure it should approach the K_5/K_6 ratio. Correction for the presence of XeF_4 (and XeF_2) in these samples has not been made. The corrections would be most important at the lowest pressures tabulated and result in the effective value of r being increased. This would in turn lead to smaller values of the K_5/K_6 ratio. The most reliable value of K_5/K_6 that one can deduce at this time from this work would be taken from the four high-pressure values. This then gives $K_5/K_6 = 1.9 \pm 0.6$ atm.$^{1/2}$.

Since XeF_5 will have an odd number of electrons, we attempted to observe its ESR spectrum to confirm this. The sample studied contained mainly XeF_5 and XeF_6. It was condensed directly into a baked-out cylindrical quartz tube, with extreme care being taken to prevent the sample from warming, and sealed off. Visual inspection of the sample at the temperature of liquid nitrogen at first gave the impression that it was colorless. However, closer inspection revealed a faint yellow color. This faint color was observed to fade rapidly upon warming even slightly. The ESR spectrum was looked for by H. C. Heller, T. Cole, and J. J. Lambe of this laboratory. Careful scans at high sensitivity over a wide range of g values were made at liquid-nitrogen and -helium temperatures. A spin-resonance signal was not observed. There are a variety of reasons why an ESR signal was not observed that need not be discussed here. This result, although negative, is therefore not decisive in determining whether or not the sample studied contained a paramagnetic species.

In view of the uncertainty surrounding the identity of XeF_5, a quantitative study of a batchwise fractional distillation of a mixture of XeF_5 and XeF_6 was undertaken. In this way we hoped to establish that the absorption at 520 cm.$^{-1}$ was definitely associated with a less volatile species and did not vary in intensity for a spurious reason because of reactivity. Furthermore, we used as our starting material a compound that had a composition corresponding to $r = 5.86$ and that therefore contained about 14 per cent XeF_5 and 86 per cent XeF_6. Additionally small amounts of XeF_4 and $XeOF_4$ were present. If XeF_5 and XeF_6 do not form mixed crystals, we should observe no change in their relative composition during the fractionation, since each would be present in proportion to its saturated vapor pressure until it had been completely volatilized. If mixed crystals are formed, then the less volatile species would be observed to concentrate as the sample volume decreased. Thus, if the less volatile species corresponded to XeF_5 and therefore to the minor constituent of the mixture, the absorption at 520 cm.$^{-1}$ would be found to increase relative to that at 610 cm.$^{-1}$. If the 520 cm.$^{-1}$ absorption was associated with the major constituent, then very little change in the relative absorbence of 520 cm.$^{-1}$ and 610 cm.$^{-1}$ would be observed.

The mixture was kept at 0° C. and successive portions consisting of 2.5 per cent of the sample were volatilized and removed. After each removal of material, the sample was warmed to room temperature in order to help the crystals maintain a homogeneous composition. We observed that the absorbence at 520 cm.$^{-1}$ increased in intensity relative to that at 610 cm.$^{-1}$ during the experiment. When 70 per cent of the material had been volatilized, the relative absorbence of the 520 cm.$^{-1}$ band had doubled. This experiment thus confirmed that the minor constituent in the mixture, XeF_5, was associated with the 520 cm.$^{-1}$ band and was the less volatile. The absorption at 928 cm.$^{-1}$ ($XeOF_4$), while small, was observed to decrease monotonically during the experiment until it was lost in the noise. This demonstrated that $XeOF_4$ is more volatile than XeF_6. The shoulder in the spectrum corresponding to XeF_4 was observed to remain constant in intensity and suddenly disappear, suggesting that XeF_4 does not form mixed crystals with these compounds. Careful examination of the 610 cm.$^{-1}$ band profile during the experiment showed that as the 520 cm.$^{-1}$ band increased in relative intensity, the maximum at 610 cm.$^{-1}$ moved toward a higher wave number. This confirmed the observation that XeF_5 absorbs significantly near 610 cm.$^{-1}$, but at a somewhat higher wave number.

A method that we have under way that may permit the preparation of the xenon fluorides in high purity and the demonstration of their important chemical species is based on the phase rule. The $Xe\text{-}F_2$ system represents a two-component system that has ($4\text{-}p$) degrees of freedom, where p is the number of phases. If we confine our considerations to a single temperature and just have a vapor phase present, there remain two degrees of freedom. In that case the amounts of xenon and fluorine can be varied independently. If the vapor volume is small and the quantity of xenon large, a condensed phase of, say, XeF_2, could be present. This would correspond to the region in Figure 3 marked XeF_2. Figure 3 is a hypothetical phase diagram for the $Xe\text{-}F_2$ system for which log P for F_2 is plotted against the reciprocal of the absolute temperature. In the region below the first line shown in the diagram, the system maintains a single degree of freedom at a particular temperature, namely, the pressure of fluorine. If we then imagine the system at a fixed temperature with a fixed amount of xenon to which we continue to add fluorine, we will proceed vertically in the diagram through the XeF_2 region until we reach the first line. This line represents the equilibrium where XeF_2 and XeF_4 both are present as condensed phases. As more fluorine is added, the relative amount of condensed XeF_4 will increase until all of the condensed XeF_2 is converted. The pressure will then rise again until the next line in the graph is reached. The successful use of this method requires that the temperature be high enough to insure a mobile equilibrium but that the various compounds be below their critical temperatures and that there is no mutual solubility of these compounds in the condensed phases.

Thus far, we have made only a preliminary test of this procedure at 250° C. In this experiment the reactor was loaded with xenon fluorides from a series of previous preparations. It was estimated that if all of the xenon compounds were in the vapor phase, their pressure would be 18 atm. Equilibrium pressures of fluorine of 170 and 40 atm. of fluorine were used. The qualitative results obtained were not very different from similar experiments at lower xenon pressures. Nothing conclusive can be decided about these preliminary results. The work is continuing.

At low fluorine pressures the major species appear to be XeF_2 and XeF_4. In a single experiment at 250° C. we have measured the equilibrium constants, K_2 for the formation of XeF_2 and K_4 for the formation of XeF_4. These results are

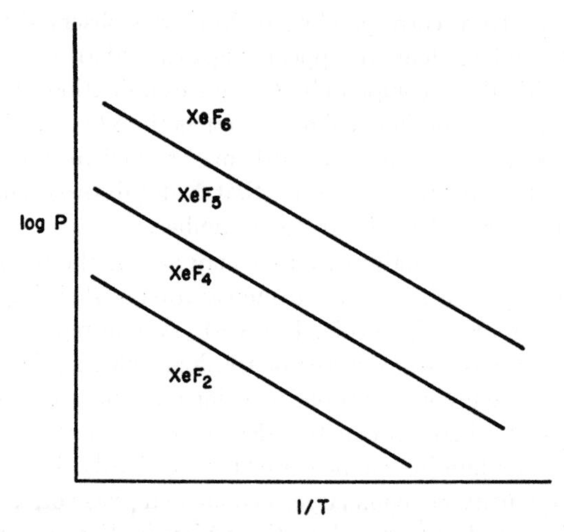

FIG. 3.—Hypothetical phase diagram for the Xe-F_2 system. The logarithm of the fluorine pressure is plotted against the reciprocal of the absolute temperature. The lines represent conditions necessary for the presence of two condensed phases. In the areas between the lines, only one condensed phase is present.

preliminary but should be good to the stated uncertainty. We find $K_2 = 200 \pm 20$ atm.$^{-1}$ and $K_4 = 639 \pm 200$ atm.$^{-2}$. The final pressure of fluorine was 21 mm. and r was 2.165 in this experiment.

For XeF_2, the free energy of formation at 250° C., with everything in the ideal gaseous state at 1 atm., is then -5510 cal. mole^{-1}. The value of $\Delta S°$ for the formation of XeF_2 at this temperature was calculated to be -26.96 cal. mole^{-1} deg.$^{-1}$. In this calculation the entropy of fluorine was taken from Cole, Farber, and Elverum [8], the value of 2.00 A. was used for the Xe-F distance [9], and for the three fundamental vibrational frequencies, $\nu_1 = 515$, $\nu_2 = 213.2$, and $\nu_3 = 555$ cm.$^{-1}$ [9].

The heat of formation of XeF_2, at 250° C. with everything in the ideal gaseous state at 1 atm., is then $-19{,}400$ cal. mole^{-1}. The value for the heat of formation at 25° C. will be about the same.

Pitzer [10] has suggested that the heat of formation of XeF_2 should fall between the values of ΔH for the reactions:

$$ClF(g) + F_2(g) = ClF_3(g)\,, \tag{10}$$

and

$$BrF(g) + F_2(g) = BrF_3(g)\,, \tag{11}$$

with the reactants in the ideal gaseous state at 1 atm. These values are $-25{,}000$ and $-48{,}000$ cal. mole^{-1} [11], respectively. The value of $-19{,}400$ cal. mole^{-1} that we have obtained is close to that for the ClF_3 dissociation but is not intermediate between the values for ClF_3 and BrF_3.

Preliminary equilibrium studies at temperatures lower than 250° C. reveal qualitatively that the XeF_5/XeF_6 ratio is higher than at 250° C. This is consistent with $K_5 > K_6$. If the heat of formation of XeF_5 is more negative than that for XeF_6 near 200° C., raising the temperature will favor the formation of XeF_6. At higher temperatures, entropy effects will reverse this behavior. Ultimately, the lower xenon fluorides will be the important species even at high pressure.

The data presented here are preliminary but should be of interest to others working in this field. If the new xenon fluoride is XeF_5, as our present evidence suggests, this will be an unusual result.* An odd xenon fluoride has been excluded by Allen [12] and Allen and Horrocks [13] on theoretical grounds. However, the same arguments were used by these authors to preclude XeF_6. Since a number of independent studies are in agreement on the existence of XeF_6, these theoretical arguments ruling out the possibility of an odd xenon fluoride need not be taken as conclusive.

Another result of this work is that it suggests that some studies with binary xenon fluorides may have been made with impure materials. Thus, none of the other research with XeF_6 [5, 6, 7] mentions the presence of XeF_5 as a major impurity. There is, therefore, some danger that the properties of an XeF_5-XeF_6 mixture will be mistaken for that of pure XeF_6. The direct synthesis of the lower fluorides will also, in general, lead to the production of mixtures.

Acknowledgments.—We gratefully acknowledge assistance from John L. Parsons in the infrared spectroscopy and from Jack S. Ninomiya in the chemical operations.

* Since this work was reported, we have refluorinated a xenon fluoride mixture in a vessel of small vapor volume and at a high pressure of fluorine and have obtained a sample of relatively pure XeF_6. This sample was further purified by fractional distillation in a closed system and monitored by infrared spectroscopy. By far the largest fraction of the material showed a spectrum similar to curve *A*, Figure 2. This material was subsequently analyzed chemically and gave the formula $XeF_{5.96}$; the vapor density was measured and gave a molecular weight of 245.5 compared to the formula weight for XeF_6 of 245.3. These results invalidate some of the qualitative evidence for the existence of XeF_5 that we have reported. The equilibrium studies of the Xe-F system continue to give inconclusive and inconsistent results.

REFERENCES

1. H. H. Claassen, H. Selig, and J. G. Malm, *J. Am. Chem. Soc.* **84**, 3593 (1962).
2. C. L. Chernick *et al.*, *Science* **138**, 136 (1962).
3. D. F. Smith, *J. Chem. Phys.* **38**, 270 (1963).
4. E. E. Weaver, B. Weinstock, and C. P. Knop, *J. Am. Chem. Soc.* **85**, 111 (1963).
5. J. Slivnik *et al.*, *Croat. Chem. Acta* **34**, 253 (1962).
6. J. G. Malm, I. Sheft, C. L. Chernick, *J. Am. Chem. Soc.* **85**, 110 (1963).
7. F. B. Dudley, G. Gard, and G. H. Cady, *J. Inorg. Chem.* **2**, 228 (1963).
8. L. G. Cole, M. Farber, and G. W. Elverum, Jr., *J. Chem. Phys.* **20**, 586 (1952).
9. P. A. Agron *et al.*, *Science* **139**, 842 (1963).
10. K. S. Pitzer, *ibid.* 414 (1963).
11. E. H. Wiebenga, E. E. Havinga, K. H. Boswijk, *Advan. Inorg. Chem. Radiochem.* **3**, 133 (1962).
12. L. C. Allen, *Science* **138**, 892 (1962).
13. L. C. Allen and W. De W. Horrocks, *J. Am. Chem. Soc.* **84**, 4344 (1963).

XENON HEXAFLUORIDE

Francis B. Dudley,* Gary L. Gard, and George H. Cady
University of Washington

The recent synthesis of xenon tetrafluoride [1, 2] by heating xenon-fluorine mixtures at 400° in a sealed nickel tube suggested that pressure-temperature measurements might indicate the minimum temperature at which combination occurs. It also was considered that higher fluorides might be capable of existence at temperatures below 400°. Xenon-fluorine mixtures were, therefore, introduced into a 1500 cc. prefluorinated nickel vessel and the pressure-temperature dependence followed from room temperature up to 720° K. The mixtures followed ideal gas behavior up to 390° K., when a pressure decrease began to occur.

The pressure decreased rather rapidly in the temperature range 480° to 520° K. until the total decrease was about 2.5 times the pressure that would have been due to xenon alone. The temperature then was raised gradually; above 560° K. the rate of increase in pressure with temperature was abnormally large, suggesting the dissociation of a molecular species. At about 710° K. the pressure *vs.* temperature curve merged into that expected for a mixture of F_2 with XeF_4. At 400° most of the xenon apparently would have been present as XeF_4.

As pressure would be expected to favor the formation of higher fluorides of xenon, three fluorine-xenon mixtures with mole-ratios varying from 4:1 to 20:1 were sealed in prefluorinated nickel tubes fitted with Hoke needle valves and brass 10/30 connections. These were heated at 500° K. for from 1 to 10 days. Estimated pressures for runs I, II, and III were, respectively, 500, 150, and 25 atmospheres. They were then quenched in cold water, and the excess fluorine was removed by pumping at liquid nitrogen temperature. The tube was warmed to room temperature and later pumped while at −78° C. The gain in weight of the nickel reaction tubes permitted a calculation of the empirical formula of the xenon fluoride formed under the reaction conditions.

Samples from the reaction tubes were condensed in glass bulbs under vacuum transfer conditions, forming a white solid. After these bulbs had warmed to

* Permanent address: University of New England, Australia.

This chapter is essentially the same as the authors' earlier communication (*Inorg. Chem.* 2, 228 [1963]).

The research reported in this chapter was supported in part by the Office of Naval Research.

room temperature, the gas phase had a pale greenish-yellow color which faded away in about five minutes. The remaining solid slowly became a liquid.

A sample was allowed to react either with water or potassium iodide solution by adding the air-free liquid to a bulb. After reaction with water, the amount of xenon which could be removed by pumping was small, but the solution had powerful oxidizing properties—much more so than hydrogen peroxide solutions of comparable strength. The solution failed to give the typical titanium–hydrogen peroxide test. Such solutions were analyzed for fluoride and hydrogen ions and for oxidizing equivalents. On the other hand, the reactions with potassium iodide liberated iodine. Gas was evolved, and xenon was recovered nearly quantitatively. Some oxygen also was released in this reaction. After titration of free iodine, potassium iodate was added, but no further iodine was liberated, indicating the absence of free acid. The inability to recover xenon by pumping after hydrolysis of samples suggested that the xenon was bound in solution, possibly as a xenic acid, $Xe(OH)_6$. This behavior is similar to that reported for xenon tetrafluoride [1, 2]. Such an acid, if it existed, must have been very weak, because it was not neutralized when titrating the solution with 0.1 *N* sodium hydroxide to the phenolphthalein end-point. A graph of pH *vs.* volume of added base failed to indicate a wave corresponding to the neutralization of an acid in the pH region 8 to 11.

In the fluoride analyses the sodium alizarin sulfonate indicator was decolorized while the solution contained oxidizing power. It was, therefore, necessary to add up to four times the normal amount of indicator solution before carrying out the titrations.

The observed behavior with water and aqueous potassium iodide solutions is consistent with the equations:

$$XeF_6 + 6H_2O \rightarrow 6HF + [Xe(OH)_6]? ,$$

$$[Xe(OH)_6]? + 6HF + 6KI \rightarrow Xe + 6H_2O + 3I_2 + 6KF .$$

With water, side reactions yielding a small amount of oxygen were observed.

The ratio of xenic acid to oxygen produced was variable as indicated by the various numbers of oxidizing equivalents in solution formed per mole of XeF_6 hydrolyzed.

a) Composition of higher xenon fluoride by synthesis. (Theoretical per cent F for XeF_6 = 46.5.)

Wt. of xenon taken, gm.	0.409	0.455
Wt. of fluoride formed, gm.	0.758	0.826
% fluorine by weight increase	46.05	45.0

b) Xenon recovered after reaction with KI solution. (Theoretical for Xe in XeF_6 = 53.5 per cent.)

Wt. of sample, gm.	0.0607	0.2060
Wt. Xe recovered, gm.	0.0327	0.1093
% Xe	53.8	53.1

c) Analysis of higher xenon fluoride by determination of F^- (by thorium nitrate titration) and of H^+; determination of oxidizing equivalents remaining after hydrolysis.

Batch of xenon fluoride	Wt. of sample, gm.	% fluorine based on		Equiv. ox. agent in soln. per mole of XeF_6 used
		F^- detn.	H^+ detn.	
I........	0.1260	44.5	47.0	5.25
	.0392	46.9	47.2	5.31
	.0525	45.4	46.3	5.78
II........	.138	45.9	45.1	5.45
	.197	43.4	46.1	5.72
III........	0.1023	45.6	47.2	5.54

REFERENCES

1. C. L. Chernick *et al.*, *Science* **138,** 136 (1962).
2. H. H. Claassen, H. Selig, and J. G. Malm, *J. Am. Chem. Soc.* **84,** 3593 (1962).

ON THE SYNTHESIS OF HIGHER XENON FLUORIDES

J. Slivnik, B. Volavšek, J. Marsel, V. Vrščaj,
A. Šmalc, B. Frlec, and A. Zemljič
Nuclear Institute Jožef Stefan

A splendid accomplishment of the Argonne National Laboratory investigators [1] in the synthesis of XeF_4 initiated its preparation in our laboratory (Plate I). We used somewhat different conditions. An initial mixture of one volume of xenon at 1 atm. at 20° C. and three volumes of fluorine was heated up to 400° C. The analysis of the product in a mass spectrometer showed traces of higher fluorides of xenon up to XeF_8 [2]. This observation resulted in the preparation of XeF_6, as already reported [3]. The comparison of our results was in a good agreement with the results of three other independent syntheses [4, 5, 6]. However, certain differences were noted: (*a*) Our XeF_6 melted at a temperature some 20° C. lower. (*b*) The Xe/F ratios of our products were as high as 6.4 in some cases. Since we carried out the reaction under higher pressure and at higher temperature than the other investigators, the possibility of the formation of higher fluorides was suggested.

Accordingly, the synthesis was repeated [3]. A gaseous mixture, 16 volumes of fluorine gas and one of xenon was heated to 620° C., where the pressure was estimated to be 200 atm. The reaction vessel was slowly cooled down to $-78°$ C., and the excess fluorine was pumped off through the liquid-nitrogen trap. A yellow product condensed in the trap. The vapor pressure of this product was considerably lower than that of fluorine, since the product was retained in the trap even after prolonged pumping with the mercury diffusion pump.

The product is not stable in glass. At room temperature the yellow color faded in a few minutes.

The product was hydrolyzed in a weak NaOH solution. The solution turned yellow; this color was stable for some hours. The fluorine content was determined and the rest was assumed to be xenon. The F/Xe ratio was 7.9 ± 0.3. An attempted analysis in a mass spectrometer was not successful. A very strong xenon spectrum was observed; however, the presence of so much unreacted xenon seems highly improbable in view of the conditions of synthesis.

To investigate the synthesis more closely, continuous pressure measurements of the reaction mixture were made during the heating period.

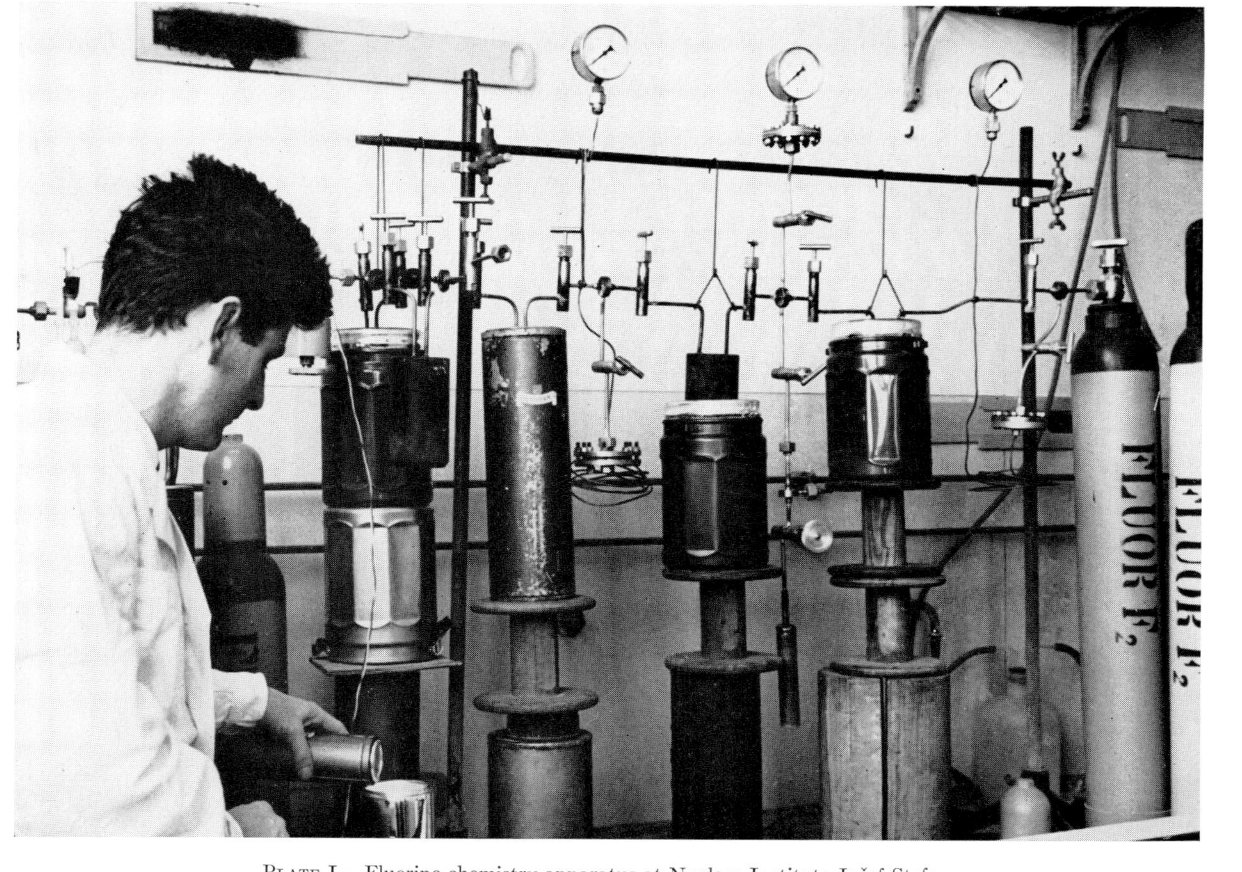

PLATE I.—Fluorine chemistry apparatus at Nuclear Institute Jožef Stefan

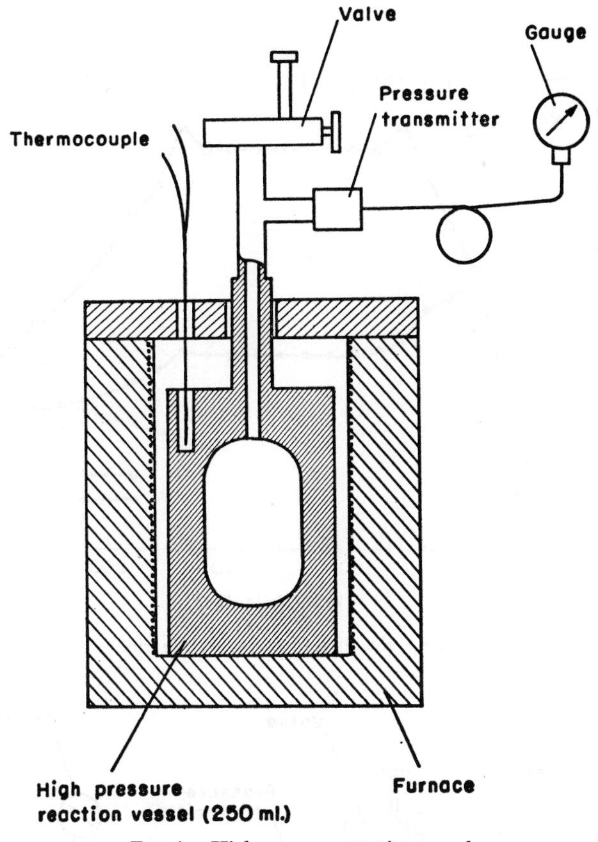

FIG. 1.—High-pressure reaction vessel

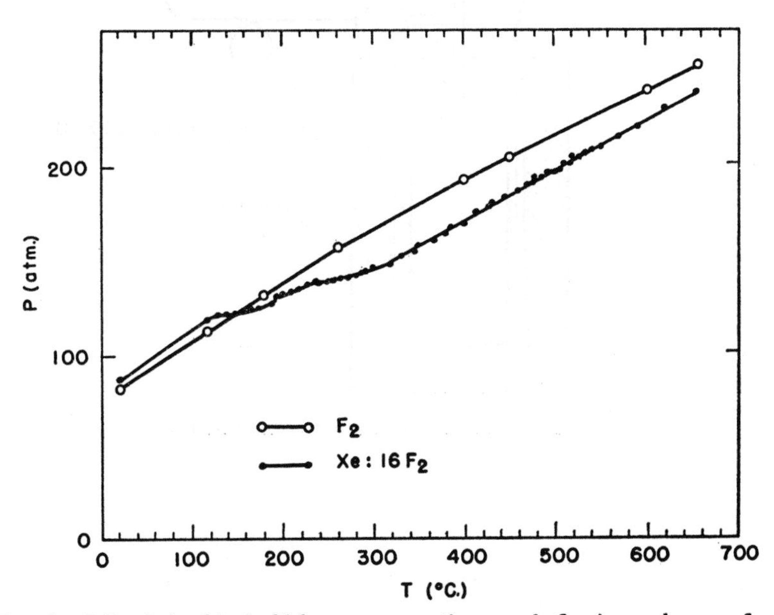

FIG. 2.—P-T relationship in high-pressure reaction vessel, fluorine and xenon-fluorine mixtures.

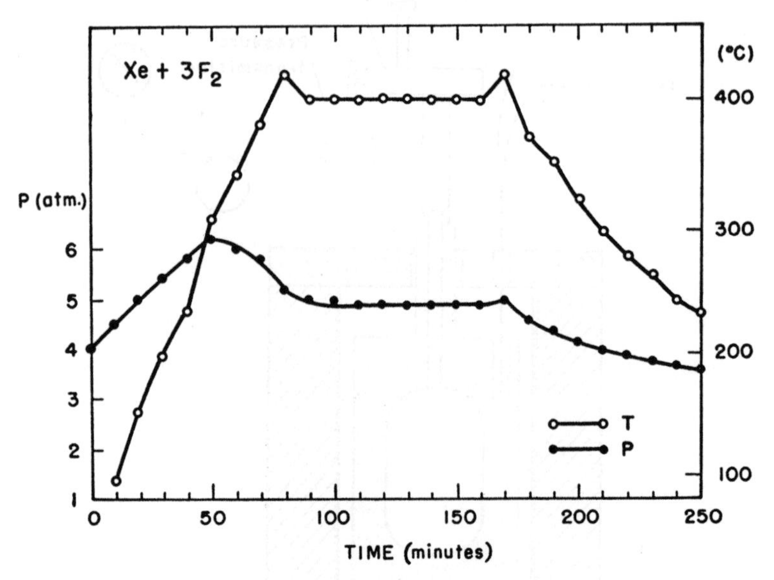

FIG. 3.—Reaction of xenon and fluorine

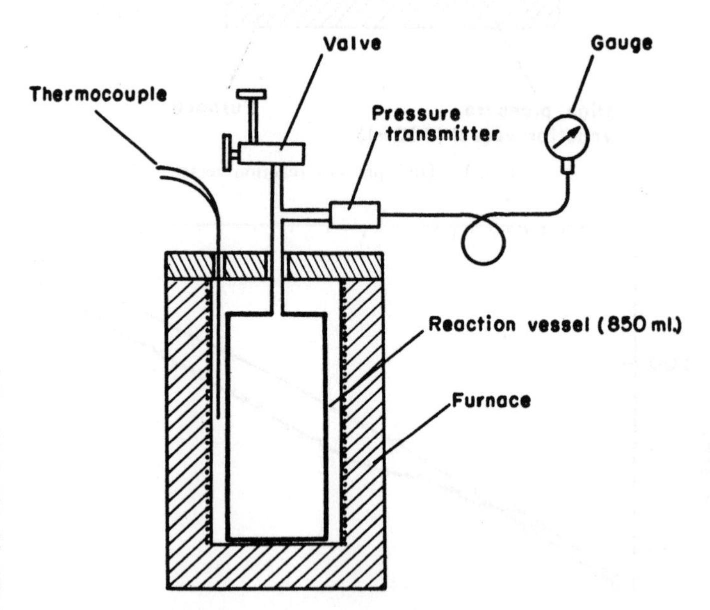

FIG. 4.—Moderate-pressure reaction vessel

As a first step, 0.84 moles of fluorine gas were heated in a 254 ml. reaction vessel and the pressure measured (Figs. 1 and 2). Upon cooling, 0.052 moles of xenon were added and the procedure was repeated. The curve shows two distinct reaction steps. In addition, at 325° C. more fluorine was consumed than would be required for the synthesis of XeF_6. The simple calculations were made from the observed data, as follows: initial xenon pressure, 5.0 atms.; initial fluorine pressure, 81.0 atms.; pressure of fluorine alone at 320° C., 171.5 atms.; pressure of the fluorine-xenon system at 320° C., 149.0 atms.; decrease of fluorine pressure, 32.6 atms.—i.e., the decrease of fluorine pressure was 3.23 times the pressure of xenon, corresponding to the product composition of "$XeF_{6.5}$" or to a mixture of 75 per cent XeF_6 and 25 per cent XeF_8.

Figure 3 shows the results of a study of the reaction of xenon and fluorine under the less strenuous conditions employed in our early experiments. A vessel 850 ml. in volume (Fig. 4) was employed and the initial xenon pressure was 1 atm., and fluorine pressure 3 atm. The observed temperature and pressure measurements suggested that the synthesis of XeF_4 was completed in about an hour as the temperature reached 390° C.

REFERENCES

1. H. H. Claassen, H. Selig, and J. G. Malm, *J. Am. Chem. Soc.* **84,** 3593 (1962).
2. J. Slivnik *et al.*, *Croat. Chem. Acta* **34,** 187 (1962); J. Marsel and V. Vrščaj, *ibid.* 191 (1962).
3. J. Slivnik *et al.*, *ibid.* 253 (1962).
4. J. G. Malm, I. Sheft, and C. L. Chernick, *J. Am. Chem. Soc.* **85,** 110 (1963).
5. E. E. Weaver, B. Weinstock, and C. F. Knop, *ibid.* 111 (1963).
6. F. B. Dudley, G. Gard, and G. H. Cady, *Inorg. Chem.* **2,** 228 (1963).

FLUORINE EXCHANGE BETWEEN XENON HEXAFLUORIDE AND GASEOUS FLUORINE

IRVING SHEFT AND H. H. HYMAN
Argonne National Laboratory

A study of the rate of exchange of F^{18} between fluorine gas and metal hexafluorides has provided information useful in characterizing the binding power of the central atom toward fluorine and systematizing the chemical behavior of this group of compounds [1]. The nature of the intermediate complex and a measure of the bond energies involved in its formation can be deduced from the dependence of the rate of exchange on fluorine concentration and on temperature.

EXPERIMENTAL

The xenon hexafluoride used in these experiments was either prepared *in situ* as previously reported [2] or was measured tensimetrically into the exchange reactor after external separation and purification. The fluorine was obtained commercially and purified by distillation. The impurities remaining were mainly air, silicon tetrafluoride, and carbon tetrafluoride, totaling about 0.1 per cent by volume.

Fluorine-18 was prepared by the direct irradiation of fluorine gas in a nickel container in the beam of a linear accelerator. A tungsten converter target was used; the preparation involves the two nuclear reactions $F^{19}(\gamma, n)F^{18}$ and $F^{19}(n, 2n)F^{18}$. Fluorine-18 is a positron emitter with a 110-minute half life. It is detected by means of the resulting 0.5 mev. annihilation gamma radiation. The counting equipment used, a single channel analyzer and scintillation crystal, has already been described [3].

A metal vacuum line of the sort usually used for work with fluorine and previously described [4] was used to carry out these experiments. A nickel exchange reactor $\frac{5}{8}$ in. wide, $1\frac{1}{2}$ in. long was connected to a modified right-angle Hoke 411 valve. The valve was extended and the seat of the valve was lowered to about $1\frac{1}{2}$ in. below the body of the valve. This permitted the entire exchange volume to be uniformly heated inside the furnace and to be entirely within the well ($\frac{11}{16}$ in. by 3 in. deep) of the crystal for more consistent counting. A thermocouple well in the reactor was used to measure the temperature of the exchanging gases. The counting tubes and exchange reactors were attached to the vacuum line with flexible Kel-F tubing to permit these tubes to be put alternately into the furnace and into the counting well of the crystal.

PROCEDURE

Xenon hexafluoride was prepared directly in the exchange reactor or was measured tensimetrically into it after external preparation and purification. Fluorine-18 was transferred from its container immersed in liquid oxygen into the reactor immersed in liquid nitrogen. The γ-ray energy spectrum was monitored and the disintegration rate followed for a time sufficient to ensure radiochemical purity. The reactor was heated and maintained at the desired temperature for a predetermined time, then quickly cooled to room temperature. The

TABLE 1

XeF_6-F_2 EXCHANGE*

F_2		FRACTION EXCHANGE	RATE $\times 10^3$ (MOLE LITER^{-1} MIN.$^{-1}$)
(mmoles)	(mole liter^{-1})		
3.91	1.016	0.121	1.22
8.40	2.182	.269	3.02
5.77	1.499	.325	3.78
6.78	1.761	.353	4.21
2.22	0.577	.185	1.86
1.06	0.275	.085	0.73
7.66	1.990	.385	4.72
3.93	1.021	.323	3.68
4.22	1.096	0.056	0.078†

* 0.020 mole liter^{-1} (0.077 mmoles) XeF_6 at 150° C. for two hours.
† 0.014 mole liter^{-1} (0.054 mmoles) XeF_6 at 100° C. for one hour.

decay of the fluorine-18 was again followed. The exchange reactor was cooled in liquid nitrogen to condense the hexafluoride, and the fluorine transferred to a second counting container cooled in liquid helium. The separated fluorine and hexafluoride were then counted. The fluorine was expanded into a calibrated volume and its pressure measured to determine the amount used in the experiment.

RESULTS AND DISCUSSION

The results of the exchange of 0.020 moles/liter XeF_6 as a function of fluorine concentration at 150° C. are shown in Table 1.

Figure 1 shows the rate of exchange to be a linear function of fluorine concentration and to extrapolate back through zero. This behavior is consistent with an associative mechanism with no contribution from dissociation of xenon hexafluoride. In a single exchange between 0.014 moles/liter XeF_6 and 1.10 moles/liter F_2 at 100° C., a rate of exchange of 0.078×10^{-3} mole liter^{-1} min.$^{-1}$ was obtained. An activation energy of 20 kcal/mole is calculated from these results. This value is the same as the one obtained for the UF_6-F_2 exchange [1], where the associative nature is more firmly established by the existence of UF_8 groups, at least in solid compounds [3, 5]. The activation energy of the IrF_6-F_2 exchange [1], which involves a dissociative mechanism, is about 12 kcal/mole. Although the results of the XeF_6-F_2 exchange experiments do not

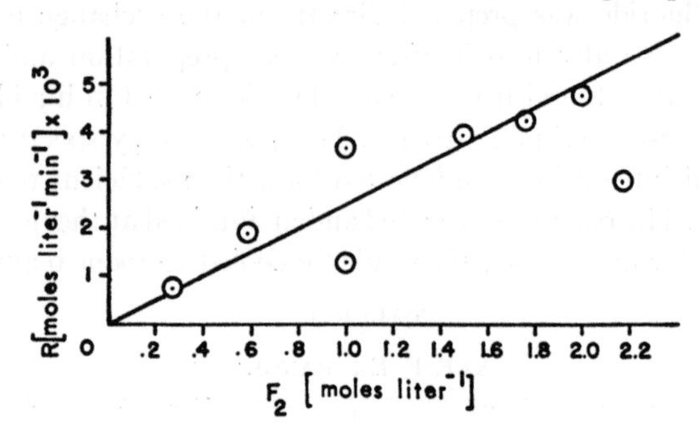

FIG. 1.—Exchange of XeF_6 and F_2. XeF_6 = 0.020 mole liter^{-1}. T = 150° C.

completely rule out an association complex with fluorine atoms, the data are best explained by an exchange mechanism involving a complex with fluorine molecules.

REFERENCES

1. I. SHEFT and H. H. HYMAN, Abstracts of Papers Presented at Am. Chem. Soc. Meeting, Los Angeles, March 31, 1963. P. 57*P*.
2. J. G. MALM, I. SHEFT, and C. L. CHERNICK, *J. Am. Chem. Soc.* **85,** 110 (1963).
3. I. SHEFT, H. H. HYMAN, R. M. ADAMS, and J. J. KATZ, *ibid.* **83,** 291 (1961).
4. I. SHEFT and J. J. KATZ, *Anal. Chem.* **29,** 1322 (1957).
5. J. G. MALM and H. SELIG, Abstracts of Papers Presented at Am. Chem. Soc. Meeting, Atlantic City, September 9, 1962. P. 40*N*.

SOME PROPERTIES OF THE XENON FLUORIDES

A. J. EDWARDS, J. H. HOLLOWAY, AND R. D. PEACOCK
University of Birmingham

It has recently been shown that xenon tetrafluoride is formed when xenon is allowed to diffuse into a fluorine stream and the mixture of gases allowed to pass through a heated nickel tube [1]. In this reaction no other fluoride appears in significant quantities as long as a sufficient excess of fluorine is maintained, and the addition of a little nitrogen as carrier gas does not affect the reaction.

We carried out preliminary experiments, using a flow apparatus in which oxygen was previously mixed with xenon and found that the reaction takes a different course. Xenon tetrafluoride is still formed as a minor product, but the major product is xenon difluoride. This appears at the mouth of the reaction tube. In addition, a colorless, more volatile material passes over into the upper part of the first cold trap (−72°). This material melts at about 90° and is very volatile at the melting point. We estimate the boiling point to lie near 115°. It is possible that this substance is an oxyfluoride (possibly $XeOF_2$) [see Grosse *et al.*, p. 78, this volume] and that its decomposition in the hot reaction tube leads to the difluoride. At 25° this volatile phase does not attack glass and has been stored for weeks at −25° without decomposition. Vessels containing xenon difluoride, on the other hand, soon show a gas pressure. Whether this is due to attack on the glass or to spontaneous decomposition has not been settled.

Certain chemical properties of the xenon fluorides are also under investigation in these laboratories, and we report here some preliminary results.

Xenon tetrafluoride does not combine at temperatures up to 200° with sodium fluoride or with potassium fluoride. No apparent reaction occurs with boron trifluoride from 25° to 100°. However, xenon tetrafluoride combines with tantalum pentafluoride when the two solids are melted together. The complex probably has the composition $XeTaF_9$; it is diamagnetic, and the X-ray powder pattern differs from that of TaF_5. The cream-colored solid melts at 81° to 82° to a pale straw-colored liquid, which is mobile at 87°. The complex sublimes unchanged at 70° under a high vacuum.

Xenon tetrafluoride dissolves in iodine pentafluoride on warming to give a colorless solution from which the tetrafluoride can be recovered unchanged. Addition of potassium fluoride to the solution yields only KIF_6.

Xenon tetrafluoride dissolves in anhydrous ether, with gas evolution, to give a solution with strong oxidizing properties. This solution has the same smell, reminiscent of osmium tetroxide, as an aqueous solution of the tetrafluoride. The addition of pyridine in ether results in a white precipitate, which IR measurements suggest to be a mixture, and which does not appear to contain xenon. This precipitate is not formed quantitatively, and the supernatant liquid retains oxidizing properties. The addition of triphenylphosphine in ether to an ethereal solution of xenon tetrafluoride also results in a white precipitate which redissolves at once. If the clear liquid is allowed to stand in a moist atmosphere, triphenylphosphine oxide crystallizes out.

These observations, which are incomplete, show that the tetrafluoride is a poorer Lewis acid than iodine pentafluoride and that it is a Lewis base only under special conditions. We have not proved that compounds are not formed initially with pyridine and with triphenylphosphine; if they exist, they are very reactive. The lack of reactivity of XeF_4 towards boron trifluoride suggests that π-bonding plays an important part in complex formation. We suggest that in $XeTaF_9$ bonding occurs through a fluorine bridge rather than through a xenon-tantalum link.

Note added in proof.—The tantalum complex has been shown to have the composition $XeF_2 \cdot 2TaF_5$. A similar complex is formed with antimony pentafluoride.

Acknowledgments.—We thank Imperial Chemical Industries for the loan of the necessary fluorine cell and the Department of Scientific and Industrial Research for a fellowship and maintenance grant.

REFERENCE

1. J. H. Holloway and R. D. Peacock, *Proc. Chem. Soc.* (London) **389** (1962).

PREPARATION OF RARE-GAS FLUORIDES AND OXYFLUORIDES BY THE ELECTRIC-DISCHARGE METHOD AND THEIR PROPERTIES

A. G. Streng, A. D. Kirshenbaum, L. V. Streng, and A. V. Grosse
Research Institute of Temple University

Abstract

A number of noble-gas compounds have been prepared with the aid of an electric discharge: with xenon and fluorine alone, XeF_4 and XeF_6; with krypton and fluorine, KrF_4; with xenon and oxygen difluoride, $XeOF_2$. A product containing one or more compounds having the approximate formula $XeOF_3$ was prepared by heating Xe and OF_2.

Bartlett's synthesis of $XePtF_6$ [1] disproved that the rare gases were inert and started an enormous amount of research on the chemistry of the rare gases. His results were followed rapidly by reports from Argonne Laboratory's scientific team that xenon fluorides can be prepared from xenon and fluorine [2, 3].

The methods used to prepare xenon fluorides at Argonne and other laboratories [2, 3, 4, 5, 6] consisted of heating the rare gas with fluorine in a nickel or stainless steel container.

XENON FLUORIDES

We have prepared XeF_4 and XeF_6 by the electric-discharge method, making use of the apparatus and experience gained in our work with the thermally unstable oxygen fluorides O_2F_2, O_3F_2, and O_4F_2 [7, 8, 9]. The apparatus consisted of a Pyrex reaction vessel 6.5 cm. in diameter. The prefluorinated copper electrodes were 2 cm. in diameter and 7 cm. apart. A schematic diagram of the apparatus is shown in Figure 1.

The reaction vessel was cooled to 195° K. (−78° C.) in a dry ice–Freon 12 bath. In preparing the XeF_4 [10], a gas mixture containing one volume of xenon and two volumes of fluorine was passed into the electric discharge and was converted quantitatively to XeF_4. For XeF_6, the gas mixture consisted of one volume of xenon and three volumes of fluorine. The amperages, voltages, and pressures used are tabulated in Table 1. In a typical XeF_4 run 7.10 mmoles of xenon and 14.20 mmoles of fluorine gas were fed into the reaction vessel in 3.5 hours and yielded 1.465 gm. of XeF_4. A typical preparation of XeF_6 yielded 264 mg. of XeF_6 in six hours with no residual xenon.

The research reported in this chapter was supported by the Office of Naval Research.

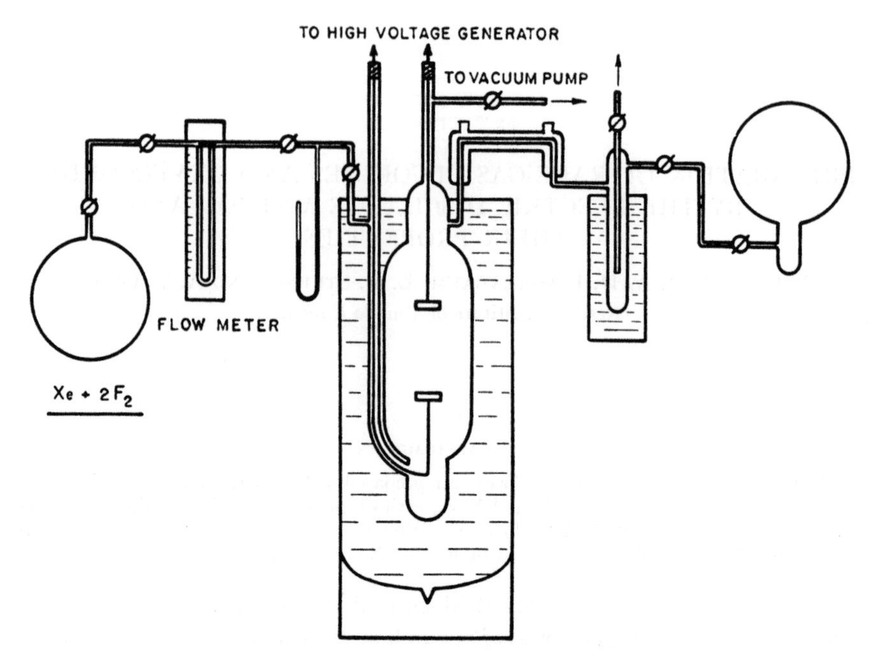

FIG. 1.—Electric-discharge apparatus

TABLE 1

CONDITIONS FOR PREPARING XeF_4, XeF_6, KrF_4, AND XENON OXYFLUORIDES BY THE ELECTRIC-DISCHARGE METHOD

Compound	Ratio of Rare Gas to F_2 or OF_2	Maximum Pressure (mm.)	Volts	Milliamps	Temperature (° K.)	Flow (cc/hr) (NTP)
XeF_4.	1:2	15	1100 to 2800	12–31	195	136
XeF_6.	1:3	16	1500 to 2000	25–30	195	16
KrF_4.	1:2	12	700 to 2200	24–37	85	125
Xenon Oxyfluorides.	1:1	60	1200 to 3600	12–32	195	130

The advantages of the electric-discharge method over the thermal methods are that the electric-discharge method is a continuous one and quantitative; it requires neither the use of excess fluorine nor an application of high pressure.

The XeF_4 prepared by the electric-discharge method formed clear colorless crystals identical in appearance with those prepared by the thermal method (see Plate I). The XeF_4 was proven both by synthesis ($Xe + 2F_2 \rightarrow XeF_4$) and by analysis of the product (Xe/F = 1:4).

The XeF_6 was obtained in the form of colorless crystals, and is stable at room temperature. It decomposes or reacts slightly with glass at ~50° C. The formula was confirmed both by synthesis,

$$Xe + 3F_2 \rightarrow XeF_6\,, \tag{1}$$

and by chemical analysis (Xe/F = 1:6). The vapor pressure of the XeF_6 was found to be 6.5 ± 1.0 mm. Hg at 0° C. and 23.0 ± 1.0 mm. Hg at 20.6° C. These data are in good agreement with the results obtained by Malm and coworkers [5] (7.5 mm. at 0° C., 30 mm. at 25° C.) and by Weaver *et al.* [6] (6 mm. at 0° C. and 27 mm. at 20° C.). The vapor-pressure equation for XeF_6 evolved from the above experimental data is

$$\log P_{\text{mm.}} = 8.155 - 1900/T° \text{ K.}\,,$$

with a $\Delta H_{\text{sub}} = 9150$ cal/mole.

KRYPTON TETRAFLUORIDE

After the discovery of xenon and radon fluorides [3], the obvious question was: Does krypton, the lower analog of these two inert gases, form similar compounds? The answer is yes.

It was first found that gaseous krypton and fluorine do not form detectable amounts of krypton fluorides when they are heated at 400° to 500° C. in a nickel vessel. A mixture of krypton (one part) and fluorine gas (five parts) was heated for one hour, with negative results. Under the same conditions XeF_4 is formed from xenon and fluorine [2]. Similarly negative results were obtained when a mixture of krypton and fluorine was irradicated with ultraviolet light at −60° C. [11]. Furthermore, when Bartlett [12] tried to prepare $KrPtF_6$ by reacting Kr and PtF_6, no oxidation took place at temperatures up to 50° C.

However, using the electric-discharge method described above, we were able to produce krypton tetrafluoride [13]. Using the same apparatus and electrical conditions (see Table 1), a mixture of one volume of krypton and two volumes of fluorine (to within ±0.1 per cent) was admitted, at a pressure of from 7 to 12 mm. Hg, to the discharge vessel. The vessel was kept at 84° to 86° K. (−187° to −189° C.) by mixtures of liquid oxygen and nitrogen. When higher temperatures were used, poorer or no yields of KrF_4 were obtained. In a successful experiment 500 cc. of the mixture of krypton and fluorine at NTP was completely converted to 1.15 gm. of KrF_4 in four hours. The rate of conversion depends on the surface condition of the copper electrodes. In the course of time the electrodes become covered with a white layer of CuF_2. With less efficient electrodes, the production rate may go down to a tenth of the rate just indicated.

The KrF_4, a white solid deposited on the glass wall of the discharge vessel between the electrodes, is readily sublimed at −30° (243° K.) to −40° C., or even at 0° C. into a glass storage vessel containing some dry potassium fluoride powder (as a getter for hydrogen fluoride). The composition of the compound was proven by synthesis,

$$Kr + 2F_2 \rightarrow KrF_2\,, \qquad (2)$$

and from analysis on heating (as mentioned below). Like XeF_4, KrF_4 forms beautiful, transparent, colorless crystals as shown in Plate II.

TABLE 2

VAPOR PRESSURE OF SOLID KrF_4
[$\log_{10} P_{mm.} = 8.531 - 1930/T°$ K.]

Temperature (° K.)	Vapor Pressure (mm.)	
	Experimental	Calculated from Equation
293.2	90	90
273.2	26	29.7
	30	
250.4	6	6.75
	7	
243.2	4	4.0
	5	

Krypton tetrafluoride is thermally much less stable than xenon tetrafluoride. However, at −78° C. (195° K.), it can be stored for months without decomposing. In a polychlorotrifluoroethylene (Kel-F) tube with copper valves, at about 20° C., approximately one-tenth of the KrF_4 present decomposed in one hour. These figures are given for purposes of orientation only. Deposit of CuF_2 on the surface of the valves may have catalyzed the decomposition. At 60° C. the decomposition is rapid; KrF_4 decomposes into its elements—that is,

$$KrF_4 \rightarrow Kr + 2F_2\,, \qquad (3)$$

as determined by analysis. The fluorine gas is determined by reaction with mercury in a buret, and the residual krypton gas is determined by volumetric analysis. This rapid decomposition at higher temperatures explains why previous attempts to prepare krypton tetrafluoride had been unsuccessful.

The vapor pressure of solid KrF_4 was determined by determining, and subtracting from the total pressure, the amounts attributed to decomposition. The values obtained are tabulated in Table 2 and compared with the data on XeF_6

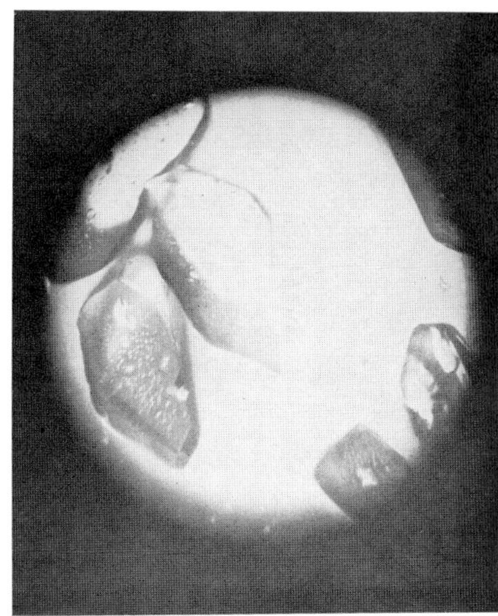

PLATE I.—Crystals of xenon tetrafluoride (*left*) and oxyfluoride (*right*)

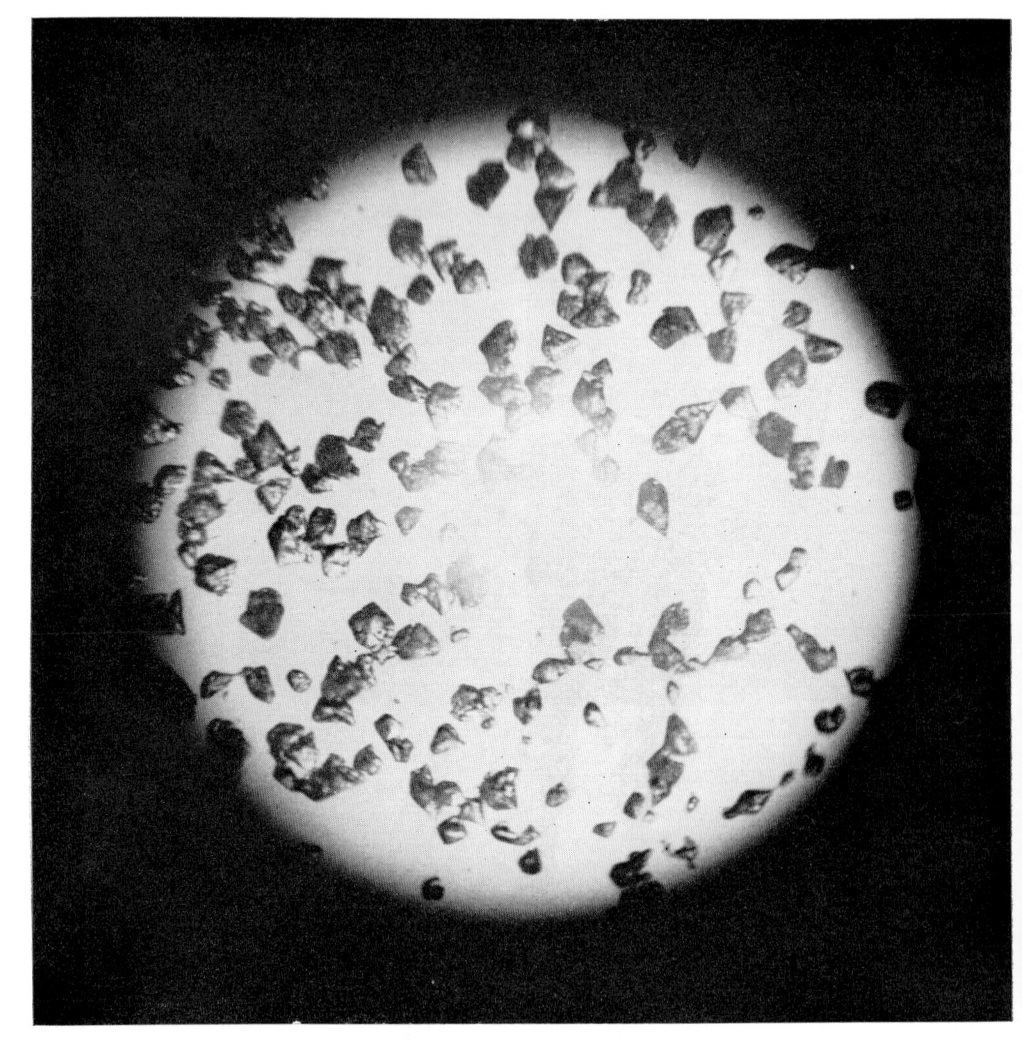

PLATE II.—KrF_4 crystals

of the Argonne Laboratory [5], Ford Laboratory [6], and our own laboratory in Figure 2. The vapor-pressure equation for solid KrF_4 is

$$\log_{10} P_{\mathrm{mm.}} = 8.531 - 1930/T^{\circ}\,\mathrm{K}.$$

with a ΔH_{sub} of 8840 $\pm\sim$300 cal/mole. The curve is practically parallel to the vapor-pressure curve of XeF_6. The extrapolated sublimation temperature (at pressure of 760 mm. Hg) is approximately 70° C.

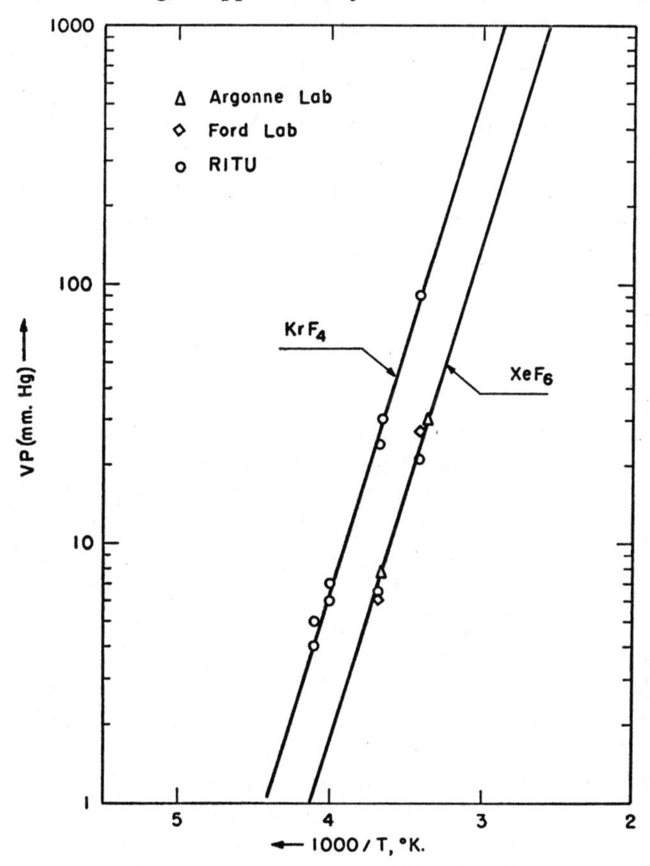

FIG. 2.—Vapor pressure of KrF_4 and XeF_6

XENON OXYFLUORIDES

It was reported by the Argonne Laboratories [2, 3] that traces of $XeOF_3$ and $XeOF_4$ were detected by mass spectrometer due to impurities of oxygen in the fluorine gas used in preparing XeF_4. No other oxyfluorides were reported and no visible quantities of oxyfluorides were prepared. At the Research Institute of Temple University, however, we have prepared appreciable quantities of oxyfluorides by both the electric-discharge and the thermal methods by reacting Xe with oxygen difluoride (OF_2) [14].

The electric-discharge apparatus used for the reaction between xenon and oxygen difluoride is the same as that described previously. A 1:1 volume mixture of xenon and oxygen difluoride was admitted to the electric-discharge vessel cooled to 195° K. in a dry ice–Freon 12 bath with an average velocity of about 130 cc/hr (at NTP). The pressure maintained in the reaction vessel was 3–62 mm. Hg; the discharge current varied from 12 to 32 milliamperes and the voltage from 1200 to 3600 volts. In a typical run 770.7 mg. xenon and 317.0 mg. oxygen difluoride were converted to 664.3 mg. of transparent crystals (=61 per cent by weight of the Xe-OF_2 mixture) with some unreacted gas. No O_2F_2 formation was observed. If the improbable assumption is made that all of the original F-content of the OF_2 reacted to XeF_4 only, the maximum weight of the XeF_4 would be 609 mg. Thus, the difference, i.e., 664 − 609 mg. or 55 mg. of oxygen, should be present in the product. The product, colorless crystals, were stable at room temperature. The compound was analyzed by reacting it with mercury to give Hg_2F_2, Xe, some Hg_2O, and some O_2. The amount of fluorine was determined by reacting the Hg_2F_2 with a known amount of NaOH as follows:

$$2NaOH + Hg_2F_2 \rightarrow 2NaF + Hg_2O + H_2O\,. \qquad (4)$$

The excess NaOH was titrated with HCl. The free oxygen was removed from the gas with pyrogallol and the residual xenon then determined volumetrically. The total combined oxygen was then determined by difference. The crystals obtained from the electric-discharge method had the empirical formula $XeOF_2$.

Xenon oxyfluorides were also prepared by the thermal method [14]. When a 1:1 molar mixture of xenon and oxygen difluoride were heated in a nickel tube for two hours at 200° C., no noticeable reaction occurred. When the mixture was heated to 300–400° C., colorless, transparent crystals were formed. These crystals were stable at room temperature. They can be readily re-sublimed and grown to large single crystals >2 mm. in length. These crystals are shown in Plate I. In a typical preparation, 220 mg. of xenon and 90 mg. of oxygen difluoride (1:1 molar mixture) were heated to 300° C. for $4\frac{3}{4}$ hours. After the reaction, 73 mg. xenon, 9.4 mg. of oxygen and a trace of fluorine were found in the unreacted gas along with 225 mg. of colorless crystals. The material balance was as follows: before the reaction, 310 mg.; after the reaction, 308 mg. The oxygen content in the crystals, by difference was 19.5 mg. or 8.57 per cent oxygen by weight. Thus, 54 per cent of the oxygen of the OF_2 is contained in the crystals. Whenever the crystals, prepared by heating to 300–400° C., were analyzed, they gave the following empirical formula:

$$Xe_{1.2}O_{1.1}F_{3.0}\,.$$

This may be a mixture of xenon fluoride and/or oxyfluorides.

In order to see whether a mixture of $\frac{1}{2}O_2$ and F_2 would reaction as OF_2, an experiment was made using a mixture in the molar ratio of $Xe + \frac{1}{2}O_2 + F_2$.

This mixture was heated in the same nickel tube at 400° C. for three hours. Only 18 per cent of the original oxygen reacted compared to 54 per cent when OF_2 was used. Thus, a mixture of $\frac{1}{2}O_2 + F_2$ can be substituted for OF_2, but the oxygen will be less readily incorporated into the product.

CHEMICAL REACTIONS WITH XENON TETRAFLUORIDE AND OXYFLUORIDES

Reaction with H_2O.—The only chemistry reported on the xenon fluoride has been the reaction of XeF_4 and XeF_6 with H_2O or dilute alkali to yield xenic acid [2, 4, 15, 16, 17]. When dried, the XeO_3 is reported to have a tendency to detonate very readily. We have found that the xenon oxyfluorides also form xenic acid on hydrolysis.

In our studies it was observed that xenic acid (or a precursor) is volatile and can be distilled with water vapor. It was found that an aqueous solution can be distilled under vacuum from one leg (at 0°–10° C.) of a λ tube to another at liquid oxygen temperature and still contain the xenic acid.

Barium xenate has been prepared in our laboratories as a white microcrystalline powder. Its methods of preparation, composition, and properties are being investigated and will be reported at a future time.

Reaction with SF_4.—Studies are under way on the reactivity of XeF_4 with SF_4. Preliminary results indicate that when 1:1 molar mixture of the two is kept at 23° C. at a starting pressure of 360 mm. Hg, they react slowly. After twenty hours 20 per cent of the SF_4 was oxidized to SF_6 and 20 per cent of the XeF_4 was reduced to Xe according to the equation

$$XeF_4 + SF_4 \rightarrow Xe + SF_6 . \qquad (5)$$

Reaction with O_2F_2.—Preliminary studies show that XeF_4 readily reacts with O_2F_2. When a 1:1 molar mixture was kept between 140° to 195° K., they reacted completely in one hour proceeding mainly according to the equation

$$XeF_4 + O_2F_2 \rightarrow XeF_6 + O_2 . \qquad (6)$$

Some small quantities of other unidentified products were also found.

REFERENCES

1. N. Bartlett, *Proc. Chem. Soc.* **218** (1962).
2. H. H. Claassen, H. Selig, and J. G. Malm, *J. Am. Chem. Soc.* **84,** 3593 (1962).
3. C. L. Chernick *et al., Science* **138,** 136 (1962).
4. F. B. Dudley, G. Gard, and G. H. Cady, *Inorg. Chem.* **2,** 228 (1963).
5. J. G. Malm, I. Sheft, and C. L. Chernick, *J. Am. Chem. Soc.* **85,** 110 (1963).
6. E. E. Weaver, B. Weinstock, and C. P. Knop, *ibid.* **85,** 111 (1963).
7. A. D. Kirshenbaum and A. V. Grosse, *ibid.* **81,** 1277 (1959).
8. A. D. Kirshenbaum, A. V. Grosse, and J. G. Aston, *ibid.* **81,** 6398 (1959).
9. A. V. Grosse, A. G. Streng, and A. D. Kirshenbaum, *ibid.* **83,** 1004 (1961).
10. A. D. Kirshenbaum, L. V. Streng, A. G. Streng, and A. V. Grosse, *ibid.* **85,** 360 (1963).
11. J. L. Weeks, C. L. Chernick, and M. S. Matheson, *ibid.* **84,** 4613 (1962).

12. N. Bartlett, *Chem. Eng. News 1963* **36** (4 Feb. 1963).
13. A. V. Grosse, A. D. Kirshenbaum, A. G. Streng, and L. V. Streng, *Science* **139**, 1047 (1963).
14. A. G. Streng, A. D. Kirshenbaum, and A. V. Grosse, "Addition and Substitution Products of Oxygen Fluorides," Third Annual Progress Report for the Office of Naval Research, Contract NONR 3085(01); Research Institute of Temple University, January 15, 1963.
15. N. Bartlett and R. Rao, *Science* **139**, 506 (1963).
16. S. M. Williamson, *ibid.* 1046 (1963).
17. D. F. Smith, *J. Am. Chem. Soc.* **85**, 816 (1963).

THE SYNTHESIS OF XENON COMPOUNDS IN IONIZING RADIATION

D. R. MacKenzie and R. H. Wiswall
Brookhaven National Laboratory

Abstract

Xenon and fluorine form compounds when irradiated with γ-rays or energetic electrons. At and above room temperature a mixture of XeF_2 and XeF_4 is formed; at a temperature of about $-35°$ C., XeF_2 is the only product. Utilization of absorbed energy is efficient since initial G-values for consumption of xenon are in the range from 5 to 15 atoms per 100 ev. absorbed. XeF_4 is relatively stable to γ-radiation, with an initial G-value of 0.6 to 1.8 at 45° C., depending on the decomposition products. Krypton and fluorine have not been made to react in an electron beam at temperatures around $-130°$ C.

INTRODUCTION

Ionizing radiation is known to act as a catalyst for exothermic reactions which are ordinarily prevented from taking place at room temperature by high activation energy. Two examples of gas-phase reactions of this sort are the formation of ammonia from its elements and the oxidation of SO_2. It seemed appropriate, therefore, to irradiate mixtures of rare gases and elements with which they might form compounds. Most of the work so far has involved xenon and fluorine, which readily form compounds under the influence of γ-rays and energetic electrons.

Three sources of radiation have been used: high-intensity γ-rays from Co^{60}, an electron Van de Graaff, and the Brookhaven graphite reactor. Positions in the Co^{60} γ-facility gave fields of up to 6.5×10^6 r/hr, measured with the Fricke dosimeter. Such fields are sufficient to give a slow but easily measurable reaction in a gaseous system for which G is of the order of 1 to 10, i.e., 1 to 10 molecules transformed per 100 ev. of energy absorbed. Much useful information was obtained in the γ-facility, but it had two disadvantages. One was a very restricted experimental volume; apparatus was confined to a cylindrical space about 12 in. long and 1 or 2 in. in diameter 10 ft. below the shielding water surface. The other was the difficulty of using a flowing system with a cold trap outside the radiation zone. Use of an electron Van de Graaff permitted a more flexible experimental setup, and this machine was used in several runs in which gases were circulated. It produced an electron beam of 20 microamps at 1.5 mev. Of course, only a small fraction of the beam energy was absorbed in the gaseous systems being studied; complete absorption of such a beam would require a path length of several meters, depending on the gas pressure used.

STATIC IRRADIATIONS

A positive effect was found in the first experiment, in which a mixture of about 3 parts fluorine to 1 part xenon at a total pressure of 3 atmospheres was γ-irradiated in a prefluorinated nickel can provided with a pressure gauge. The pressure dropped initially about 1 p.s.i. per day, and a small quantity of colorless

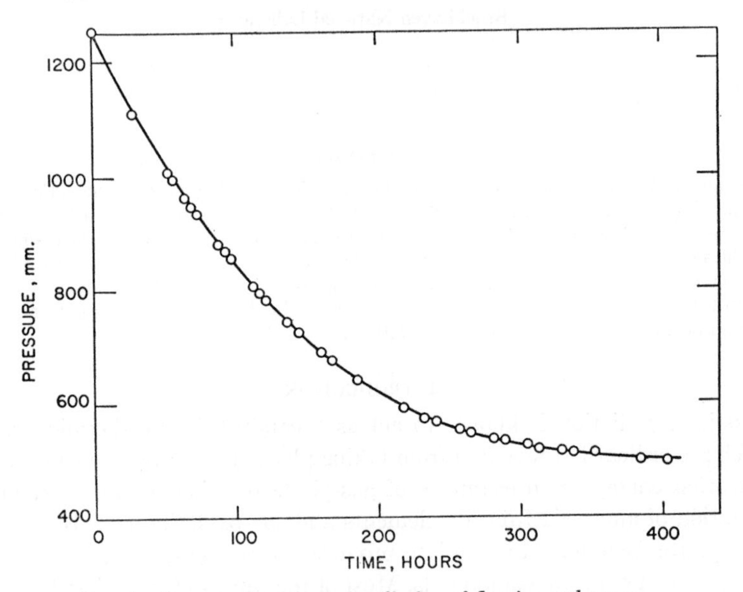

FIG. 1.—Pressure decrease in irradiation of fluorine and xenon

crystalline solid was recovered from the can by sublimation after the unreacted gas had been pumped off. The behavior of this material resembled that of the XeF_4 described by C. L. Chernick *et al.* [1]. Two later experiments were carried out in improved apparatus with small dead volume between reaction vessel and pressure gauge. In the first of these no attempt was made to cool the nickel vessel, which was maintained at 64° C. by γ-heating of the vessel and the irradiation tube. In the second, water-cooling was used to keep the vessel close to room temperature (Plate I).

Figure 1 shows the pressure-time curve obtained in the experiment using water-cooling. From the initial rate of pressure drop one can calculate a radiation yield, provided the composition of the product or the change in composition of the gas mixture is known. At the end of each run, portions of the solid product were analyzed by hydrolyzing a weighed amount in sodium hydroxide solution and determining fluoride by thorium titration. The results gave a composition

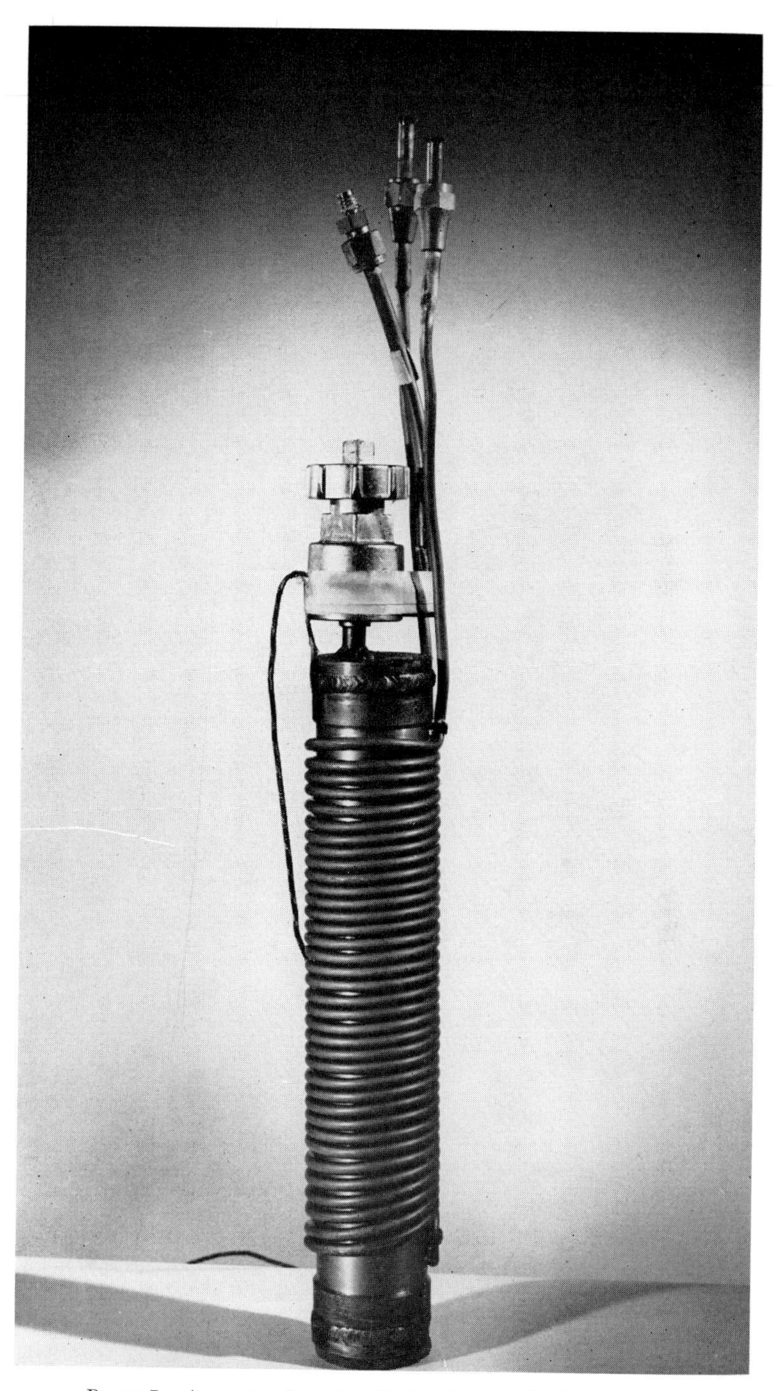

PLATE I.—Apparatus for γ-irradiation of xenon-fluorine mixtures

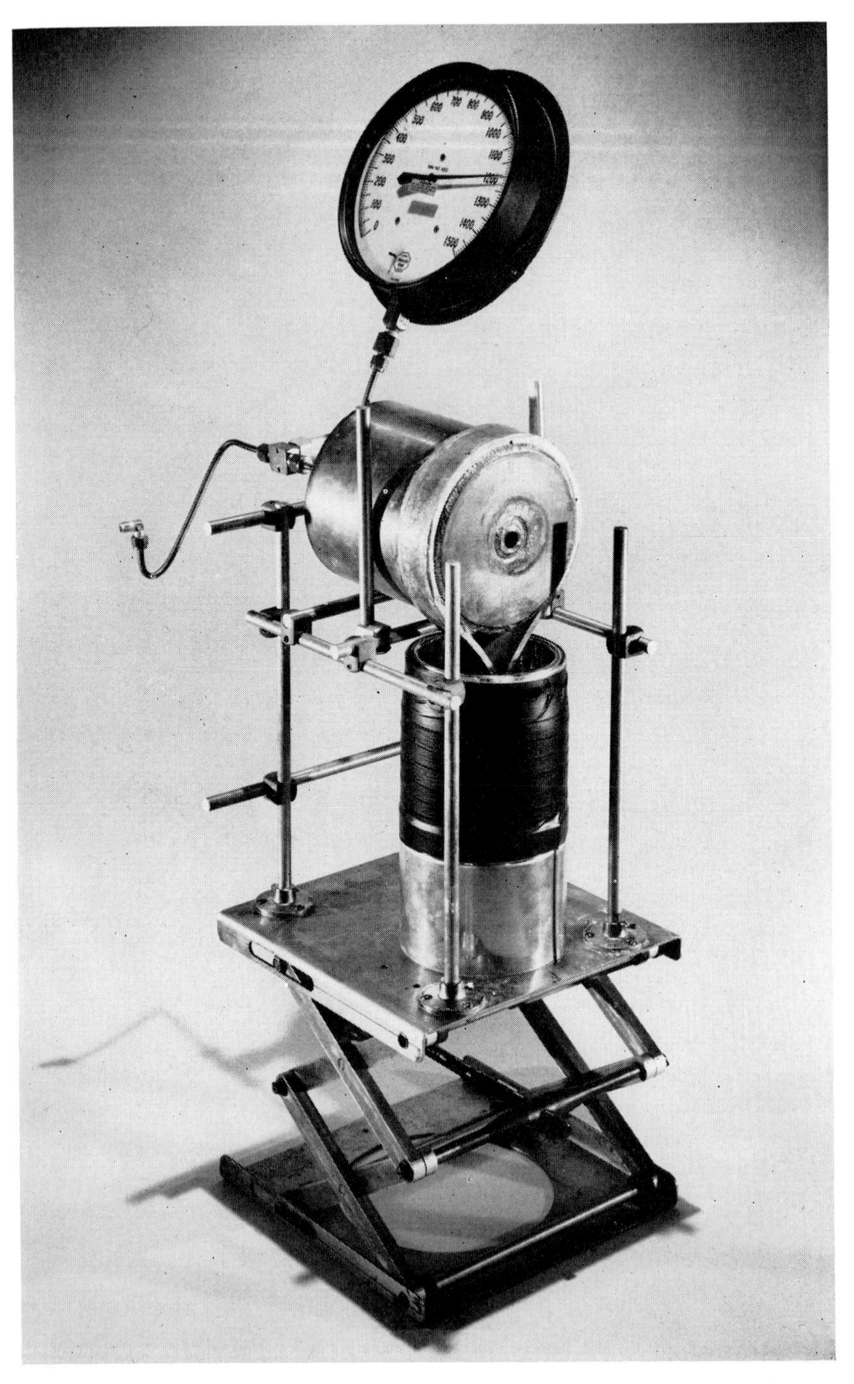

PLATE II.—Apparatus for irradiation of xenon-fluorine mixtures in the Van de Graaff electron beam.

corresponding to $XeF_{3.0}$ in run II and $XeF_{2.33}$ in run III; presumably a mixture of XeF_2 and XeF_4 formed [2]. Assuming that the initial and over-all reactions are the same, the initial value of G is 5.4 atoms of xenon consumed per 100 ev. absorbed in the reaction mixture in run II and 14.6 in run III. The conditions of the two runs differed slightly (Table 1). The partial pressures were 9.2 p.s.i. Xe and 24.0 p.s.i. F_2 in run II, and 387 mm. Hg Xe and 871 mm. Hg F_2 in run III. The reaction temperatures were 64° C. and 27° C., respectively.

The data of Figure 1 do not lend themselves to easy analysis. Only where the initial reactant ratio is the same as in the solid product can a simple function be hoped for. One deduction from this curve and from analysis of the residual gas seems legitimate, however: the pressure will evidently level off before the xenon is exhausted, though not much before. (It had not quite leveled off with 94 per

TABLE 1

IRRADIATIONS OF $F_2 + XE$

(Co^{60} γ)

	Without Cooling	With Cooling	1.5 Mev. Electrons
Volume of vessel, cc.	330	330	3050
Temperature, ° C.	64	27	~−35
Duration of run, hrs.	230	404	4.5
Initial pressure	33.2 p.s.i.a.	1258 mm. Hg	1242 mm. Hg
Final pressure	25.9 p.s.i.a.	496 mm. Hg	1217 mm. Hg
Initial ratio F_2/Xe	2.6	2.25	2.08
Final ratio F_2/Xe			
Gas phase	....	20.3	....
Solid	1.5	1.16	1.00
Wt. Xe fluoride formed, gm.	0.31	1.26	0.58
Fraction of initial Xe used up	0.22	0.94	0.02
Initial $G_{(-Xe)}$	5.4	14.6	....

cent of the xenon used up.) That is, the solid product is rather slowly decomposed by radiation, and a steady-state pressure is reached only when the xenon is nearly gone. This was shown when the solid product was re-irradiated for three days after pumping off the residual gas. Only about 1 mm. pressure developed in that time.

This phenomenon was further studied with pure XeF_4, approaching the steady state from the other side. A sample of 960 mg. of XeF_4 prepared by thermal combination of the elements at 400° C. was exposed to 420 megarads γ-radiation in an evacuated nickel can at 45° C. Figure 2 shows the increase in pressure with time. Figure 3 is a log-log plot of the data. These fit the empirical equation $P = 0.47t^{0.58}$ with P in millimeters and t in minutes. From the initial slope, one calculates an initial G-value of 1.8 molecules XeF_4 converted to XeF_2 and F_2, or 0.6 molecules XeF_4 converted to Xe and $2F_2$. Whichever the mode of decomposition may be, the stability of XeF_4 is remarkably good for a covalent compound. Water and most organic compounds show higher G-values.

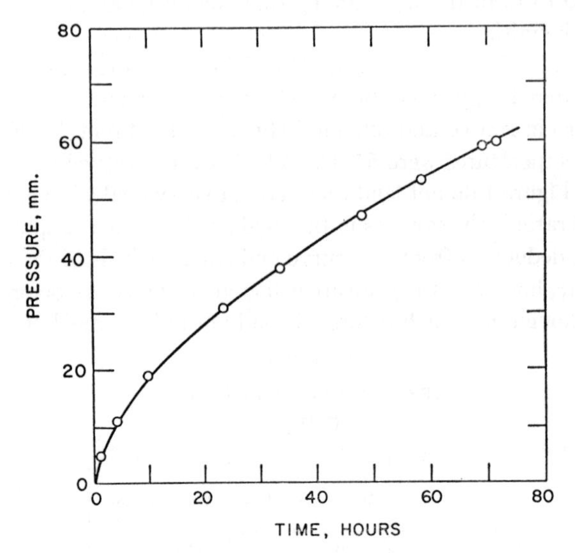

FIG. 2.—Pressure increase in XeF_4 radiolysis

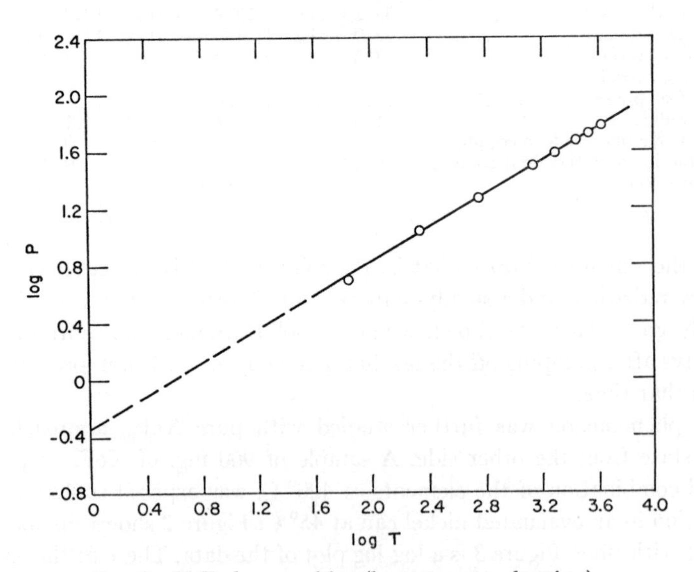

FIG. 3.—XeF_4 decomposition (log pressure *vs.* log time)

The behavior of a Xe-F_2 mixture in the Van de Graaff beam was examined in a special cold-wall apparatus which had been developed with the hope of isolating some of the more fugitive rare-gas compounds. It consisted of a cylindrical nickel vessel about 6 in. in diameter and 7 in. long (Plate II). The electron beam entered one end, the "front," through a 0.005-in. nickel window mounted on the inner end of a $\frac{3}{4}$-in. (inside diameter) re-entrant tube which projected about $1\frac{1}{4}$ in. into the vessel. Such an arrangement should diminish the radiation intensity at the front wall of the vessel, thus permitting the accumulation of product there. Such accumulation was further induced by preferentially cooling the front end by fastening a heavy copper strap tightly around that end of the cylinder; the ends of the strap were dipped into a dry-ice bath or (for krypton runs) liquid nitrogen.

A xenon-fluorine mixture, in which the initial partial pressures were 395 mm. Xe and 823 mm. F_2, was given an electron irradiation for four and a half hours. Conditions are given in Table 1 for comparison with the γ-irradiations. Temperature was not uniform over the large reaction vessel, but for the most part the surface temperature was between $-30°$ and $-40°$ C. Weight of product was 0.58 gm., corresponding to about 2 per cent of the starting material. Chemical analysis showed it to be practically pure XeF_2. The relative purity of the XeF_2 was shown qualitatively by the complete lack of yellow product formed during hydrolysis and by the slow reaction with NaOH solution. XeF_4 reacts rapidly with NaOH solution, forming a bright-yellow substance which subsequently disappears.

Attempts were made to induce a reaction between krypton and fluorine in experiments similar to those which yielded xenon-fluorine compounds. Neither in the γ-field nor in the electron beam in the presence of a cooled wall ($-130°$ to $-140°$ C.) was there any sign of reaction.

CIRCULATING SYSTEMS

Apparatus was devised in which a fast-moving stream of gas mixture could be exposed to radiation—most conveniently a Van de Graaff beam—and then passed immediately over a cold surface. For mixtures of rare gases with oxygen or chlorine, a glass system was used with a modified Neptune Dynapump. Velocities of about 80 ft/min past the electron beam and an interval of one-half second between beam and cold trap were attained. For mixtures containing fluorine, a copper-nickel system was used, and circulation was provided by a special piston pump designed for use with fluorine and provided by the Oak Ridge Gas Diffusion Plant. The pump could operate at any speed from 1 to 40 l/min. Gas velocities of 20 and 30 l/min were used, with an interval of approximately one-quarter second between beam and trap.

In the first of these systems, mixtures of xenon and oxygen and xenon and chlorine were irradiated without apparent result. In the metal fluorine circulating system, two irradiations of krypton-fluorine mixtures were carried out, both with F_2/Kr ratios about 1.2 and total pressures of 1.5 and 1.6 atmospheres, respectively. Temperature of the trap was kept around −130° C. A small amount of liquid product was obtained in each run (measured as 10 mg. in the second one), but it showed only about 15 per cent fluorine and no inert gas when hydrolyzed with NaOH solution.

ANALYTICAL PROCEDURES

Our standard method of analyzing inert-gas fluorides has been to hydrolyze a known weight of compound with from 5 to 10 per cent NaOH solution and determine fluoride by titration with standard thorium nitrate.

Representative samples cannot be taken from a mixture of XeF_2 and XeF_4 by sublimation because of their different volatilities. One way to obtain a representative sample is to take all the material at once. But when sample weights are large, it is difficult to carry out the hydrolysis smoothly, and there is always the danger of spoiling the determination, thus losing all the material. To get

TABLE 2

ANALYTICAL RESULTS WITH XENON FLUORIDES

Compound	Weight Taken, mg.	Weight Fluorine Found, mg.	F/Xe Ratio
XeF_2.......	35.8	7.70	1.90
	37.2	8.45	2.04
XeF_4.......	30.0	10.80	3.90
	36.9	13.50	3.99

around this difficulty, we have handled our products from irradiations in an inert-atmosphere dry box, achieving thorough mixing by grinding with a glass mortar and pestle. Metal mortar and pestles would probably be preferable, but so far we have noticed no etching of the glass. Material is stored in quartz weighing-bottles with only a slight loss of weight through evaporation over several weeks.

For analysis, a sample of from 25 to 50 mg. is weighed into a small glass tube with a side arm and a rubber septum fitted into the top. After removal from the lock of the dry box, the side arm is quickly fitted into the end of a piece of Tygon tubing connected to a delivery tube dipping into 5 to 10 per cent NaOH solution. Approximately 5 cc. of NaOH solution is injected by syringe into the sample tube through the septum. The collected NaOH solutions and rinsings are then made up to volume for titration. Examples of results obtained with samples of relatively pure XeF_2 and XeF_4 are given in Table 2.

The absolute values obtained in the fluorine analysis are correct to within the limits of precision of the method. The precision attainable in the rather difficult thorium titration leaves something to be desired; however, results were good to within ±4 per cent and usually much less.

DISCUSSION

Results with ionizing radiation.—Krypton fluorides have been reported to be formed at −195° C. in an electric discharge [3] and at −253° C. in an inert-gas matrix irradiated with UV light [see Pimentel *et al.*, p. 103, this volume]. Grosse *et al.* report [see p. 75, this volume] that the yield of compound dropped off drastically as the temperature increased and was negligible at −80° C. In both our static system and our flow system, one would expect a small amount of product to form from electron irradiation at −130° C. That we were unable to isolate any krypton compound from either system probably indicates a need for lower temperature operation. The liquid product obtained from the flow system may have been $XeOF_4$ formed from small amounts of xenon impurity in the krypton, since $XeOF_4$ liberates only a small fraction of its xenon in NaOH hydrolysis.

The products of the radiolysis of our F_2-Xe mixtures, the compositions of which are given in Table 1, are comparable to those obtained in photochemical experiments at Argonne National Laboratory. Weeks, Chernick, and Matheson [4] reported that only XeF_2 was formed in the photochemical reaction regardless of the F_2/Xe ratio. In these experiments a flow system was used with a trap at −78° C. immediately downstream from the reaction cell. When the reaction was carried out with no coolant, XeF_4 was formed as well [see Matheson and Weeks, p. 92, this volume].

It thus appears that in our experiments the temperature of the reaction vessel is the determining factor in product composition. In the electron irradiation the temperature of the walls of the reaction vessel and the gas was well below 0° C. Thus, XeF_2 formed would be in the solid state almost entirely and "frozen out" on the walls where it would undergo little or no reaction. In the uncooled γ-irradiation, the vessel temperature was 64° C.—a temperature at which the vapor pressure of XeF_2 is appreciable and a reasonable fraction would be in the vapor phase and subject to further reaction. This was the experiment in which the reaction product contained most XeF_4. The other γ-irradiation experiment, with intermediate temperature, gave a product with intermediate F_2/Xe ratio.

Comparison with other methods of synthesis.—The small extent of absorption of ionizing radiation by gases at normal pressure probably makes this type of synthesis the least efficient in terms of energy utilization. Certainly electric discharge is relatively very efficient as regards energy deposition in gases. However, with intense γ-ray sources experiments of a week's duration or less will give reasonable yields in synthesis and will also provide kinetic data with which it should be possible to develop a reaction mechanism.

The obvious type of radiation to permit efficient energy absorption is heavy charged particles such as α's or fission fragments. This type is particularly desirable for flowing systems where the fraction of material under irradiation is very small. It is planned to use at least α-particle radiation with krypton systems.

There is no reason why krypton fluoride or fluorides should not form in the presence of ionizing radiation, and the only problem should be to remove compounds out of the irradiation zone rapidly enough.

Acknowledgments.—We wish to express our appreciation to A. O. Allen and H. A. Schwarz for the use of their electron Van de Graaff and for helpful suggestions regarding conduct of the electron-irradiation experiments.

REFERENCES

1. C. L. Chernick *et al.*, *Science* **138,** 136 (1962).
2. D. R. MacKenzie and R. H. Wiswall, *J. Inorg. Chem.*, in press.
3. A. V. Grosse, A. D. Kirshenbaum, A. G. Streng, and L. V. Streng, *Science* **139,** 1047 (1963).
4. J. L. Weeks, C. L. Chernick, and M. S. Matheson, *J. Am. Chem. Soc.* **84,** 4612 (1962).

PHOTOCHEMISTRY OF THE FORMATION OF XENON DIFLUORIDE

James L. Weeks and Max S. Matheson
Argonne National Laboratory

Abstract

Xenon and fluorine combine photochemically when irradiated with near ultraviolet light in the fluorine absorption band. If the product is continuously trapped out, essentially pure XeF_2 is produced. Fluorine and XeF_2 will further react photochemically to produce XeF_4. Quantum yields for XeF_2 formation are of the order of 0.3 to 0.7, being somewhat lower at higher pressures, especially at higher pressures of fluorine. The mechanism of formation very probably involves fluorine atoms. A transient intermediate, perhaps XeF, has been detected in flash photolysis experiments. Attempts at the photochemical preparation of other rare-gas compounds have not yet been successful.

INTRODUCTION

After Claassen, Selig, and Malm [1] reported that they had prepared XeF_4 and probably a lower fluoride in a thermal reaction, it seemed probable that a photochemical reaction of xenon and fluorine could be induced. The weakness of the F-F bond (36 kcal.) [2] suggested that fluorine atoms were present during the thermal reaction at 400° C., and the continuous nature of the fluorine gas absorption band [3] (maximum 2900 A.) shows that fluorine atoms can also be produced photochemically. Preliminary experiments [4] confirmed the existence of a light-induced reaction yielding rather pure XeF_2. This work has now been extended to include studies of the effect of intensity, pressure, and the F_2/Xe ratio on the quantum yield. The quantum-yield work, some preliminary observations on transient spectra in flash photolysis, as well as attempts to prepare other rare-gas compounds photochemically are described in this paper.

EXPERIMENTAL

Xenon was obtained from several commercial suppliers; all samples were shown by mass-spectrometric analysis at this laboratory to be at least 99.9 per cent pure. Fluorine was obtained from the General Chemical Division of the Allied Chemical and Dye Corporation and was shown by reaction with mercury also to be at least 99.9 per cent pure.

Light sources used in this work were high-pressure mercury-arc lamps. An Osram HBO-500W Mercury Arc mounted in a ventilated brass box designed for proper operating temperatures was used with a direct current power supply for the quantum-yield studies. A General Electric AH-6 High-Pressure Mercury Arc in a water-cooled, fused-silica jacket with high-purity, fused-silica windows was used for the preparation of XeF_2 in gram amounts. Fused-silica lenses were used to concentrate the light in a narrow, intense beam for preparative work or to provide a fairly uniform parallel beam for quantum-yield studies.

For most of the preparative work a chemical filter solution [5] was used in a 4-cm. path length, fused-silica cell placed just behind the lens. This solution transmitted about 70 per cent of the 2500–3200 A. light from the lamp and less than 1 per cent of wave lengths outside the 2200–3400-A. region. For the more recent quantum-yield studies a Gaertner model L234-150 Quartz Monochromator was used to isolate the 3130-A. line.

For the filling of cells with xenon and fluorine and the removal of products, techniques described elsewhere in this volume [Chernick, p. 35] were employed.

The rate of reaction was followed with Monel pressure gauges of the bourdon type obtained from the Helicoid Gage Division of the American Chain and Cable Company. The type of reaction cell used is shown in Figure 1. This cell is made of nickel in a closed-loop design (the small tubing is 8 mm., inside diameter) with or without the reservoir of about 400 cc. and with a reaction chamber of about 100 cc., the latter having two vacuum-tight sapphire windows. The reservoir enabled the preparation of larger amounts of product. For "production" runs on XeF_2, a heating tape was operated at 90°–100° C. on one leg of the cell to effect gas circulation. The bottom U-bend was kept in a −78° C. bath during the irradiations in order to trap out the product. (XeF_2 has a vapor pressure of ~3–4 mm. at room temperature, but the vapor pressure is negligible at −78° C.) A water-cooling coil kept the nickel walls of the reaction chamber at about room temperature. These nickel cells were adopted after earlier experiments in quartz cells showed considerable contamination of products due to attack on the quartz.

For the intensity measurements in the quantum-yield determinations, several types of detectors were used. The uranyl oxalate actinometer was employed in the preliminary quantum-yield experiments where a chemical filter solution was used to limit the wave-length region. In the later measurements of quantum yields a calibrated photocell was used to monitor light intensities. The photocell was calibrated at a given wave length by comparing it with a thermopile which, in turn, had been calibrated by comparison with standard carbon-filament lamps from the National Bureau of Standards. Intensities measured by use of the calibrated photocell were found to be in good agreement with those obtained using the uranyl oxalate actinometer.

The apparatus used for the flash photolysis studies reported briefly here was essentially that previously described [6], except for the use here of a Jarrell-Ash 2.25-meter Ebert Grating Spectrograph.

RESULTS

The solid, water-white crystalline product was chemically analyzed by reaction with excess hydrogen at 400° C. [1] and shown to be XeF_2. The relative proportions of xenon and fluorine reacting also were confirmed by pressure measurements made during the irradiation with and without the use of liquid oxygen to freeze down the xenon. The results of the analyses are presented in Table 1 [4]. The second column of Table 1 shows that the usual initial total pressure was nearly 1000 mm. Hg (the upper limit of the pressure gauge used for most runs) in order to produce sufficient product for analysis. Later, a gauge reading up to 4000 mm. Hg was used with the cell with the reservoir, and up to 8-gm. samples of XeF_2 were prepared. The third column shows that even though the initial mole ratio F_2/Xe was varied from 1 up to 5.2, essentially pure XeF_2 was produced (as long as the product was continuously trapped out at −80° C.).

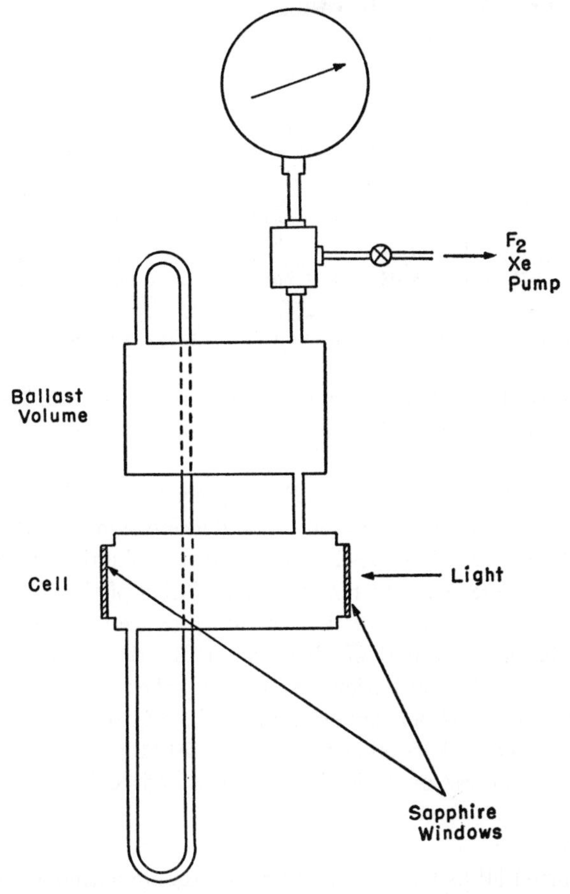

FIG. 1.—Schematic diagram of nickel-sapphire cell for preparation of XeF_2 by irradiating mixtures of xenon and fluorine with ultraviolet light.

In the fourth column, the 73 mm. remaining in run 4 is essentially xenon, if XeF_2 is produced. In runs 3 and 5 the xenon has been almost completely reacted and nearly pure fluorine gas is left. The average rate of total pressure decrease was usually about 20 mm/hr. Runs were usually stopped when a calculation based on pressure readings showed a weight of product at least sufficient for chemical analysis ($\geq$0.4 gm.). (The rate of pressure decrease falls as the reaction proceeds, due to less absorption by fluorine as its concentration decreases.) The analyzed product ratio of 0.89 found for run 1 was probably due to the use of a fused-silica cell in which silicon tetrafluoride is also formed and complicates the handling and analysis of the product.

A Bendix Time-of-Flight Mass Spectrometer [7] showed principally XeF_2 and traces of XeF_4 in the products described above. No oxyfluorides were observed. Studies on the infrared spectra of the products of some of the irradiations

TABLE 1

PHOTOLYSIS OF XE AND F_2

Run	Initial Pressure (mm. Hg)	Initial Ratio (F_2/Xe)	Final Pressure (mm. Hg)	Product Collected (gm.)	Analysis of Products (F_2/Xe)
1*....	990	2.00	447	0.31	0.89
2.....	890	2.35	546	0.73	1.04
3.....	979	5.20	640	0.61	0.98
4.....	1030	0.91	73	2.11	1.01
5.....	971	5.03	654	0.60	1.01

* Run in silica vessel; all other runs in nickel system with sapphire windows.

showed intense bands at 549 cm.$^{-1}$ and 565 cm.$^{-1}$ resulting from XeF_2, but less than 1 per cent (limit of detection) at 590 cm.$^{-1}$, where XeF_4 has a strong absorption [8].

In one experiment, about 1 gm. of XeF_2 was prepared in a nickel-sapphire cell in the usual way, and the unreacted xenon and fluorine pumped out, holding the XeF_2 in the cell at $-78°$ C. Somewhat over 1 atm. of fluorine was then added to the cell and the $-78°$ C. bath removed. Irradiation of the $XeF_2 + F_2$ mixture with a mercury arc at room temperature produced a pressure decrease (attributed to fluorine reacting) of about one-tenth the usual rate observed in XeF_2 production. The irradiation was stopped when calculations based on pressure drop showed $\sim$82 per cent conversion of XeF_2 to XeF_4. Infrared spectra on this sample showed 10–30 per cent XeF_2, 70–90 per cent XeF_4, and no XeF_6. Mass-spectrometric analysis showed XeF_2 and XeF_4, but no XeF_6 or oxyfluorides.

A preliminary value for the quantum yield of xenon reacted (in forming XeF_2) was reported [4] as 0.3, based on uranyl oxalate actinometry, in which $\phi = 0.60$ for oxalate decomposed per quantum was taken as the average for this spectral range (2500–3200 A.). It was found that a thermal reaction, resulting from the use of the heating tape, made a correction necessary. The corrected $\phi = 0.3$ was checked in another experiment in which no part of the circulating system was above room temperature. Using a more refined apparatus with monochromatic light, a number of additional quantum-yield runs at 3130 A. have been made in which the effect of initial total pressure, absorbed intensity, and initial ratio of reactants have been studied. The use of heating tape to promote gas circulation was discontinued for these additional quantum-yield experiments. These studies are not completed as yet; however, Tables 2 and 3 list the results of the investigation so far.

TABLE 2

EFFECT OF TOTAL INITIAL PRESSURE AND INTENSITY ON THE QUANTUM YIELD OF XeF_2 FORMATION AT 3130 A. $F_2/Xe \approx 1$

RUN	INITIAL PRESSURES		RATE OF PRESSURE DECREASE (mm Hg/hr)	LIGHT ABSORBED (%)	ABSORBED QUANTA/SEC. ($\times 10^{-15}$)	ϕ
	F_2 (mm. Hg)	Xe (mm. Hg)				
7	470	496	0.94	84*	50.4	0.33†
11	468	497	4.30	83*	39.6	.27
16	478	478	1.87	90	4.90	.28
21	455	482	0.71	89	1.63	.32
19	235	250	1.65	67	3.54	.35
20	118	125	1.20	37	1.84	.49
22	74	76	1.21	26	1.26	.70
26	49	50	1.00	16	0.84	0.87

* This value was determined using chemical filter solution (instead of monochromator) transmitting 70 per cent of the 2500–3200 A. range.

† This value was corrected for "dark reaction" due to heating tape.

Table 2 and Figure 2 show that the quantum yields obtained for 1:1 mixtures of xenon and fluorine decrease with increasing total pressure. The first four runs of Table 2 indicate little if any effect on the quantum yield for a thirty-fold intensity change, at least at these high pressures (~1000 mm.). The effect of intensity at lower pressures is now being investigated.

Table 3 summarizes the data presently obtained for the effect of F_2/Xe ratio on quantum yield. This table and Figure 3 show that, for either low or high constant xenon pressure, increasing the fluorine pressure decreases the quantum yield. On the other hand, as shown in Figure 4 at low fluorine pressure, increasing xenon pressure decreases the quantum yield, but contrarily, at high fluorine pressure, increasing the xenon pressure perhaps slightly increases the quantum efficiency.

Flash photolysis of a mixture of 185 mm. of xenon and 185 mm. of fluorine has shown the existence of a transient species which decays with a half life under our conditions of perhaps 20 μsec. Two series of bands, one at about

3300 A. and one at about 2500 A., have been observed. Although some features are unexplained, the appearance of the bands suggests a diatomic molecule, i.e., XeF. Pure fluorine gas at 185 mm., or 185 mm. of F_2 + 185 mm. of krypton have shown no observable transients in flash photolysis. Observation of transient spectra at higher resolution and with isotopically pure xenon are planned. (Dr. P. K. Carroll of the University of Chicago is collaborating with us in this work.)

Several unsuccessful attempts were made to find evidence of reaction in gaseous mixtures other than Xe + F_2. Krypton and fluorine were irradiated with the mercury arc in a specially designed nickel-sapphire cell at temperatures

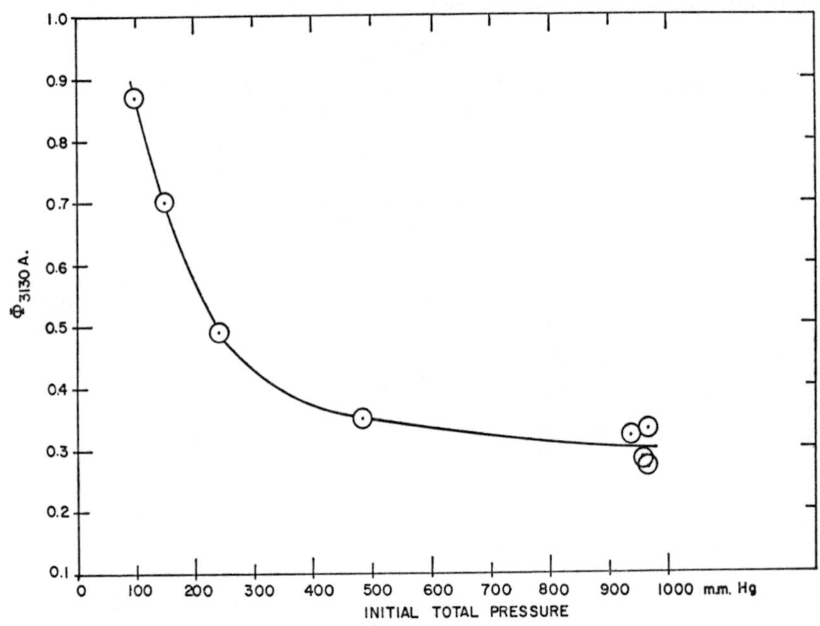

FIG. 2.—Variation of $\Phi_{3130A.}$ with pressure at constant F_2/Xe = 7

TABLE 3

EFFECT OF F_2/XE RATIO ON THE QUANTUM YIELD OF XeF_2 FORMATION AT 3130 A.

RUN	INITIAL PRESSURES			RATE OF PRESSURE DECREASE (mm Hg/hr)	LIGHT ABSORBED (%)	ABSORBED QUANTA/SEC. ($\times 10^{-15}$)	ϕ
	F_2/Xe	F_2 (mm. Hg)	Xe (mm. Hg)				
22	0.97	74	76	1.21	26	1.26	0.70
23	0.17	87	500	0.83	28	1.31	.47
25	0.089	77	903	0.55	25	1.37	.30
24	6.21	472	76	1.64	90	4.83	.25
16	1.00	478	478	1.87	90	4.90	.28
21	0.95	455	482	0.71	89	1.63	0.32

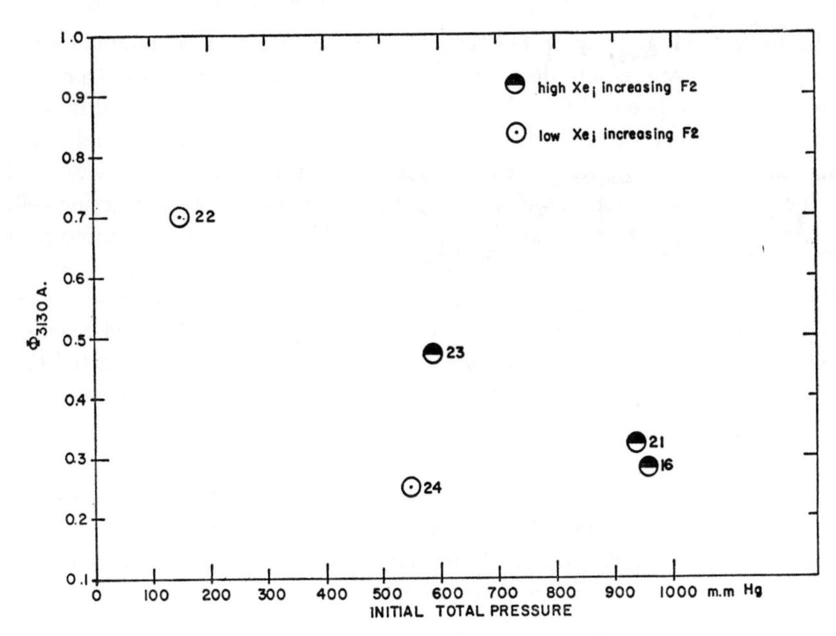

FIG. 3.—Effect of fluorine pressure on $\Phi_{3130A.}$ at a constant xenon pressure

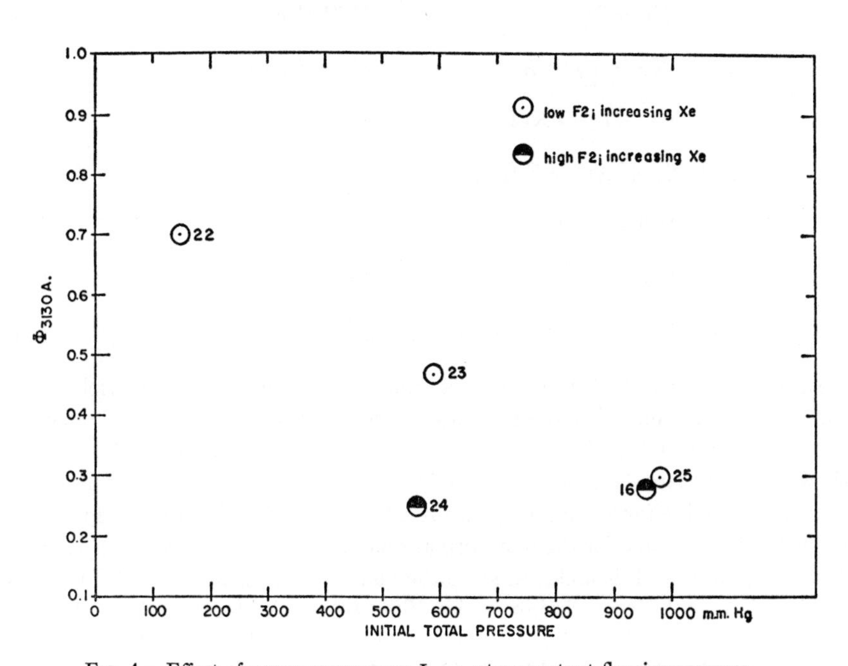

FIG. 4.—Effect of xenon pressure on $\Phi_{3130A.}$ at a constant fluorine pressure

down to that of liquid oxygen (−183° C.) with no evidence of compound formation as shown by pressure decrease or by sampling in the mass spectrometer. KrF_4 has been reported [9] as the product obtained when krypton and fluorine are reacted using an electric-discharge method. Xenon plus chlorine and radon plus chlorine were both irradiated at room temperature with a mercury arc and with convection circulation and a 0° C. bath on the U-bend of the reaction cell. No evidence of compound formation was observed in either system. A mixture of xenon + oxygen was irradiated in a high-purity silica vessel with a quartz Hanovia SC-2537 mercury resonance lamp (1849 A. intensity not measured) at room temperature and with −78° C. trapping. Again no evidence of reaction was noted, although formation of XeO_3 by other means has been reported [10, 11]. Further attempts at photochemical preparation of other rare-gas compounds are planned.

DISCUSSION

The fact that the fluorine absorption band at 2500–3500 A. is continuous and the tentative identification of XeF in flash photolysis suggest that the mechanism of XeF_2 (and perhaps of XeF_4) formation involves fluorine atoms. Possible steps in such a mechanism are listed below:

$$F_2 + h\nu \rightarrow 2F \qquad I_a, \text{ assumes } \phi_{\text{dissoc.}} = 1 \qquad (1)$$

$$2F + F_2 \rightarrow 2F_2 \qquad 2k_2(F)^2(F_2) \qquad (2)$$

$$2F + Xe \rightarrow F_2 + Xe \qquad 2k_3(F)^2(Xe) \qquad (3)$$

$$F + Xe + F_2 \rightarrow XeF + F_2 \qquad k_4(F)(Xe)(F_2) \qquad (4)$$

$$F + Xe + Xe \rightarrow XeF + Xe \qquad k_5(F)(Xe)(Xe) \qquad (5)$$

$$XeF + F \rightarrow XeF_2 \qquad k_6(F)(XeF) \qquad (6)$$

$$XeF + XeF \rightarrow XeF_2 + Xe \qquad 2k_7(XeF)^2 \qquad (7)$$

$$XeF + XeF \rightarrow 2Xe + F_2 \qquad 2k_8(XeF)^2 \qquad (8)$$

Reaction 2 is suggested by the effect of fluorine pressure on quantum yield (Fig. 3). The effect of xenon is more complex but reaction 3 may explain the effect of xenon pressure at low fluorine pressure. The steps for formation of XeF are shown by equations 4 and 5, with stabilization by a third body likely. These two steps also provide for a complex pressure effect of xenon. Either reaction 6 or 7, or both, are necessary for the final formation of XeF_2. No step involving XeF + F_2 to give XeF_2 + F is included since the quantum yields less than 1 do not as yet justify postulating a chain reaction. Step 8 provides for a back reaction to regenerate xenon and fluorine. If conditions can be found in which the quantum yield approaches 1, this would be evidence that reaction 8 is unimportant and also that a reaction of XeF + F to give Xe + F_2 need not be included. At present we need more courage or more data to face up to the formidable algebraic complexities which result when all 8 steps are included. It is hoped that further studies on intensity and pressure effects will reduce the problem of describing the mechanism to more manageable proportions. Continuing flash photolysis experiments (in co-operation with Dr. P. K. Carroll of the University of Chicago) should also help in elucidating the photochemical mechanism.

Acknowledgments.—The authors are grateful to W. M. Manning for his encouragement, to M. H. Studier and E. N. Sloth for mass-spectrometric analyses, to H. H. Claassen for infrared analyses, and especially to C. L. Chernick for his participation in the early phases of the work and for his continued co-operation. W. A. Mulac gave important technical assistance in the flash photolysis experiments.

REFERENCES

1. H. H. Claassen, H. Selig, and J. G. Malm, *J. Am. Chem. Soc.* **84,** 3593 (1962).
2. T. L. Cottrell, *The Strengths of Chemical Bonds*, p. 281. Ind. ed.; London: Butterworth Scientific Publications, 1958.
3. H. V. Wartenberg, G. Sprenger, and J. Taylor, *Z. Physik. Chem., Bodenstein-Festband*, p. 61. Leipzig: Akademische Verlagsgesellschaft, 1931.
4. J. L. Weeks, C. L. Chernick, and M. S. Matheson, *J. Am. Chem. Soc.* **84,** 4612 (1962).
5. M. Kasha, *J. Opt. Soc. Am.* **38,** 929 (1948).
6. L. I. Grossweiner and M. S. Matheson, *J. Phys. Chem.* **61,** 1089 (1957).
7. M. H. Studier and E. N. Sloth, *ibid.* **67,** 925 (1963).
8. C. L. Chernick *et al.*, *Science* **138,** 136 (1962).
9. A. V. Grosse, A. D. Kirshenbaum, A. G. Streng, and L. V. Streng, *ibid.* **139,** 1047 (1963).
10. D. F. Smith, *J. Am. Chem. Soc.* **85,** 816 (1963).
11. D. H. Templeton, A. Zalkin, J. D. Forrester, and S. M. Williamson, *ibid.* **85,** 817 (1963).

ON THE FLUORINATION OF XENON: XENON DIFLUORIDE

RUDOLF HOPPE, HARALD MATTAUCH, KARL-MATIN RÖDDER,
AND WOLFGANG DÄHNE
University of Münster

In joining the Klemm-group of Inorganic Chemistry at Münster University in 1949, one of us proposed the existence of XeF_4, which, it was thought, should result from the direct fluorination of xenon.

At that time Bode, working in the same laboratory, did not obtain KF by heating samples of KCl in fluorine but got instead a new substance (over-all composition KF_x with $x \sim 2$) showing all those qualities described later by the adjective "Fluoraktiv." This was tentatively interpreted as KF_3, and an analogous CsF_3 was proposed on the basis of the theory that a reasonable set of fluorides would include IF_7, XeF_4, CsF_3 rather than the abrupt transition IF_7, XeF_0, CsF. In those days it was impossible to get xenon in Germany, let alone xenon of sufficient purity.

Bode unfortunately prepared samples containing, in fact, compounds like $KClF_4$ (and $CsClF_4$). But, in view of the now well-established existence of fluorides of xenon, it is to be repeated, Cato-like, that CsF_3 could be obtainable. We are working on this problem.

In 1951 the heat of formation of IF_5 was determined by A. A. Woolff. It was this very heat of formation which confirmed, indirectly but immediately, the proposed existence of fluorides of xenon: From the corresponding bond energy, $E(\mathrm{I} - \mathrm{F})$ of IF_5 (64 kcal/mole) and the long-known bond energy, $E(\mathrm{Te} - \mathrm{F})$ of TeF_6 (79 kcal/mole) we extrapolated the bond energy $E(\mathrm{Xe} - \mathrm{F})$ of XeF_4 and XeF_2 (20–40 kcal/mole, at least) which indicated the *exothermic* formation of XeF_4 as well as XeF_2 from the elements. Consequently, in dealing with a lot of complex fluorides of the transition elements through the years, we have looked for the fluorination of xenon. (Attempts to prepare compounds like $KHg(III)F_4$, $K_2Au(IV)F_6$ and $KZn(III)F_4$ failed.)

THE FLUORINATION OF XENON BY DISCHARGES OF AN INDUCTION COIL

Assuming that XeF_2 and XeF_4 were the products of an exothermic reaction of xenon and fluorine, it was most reasonable to heat mixtures of xenon and excess fluorine under pressure. Steel cylinders containing fluorine under pressure are not yet available in Germany, and an attempt in the fall of 1961 to get those cylinders from the U.S.A. failed, so we prepared elementary fluorine on a laboratory scale using a small v. Wartenberg apparatus. This fluorine was carefully purified. Later on, the quartz vessel used in preparing XeF_2 showed no visible signs of being attacked by HF after being used nearly every day for months.

Due to the relatively small amounts of fluorine available at the beginning of our experiments it did not seem convenient to fill small cylinders with xenon and

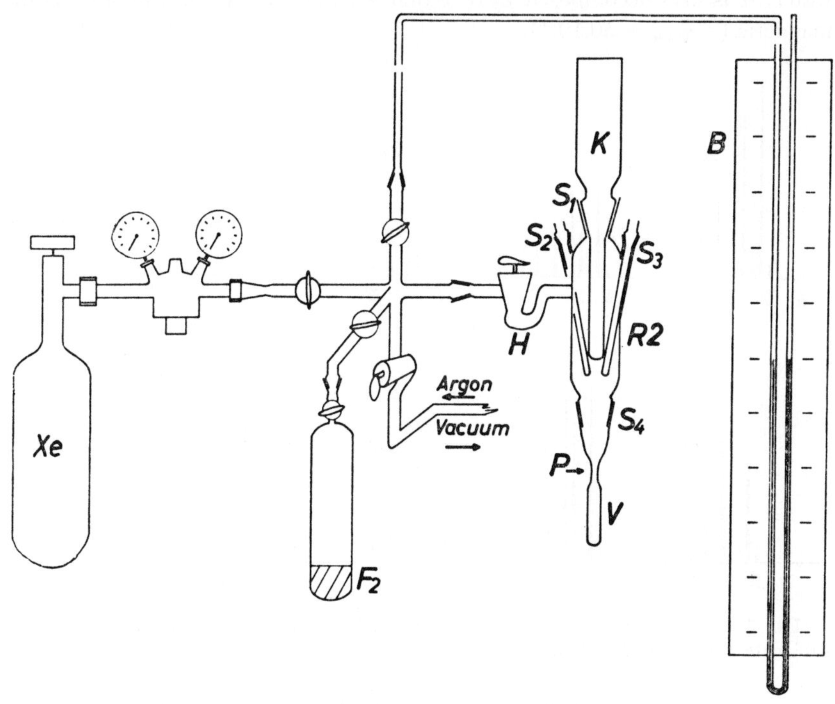

FIG. 1.—Apparatus for the synthesis of XeF_2 (schematic)

fluorine under pressure. However, there were two possibilities, in addition to heating, for activating F_2-Xe mixtures—either ultraviolet light or electric discharge using an induction coil. We chose the second possibility. The fluorine gas was mixed with xenon (99.9 per cent Xe; Linde-AG; Kr < 0.1 per cent; $N_2 < 0.01$ per cent; $O_2 < 0.001$ per cent) in the molar ratio $Xe/F_2 = 1{:}1$ to $1{:}3$, $1{:}2$ being the most frequently used molar ratio. This mixture was subjected to the discharge of an induction coil. The mixture was placed in a quartz vessel, as shown in Figure 1. The electrodes are fused into fingers projecting into this vessel. The reaction was followed by a decrease in pressure. Brilliant, glittering colorless crystals up to a millimeter in diameter condensed on a cold finger cooled with a dry ice–alcohol bath. The container was agitated and the crystals dropped off into a weighed glass trap which could be sealed after filling. This dense, glittering condensate, as well as the product of resublimation (*in vacuo* as well as under dry air), is pure XeF_2, as is proved by direct analytical determination of the F-content and by mass spectrometry (Fig. 2).

Xenon difluoride forms well-shaped colorless crystals. Weissenberg photographs suggest that it probably crystallizes in the monoclinic system. We are presently working on the elucidation of the crystal structure. It has a strong nauseating odor and is decomposed by H_2O, NH_3 water and N_2H_4 dissolved in water; it is also decomposed by reaction with acids. Xenon difluoride is diamagnetic ($-\chi_{mol} = 50.10^{-6}$).

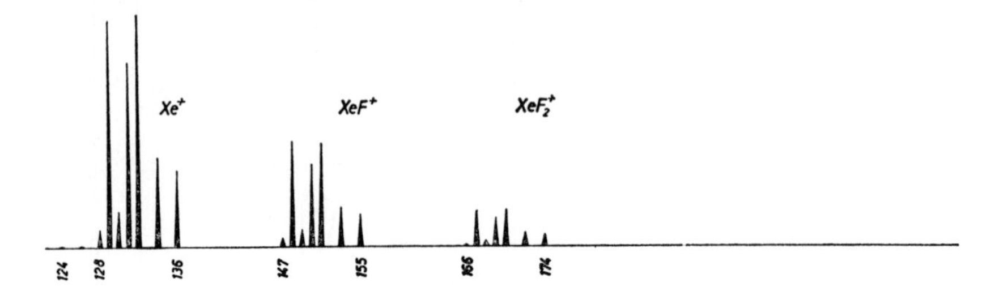

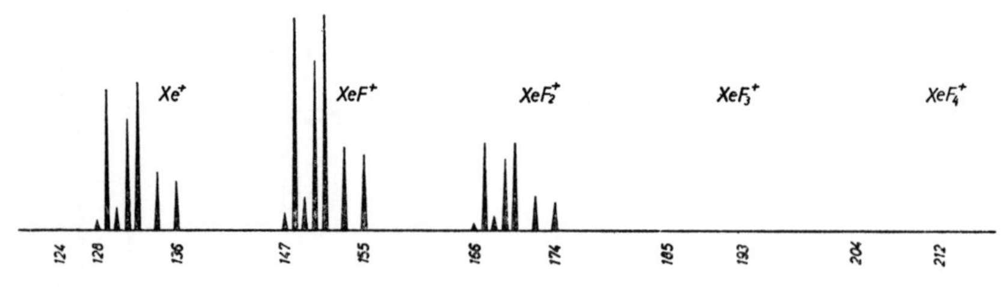

FIG. 2.—Mass spectrum of XeF_2. *Top*—primary product $E = 1/10{:}1/3{:}1/3$. *Bottom*—sublimate $E = 1/100{:}1/3{:}1/3$.

ON THE CHEMICAL BOND IN XeF_2 AND XeF_4

In 1819, Gay-Lussac discovered the interhalogen compounds ICl and ICl_3. Xe^{2+} and Xe^{4+} are isoelectronic with I^+ and I^{3+}, respectively. Hence, in a first approximation it can be assumed, that the chemical bond in ICl (or IF itself) and XeF_2 on the one hand and in ICl_3 (or much more fitting IF_3) and XeF_4 are quite similar. The existence of chemical compounds of xenon is as understandable in principle as the existence of the corresponding compounds of iodine. In discussing the existence of new compounds of xenon and other rare gases, extrapolated bond energies are to be used, leading, for example, to the possibility of exothermic formation of KrF_2, which we have probably obtained.

PREPARATION OF INERT-GAS COMPOUNDS BY MATRIX ISOLATION: KRYPTON DIFLUORIDE

J. J. TURNER* AND G. C. PIMENTEL
University of California, Berkeley

Syntheses of the xenon fluorides have been based upon various treatments of gaseous mixtures of fluorine and xenon: pyrolysis near 400° C. in a nickel tube [1–3], photolysis at −80° C. with a high-pressure mercury arc [4], or high-voltage discharge at −78° C. [5]. A high-voltage discharge through a gaseous mixture of CF_4 and xenon [6] has been used to prepare XeF_2.

Attempts to prepare other inert-gas compounds have met with mixed success. A radon fluoride was prepared by Chernick *et al.*, using pyrolysis [7], and KrF_4 by Grosse *et al.* [8], using electric discharge at 85° K. On the other hand, pyrolysis of krypton and fluorine gaseous mixtures at 400° C. failed to give any krypton fluoride [8], and no compound formation resulted from photolysis [4] of gaseous mixtures of krypton and fluorine (room temperature and −60° C.), krypton and chlorine, radon and chlorine, and xenon and oxygen. Hanlan and Pimentel sought, but were unable to find, infrared spectroscopic evidence for inert gas–BF_3 adducts using argon, krypton, and xenon at 20° K. [9].

It is timely to report on our preparation of inert-gas compounds using the matrix-isolation technique [10]. This technique was developed for the express purpose of preparing and characterizing reactive and relatively unstable species. The basic idea is to suspend the species in an inert solid matrix and investigate its spectroscopic properties. Inert gases and nitrogen are usual matrix materials.

The sample is prepared by quickly freezing on a cold salt window a gas mixture of a small amount of some relevant material in a large amount of matrix gas. The conditions must be carefully controlled [11]. There are several ways in which the species may be obtained *in situ:* photolysis of some well-isolated parent material (e.g., CH_3NO_2 to give HNO [12]), photolysis of a molecule to initiate reaction with the matrix (e.g., HI in CO to give HCO [13]), photolysis of a molecule to initiate reaction with another molecule in an adjacent site, both species being diluted in an inert matrix (e.g., N_2O to give O, reacting with acetylene to give ketene [14]). Linevsky [15] has also studied monomeric LiF by con-

* Harkness Fellow.

Research reported in this chapter was supported by the Petroleum Research Fund, the Air Force Office of Scientific Research, and the Commonwealth Fund.

densing it from the gas phase with large amounts of inert gases. Some of these methods are proving to be particularly useful for the halogens in view of their ease of dissociation and are applicable to the preparation of inert gas–halogen compounds.

Mixtures of fluorine and inert gases were deposited slowly upon a CsI window at 20° K. After an infrared spectrum had been recorded from 700 cm.$^{-1}$ to 230 cm.$^{-1}$ (using a Beckman IR-7 Spectrometer), the sample was irradiated at 20° K. with the focused light from an AH-4 medium-pressure mercury lamp (without its glass envelope). The infrared spectrum was then recorded again. Frequency accuracy is about ± 2 cm.$^{-1}$. In each experiment, about 4 millimoles of pure argon were deposited, prior to deposition of the sample, forming a layer of solid protecting the window from fluorine.

Fluorine-argon.—A gas mixture of fluorine and argon (4 millimoles at a mole

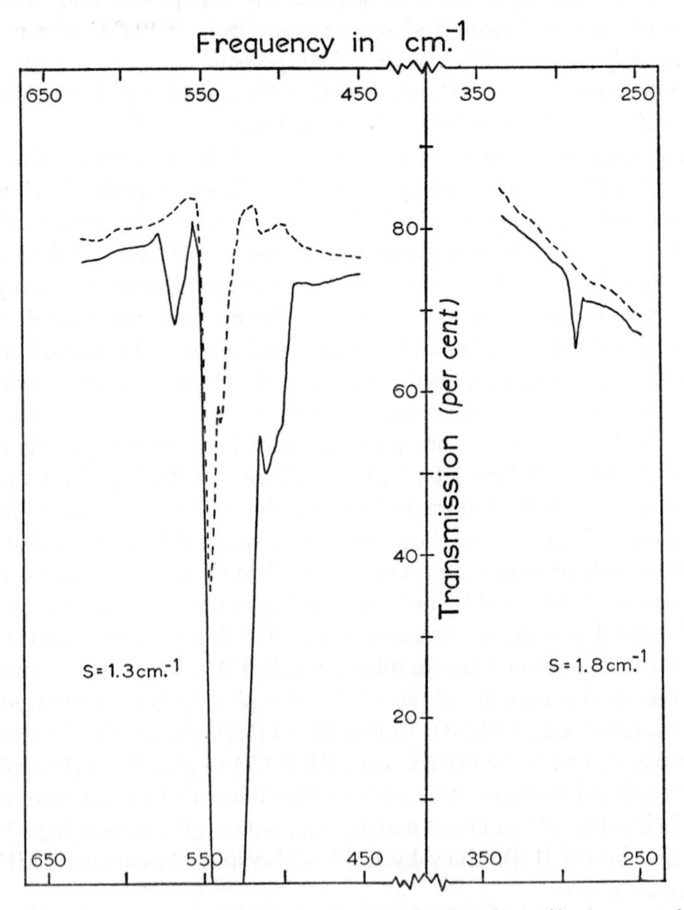

FIG. 1.—Infrared spectrum of solid sample of F_2/Xe/Ar (1:1:17). (............ after four minutes' photolysis; ——— after fourteen minutes' photolysis; S = spectral slit width.)

ratio F_2/Ar = 1:18) was deposited and photolyzed for eight hours. Six weak absorptions (optical density < 0.03) were observed, but subsequent experiments with added oxygen showed that all six bands are associated with F_2-O_2 compounds presumably resulting from oxygen impurity in the fluorine gas. Under these conditions, no evidence could be found for argon fluorides.

Fluorine-xenon-argon.—A gas mixture of fluorine, xenon, and argon (3.5 millimoles at mole ratios $F_2/Xe/Ar$ = 1:1:17) was deposited and photolyzed for four minutes. An intense absorption was observed at 547 cm.$^{-1}$, with a shoulder at 539 cm.$^{-1}$ and a weak band at ~510 cm.$^{-1}$ (Fig. 1). On further photolysis, these bands became more intense and two new features appeared. The new features, at 568 cm.$^{-1}$ and 290 cm.$^{-1}$, appeared to increase in intensity by the same factor in each successive photolysis. Diffusion experiments (in which the

TABLE 1

MATRIX ABSORPTIONS OF XENON FLUORIDES COMPARED TO GAS-PHASE FREQUENCIES

COMPOUND	VIBRATION	FREQUENCY (CM.$^{-1}$)	
		Matrix, 20 ° K.	Gas*
XeF_2..........	ν_1	~510 cm.$^{-1}$	515
	ν_2	not observed	213
	ν_3	547	555
XeF_4..........	ν_6	568	586
	ν_2	290	290

* See references 2 and 16.

sample is warmed to about 50° K. and recooled to 20° K.) show that the absorption at 539 cm.$^{-1}$ is caused by aggregates and the others by isolated molecules. In separate experiments, dilution of the mixture with argon confirms this conclusion.

Another experiment using a much larger mole ratio of F_2/Xe ($F_2/Xe/Ar$ = 5:1:100) enhanced the absorptions near 568 cm.$^{-1}$ and 290 cm.$^{-1}$, relative to a rather broad absorption near 540 cm.$^{-1}$. These experiments lead to the assignments shown in Table 1.

Fluorine-krypton-argon.—A gas mixture of fluorine, krypton, and argon (11 millimoles at mole ratios $F_2/Kr/Ar$ = 1:70:220) was deposited and photolyzed for three hours. Two absorptions were observed, a fairly narrow band (half-width about 3.5 cm.$^{-1}$) at 580 cm.$^{-1}$ with optical density 0.19 and a second band at 236 cm.$^{-1}$ with optical density 0.04 (Fig. 2). In a separate experiment (at different mole ratios) on diffusion, the band at 580 cm.$^{-1}$ broadened and shifted to about 570 cm.$^{-1}$. In analogy to the XeF_2 case, we attribute this to the formation of aggregates. The results of an experiment in which oxygen was deliberately added to the mixture eliminated the possibility that these two absorptions are associated with oxygen impurity. We believe that the absorptions at 580 cm.$^{-1}$ and 236 cm.$^{-1}$ result from KrF_2, ν_3, and ν_2, respectively.

Another experiment was performed using a larger mole ratio of F_2/Kr ($F_2/Kr/Ar$ = 5:1:20). Broad absorptions near 575 cm.$^{-1}$ and 245 cm.$^{-1}$ were observed (o.d., respectively, 0.6 and 0.1). We feel that these absorptions also result from KrF_2 with aggregation frequency shifts associated with the higher concentrations, as observed for XeF_2. Unfortunately, in this experiment spectral absorptions due to F_2-O_2 impurities obscured the region above 580 cm.$^{-1}$, the region in which the KrF_4 absorption is expected.

The frequencies assigned to monomeric KrF_2 imply the force constants shown in Table 2, contrasted there with the corresponding values for XeF_2 in an argon matrix. The notation is that used by Agron *et al.* [2].

We find it surprising that the force constants of KrF_2 are so close to those of

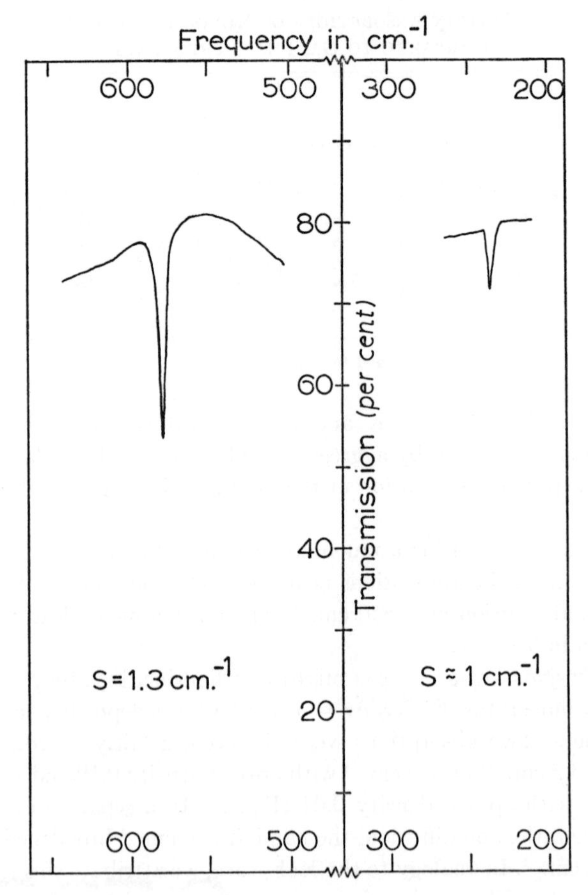

FIG. 2.—Infrared spectrum of solid sample of $F_2/Kr/Ar$ (1:70:220), after three hours' photolysis. (S = spectral slit width.)

XeF_2. It is implied that the bond energies are similar, although XeF_2 is more easily formed than KrF_2. Also, it is interesting that KrF_4 decomposes more readily than XeF_4 [8]. Further work is in progress to consider the impact of this result upon current discussions of the bonding in the inert-gas compounds.

In summary, it is clear that the matrix-isolation technique is a useful preparative method for inert-gas compounds, and it may have unique value for those that are least stable. It is significant, perhaps, that no evidence could be obtained for an argon-fluorine compound using this method.

TABLE 2

COMPARISON OF FORCE CONSTANTS OF KRF_2 AND XEF_2

COMPOUND	MDYNE/A.		
	$k_r - k_{rr}$	$k_r + k_{rr}$	$k\delta/l^2$
KrF_2..........	2.59	....	0.21
XeF_2..........	2.60	~2.91	0.20*

* Gas-phase value from reference 2.

Note added in proof.—A brief report of the preparation of KrF_2 has been published by the authors (*Science* **140,** 974 [1963]).

REFERENCES

1. H. H. CLAASSEN, H. SELIG, and J. G. MALM, *J. Am. Chem. Soc.* **84,** 3593 (1962).
2. P. A. AGRON *et al.*, *Science* **139,** 842 (1963).
3. D. F. SMITH, *J. Chem. Phys.* **38,** 270 (1963).
4. J. L. WEEKS, C. L. CHERNICK, and M. S. MATHESON, *J. Am. Chem. Soc.* **84,** 4612 (1962).
5. A. D. KIRSHENBAUM, L. V. STRENG, A. G. STRENG, and A. V. GROSSE, *ibid.* **85,** 360 (1963).
6. D. E. MILLIGAN and D. R. SEARS, *ibid.* **85,** 823 (1963).
7. C. L. CHERNICK *et al.*, *Science* **138,** 136 (1962).
8. A. V. GROSSE, A. D. KIRSHENBAUM, A. G. STRENG, and L. V. STRENG, *ibid.* **139,** 1047 (1963).
9. J. F. HANLAN, Ph.D. Thesis, University of California, 1961.
10. G. C. PIMENTEL, *Spectrochim. Acta* **21,** 94 (1958).
11. A. M. BASS and H. P. BROIDA (eds.), *Formation and Trapping of Free Radicals*, Chap. 4. New York: Academic Press, 1960.
12. H. W. BROWN and G. C. PIMENTEL, *J. Chem. Phys.* **29,** 883 (1958).
13. G. E. EWING, W. E. THOMPSON, and G. C. PIMENTEL, *ibid.* **32,** 927 (1960).
14. I. HALLER and G. C. PIMENTEL, *J. Am. Chem. Soc.* **84,** 2855 (1962).
15. M. J. LINEVSKY, *J. Chem. Phys.* **34,** 587 (1961).
16. H. H. CLAASSEN, C. L. CHERNICK, and J. G. MALM, *J. Am. Chem. Soc.* **85,** 1927 (1963).

XENON OXYTETRAFLUORIDE

CEDRIC L. CHERNICK, H. H. CLAASSEN,*
JOHN G. MALM, AND P. L. PLURIEN†
Argonne National Laboratory

ABSTRACT

Xenon oxytetrafluoride, $XeOF_4$, has been prepared by the reaction of xenon hexafluoride with water or silica. The compound is a colorless liquid with a melting point of $-28°$ and a vapor pressure of 8 mm. at 0°.

INTRODUCTION

Earlier work in this laboratory indicated the existence of xenon oxyfluorides [1, 2], and a clear, colorless liquid had been obtained when xenon hexafluoride was allowed to remain in a quartz container [3]. This liquid has now been identified as xenon oxytetrafluoride, $XeOF_4$, and has been shown to be the same material as that obtained by careful hydrolysis of XeF_6 with water in the vapor phase.

EXPERIMENTAL

Xenon hexafluoride was prepared by the method described elsewhere in this volume [see p. 37].

Reaction with silica.—Xenon hexafluoride was condensed in a quartz bulb fitted with a break-seal, and the bulb sealed while immersed in liquid nitrogen. The bulb was heated to about 50° and the XeF_6 liquid allowed to react with the quartz until the characteristic pale-yellow color had disappeared, indicating complete reaction of the hexafluoride. (About 5 gm. of XeF_6 in a 250 ml. bulb completely reacted after a couple of days.) At this stage there was considerable evidence of etching of the quartz and a colorless liquid in the bottom of the bulb. The bulb was sealed onto a vacuum line, immersed in a dry-ice–trichloroethylene bath, and opened to the line; the silicon tetrafluoride that had been produced was removed by pumping.

$$2XeF_6 + SiO_2 \rightarrow 2XeOF_4 + SiF_4 \quad (1)$$

A sample that was allowed to remain in quartz for some time after the complete reaction of the hexafluoride was found to contain a white crystalline mate-

* Permanent address: Wheaton College.

† Permanent address: Centre d'Études Nucléaires de Saclay, Seine-et-Oise, France.

rial in addition to the liquid. This material, which was highly explosive, was presumed to be xenon trioxide [4]. If this is the case, the reaction probably proceeds step-wise,

$$2XeOF_4 + SiO_2 \rightarrow 2XeO_2F_2 + SiF_4\,, \tag{2}$$

$$2XeO_2F_2 + SiO_2 \rightarrow 2XeO_3 + SiF_4\,. \tag{3}$$

Trace amounts of XeO_2F_2 have been noted in mass-spectrometric studies of some of our $XeOF_4$ samples [see Studier and Sloth, p. 47, this volume].

Hydrolysis of XeF_6.—A weighed amount of XeF_6 was hydrolyzed using the amount of water theoretically required by the reaction,

$$XeF_6 + H_2O \rightarrow XeOF_4 + 2HF\,. \tag{4}$$

The hexafluoride was condensed in a small nickel can immersed in a liquid-nitrogen bath. The required amount of distilled water was measured volumetrically and condensed into a glass tube. The tube and can were pumped down to 10^{-4} mm. Hg.; the valve to the vacuum line was closed; and the water was condensed into the can. The valve to the can was closed, and the can was warmed to room temperature. The amounts of materials taken were such that the water and XeF_6 would be completely vaporized at room temperature. After standing for approximately half an hour the can was cooled to $-78°$, and the major portion of the HF pumped away.

Purification.—The $XeOF_4$ produced by either method was subjected to the same final purification procedures. Volatile impurities were removed by repeated trap-to-trap distillations while pumping, the oxytetrafluoride fraction being collected in traps at $-78°$. The $XeOF_4$ was then separated from less volatile materials by keeping the sample at $-25°$ and pumping the $XeOF_4$ off and trapping it at $-78°$.

Analysis.—The formula has been established by reduction with hydrogen. A weighed amount of material was heated for one hour with excess hydrogen at 300° C., the following reaction taking place:

$$XeOF_4 + 3H_2 \rightarrow Xe + H_2O + 4HF\,. \tag{5}$$

After heating, the can was cooled, and the excess hydrogen pumped out through successive traps held at $-159°$ and $-195°$. The first trap caught the water and hydrogen fluoride, and the second the xenon. The contents of both traps were weighed, and the H_2O-HF mixture was then hydrolyzed in a known excess of caustic soda. From the fluoride analysis, the water content could be found by difference.

RESULTS

The analysis of the compound was as follows:

Calculated for $XeOF_4$: Xe, 58.7 per cent; O, 7.2 per cent; F, 34.1 per cent.

Found (1): Xe, 55.5 per cent; O, 7.4 per cent; F, 34.5 per cent; and (2): Xe, 59.2 per cent; O, 7.6 per cent; F, 32.0 per cent.

The formula has also been confirmed by mass-spectrometric measurements in which $XeOF_4^+$ and its degradation products have been observed [see Studier and Sloth, p. 47, this volume].

Properties.—Xenon oxytetrafluoride is a colorless, stable liquid at room temperature. It has been stored unchanged in nickel containers for several weeks. As already noted it slowly attacks quartz—finally yielding xenon trioxide. The melting point of a sample sealed in a quartz capillary was found to be $-28°$. Preliminary measurements of the vapor pressure indicate values of 8 mm. at 0° and 29 mm. at 23°.

REFERENCES

1. C. L. Chernick *et al.*, *Science* **138,** 136 (1962).
2. M. H. Studier and E. N. Sloth, *J. Phys. Chem.* **67,** 925 (1963).
3. J. G. Malm, I. Sheft, and C. L. Chernick, *J. Am. Chem. Soc.* **85,** 110 (1963).
4. D. F. Smith, *ibid.* 816 (1963).

REACTIONS OF XENON WITH CERTAIN STRONG OXIDIZING AGENTS

Gary L. Gard, Francis B. Dudley,*
and George H. Cady
University of Washington

The syntheses of the xenon fluorides [1–7] and xenon hexafluoroplatinate [8], hexafluororhodate [see Bartlett and Jha, p. 23, this volume] and hexafluororuthenate [1] suggested that other strong oxidizing agents could possibly react with xenon. Therefore, the reactions of xenon with trifluoromethyl hypofluorite

O

(CF_3OF), oxygen difluoride (OF_2), fluorine fluorosulfate (FOSF), and peroxy-

O O O

disulfuryl difluoride (FSOOSF) were studied.

O O

One set of reactions was carried out in a prefluorinated 1.5-liter nickel vessel. The experimental conditions and results are given in Table 1.

TABLE 1

Reactants y+Xe	Molar Ratio y/Xe	Minimum Temperature for Reaction (° K.)	Maximum Temperature Attained (° K.)	Products
O O FSOOSF+Xe....... O O	3:1	....	750	$Xe+O_2+SO_2F_2$
OF_2+Xe...........	1:1	560	660	?

$S_2O_6F_2$ + Xe.—The mixture of $S_2O_6F_2$ and Xe had a total pressure of 160 mm. at 300° K. As the mixture was heated, the pressure increased ideally in proportion to the absolute temperature up to about 500° K. Above this temperature the rate of increase in pressure was abnormally great because of the dissociation of $S_2O_6F_2$ into free radicals [9] according to the equation

$$S_2O_6F_2 \rightleftharpoons 2SO_3F \, .$$

* Permanent address: University of New England, Australia.

The research reported in this chapter was supported in part by the Office of Naval Research.

The observed abnormality was that expected for this equilibrium. If xenon fluorosulfate had been formed, the total pressure would have been smaller than that observed. There was no evidence for a reaction involving xenon.

$OF_2 + Xe$.—When the temperature of an equimolar mixture of oxygen difluoride and xenon having a total pressure of 104.5 mm. at 452° K. was slowly raised, the pressure increased ideally with the absolute temperature up to about 560° K. Above this temperature the pressure decreased, indicating that chemical combination was in progress. The product of the reaction was not studied. It is probably significant that decomposition of oxygen difluoride by the reaction

$$2OF_2 \rightarrow O_2 + 2F_2$$

occurs at temperatures above about 530° K. A pressure *vs.* rising temperature curve for a sample of oxygen difluoride shows an abnormal increase in pressure above 530° K. because of this dissociation. When xenon was present, the reaction of fluorine gas with xenon probably balanced the dissociation of OF_2 to permit the number of molecules in the gas to remain about constant from 530° to 560° K. Above this temperature the number of molecules decreased. It appears likely that xenon did not react directly with OF_2 but did react with F_2 formed from oxygen difluoride, as the pressure *vs.* temperature curves for mixtures of xenon and fluorine showed that chemical combination occurred at temperatures as low as 390° K.

Another set of reactions was carried out in prefluorinated Monel pressure tubes equipped with Hoke needle values and brass $\frac{10}{30}$ connections. The experimental conditions and results were as follows:

TABLE 2

Reactants y+Xe	Molar Ratio y/Xe	Pressure (atms.)	Temperature (° K.)	Products
CF_3OF+Xe....	2.2:1	about 250	500	XeF_2 and CF_3OOCF_3
FSO_3F+Xe....	2.2:1	about 150	450	XeF_2 and $S_2O_6F_2$

These pressures were estimated from the amounts of materials present and from the temperature and volume of the system.

$Xe + 2CF_3OF \rightarrow XeF_2 + CF_3OOCF_3$.—We distilled 0.0280 mole of CF_3OF into a prefluorinated Monel tube (volume 6 cc.) and added 0.0125 mole of xenon. The tube was put into a heated oven and kept at 220° to 250° C. for seven days. (CF_3OF decomposes reversibly into COF_2 and fluorine at temperatures greater than about 275° C.) The tube was then withdrawn from the oven and cooled quickly in a water bath. The tube was further cooled to −78° C., and the volatile product was pumped away through a trap held at liquid-nitrogen temperature. The volatile materials other than xenon were shown to be CF_3OF and CF_3OOCF_3 by infrared identification. The material remaining in the tube

was shown by chemical analysis to be xenon difluoride. Results of analyses are given in Table 3.

The mechanism of this reaction may have involved the dissociation of a molecule of CF_3OF into a fluorine atom and a $CF_3O\cdot$ free radical. The fluorine would have been available to combine with xenon, and free radicals could have combined to form CF_3OOCF_3.

$Xe + 2FSO_3F \rightarrow XeF_2 + S_2O_6F_2$.—We distilled 0.0079 mole of FSO_3F into a prefluorinated Monel tube (volume 3 cc.), and added 0.0037 mole of xenon. The tube was placed in a heated oven and kept for four days at 170° to 180° C. (FSO_3F decomposes at temperatures greater than 200° C.) The tube was withdrawn from the oven and cooled quickly in a water bath. The tube was cooled to

TABLE 3

	Experimental (per cent)	Theoretical (per cent)
Xenon	{76.3 76.5	77.6
Fluorine	21.98	22.4

TABLE 4

	Experimental (per cent)	Theoretical (XeF_2) (per cent)
Xenon	74.4	77.6
Fluorine	24.05	22.4

−14° C., and the volatile product was pumped away through a trap held at liquid-oxygen temperature. The volatile material was shown to be $S_2O_6F_2$ by infrared identification. The material remaining in the tube was shown to be mainly XeF_2 with small amounts of XeF_4. Results of analyses are given in Table 4.

This reaction may have involved dissociation of SO_3F_2 into fluorine atoms and $SO_3F\cdot$ radicals. The latter could have combined to form $S_2O_6F_2$, while the former yielded xenon fluorides.

In an attempt to prepare a xenon fluorosulfate, 0.0034 mole of XeF_2 was allowed to react with 0.0117 mole of SO_3 at 120° C. After five days the reactor was removed from the oven and quenched in water. The tube was further cooled to −78° C. and then pumped on through a trap held at liquid-nitrogen temperatures. The volatile materials were shown to be xenon and oxygen. The remaining material in the tube was found to be $S_2O_5F_2$ by infrared identification. Apparently the reaction was

$$XeF_2 + 2SO_3 \rightarrow Xe + S_2O_5F_2 + \tfrac{1}{2}O_2 .$$

REFERENCES

1. C. L. CHERNICK *et al.*, *Science* **138,** 136 (1962).
2. H. H. CLAASSEN, H. SELIG, and J. G. MALM, *J. Am. Chem. Soc.* **84,** 3539 (1962).
3. J. L. WEEKS, C. L. CHERNICK, and M. S. MATHESON, *ibid.* 4612 (1962).
4. J. G. MALM, I. SHEFT, and C. L. CHERNICK, *ibid.* 110 (1963).
5. E. E. WEAVER, B. WEINSTOCK, and C. P. KNOP, *ibid.* 111 (1963).
6. J. SLIVNIK *et al.*, *Croat. Chem. Acta* **34,** 253 (1962).
7. F. B. DUDLEY, G. L. GARD, and G. H. CADY, *Inorg. Chem.* **2,** 228 (1963).
8. N. BARTLETT, *Proc. Chem. Soc.* **218** (1962).
9. F. B. DUDLEY and G. H. CADY, *J. Am. Chem. Soc.*, in press.

RADON FLUORIDE: FURTHER TRACER EXPERIMENTS WITH RADON

P. R. Fields, L. Stein, and M. H. Zirin
Argonne National Laboratory

Abstract

It has been shown in tracer experiments that radon fluoride can be prepared by heating mixtures of radon and fluorine to 400° C. The compound is very stable and distils at 230° to 250° at a pressure of approximately 10^{-6} mm. Hg. It can be reduced with hydrogen at 500° to recover elemental radon.

Attempts to prepare a radon chloride by thermal and photochemical methods have been unsuccessful thus far. No chemical reactions have been detected in mixtures of radon and oxygen irradiated with ultraviolet light or passed through a Berthelot-type ozone generator. However, it has been observed that the high-voltage electrical discharge used to produce ozone causes some radon to be strongly fixed on Pyrex surfaces. Microwave discharges have also been found to be very effective in fixing radon on quartz and Pyrex. In experiments with a metal microwave cell, radon has been fixed on a central brass antenna and subsequently released by heating the brass to the softening point (~900°).

After the existence of fluorine compounds of xenon had been clearly demonstrated by Bartlett [1] and Claassen, Selig, and Malm [2], it was shown in tracer experiments [3] that radon also forms a stable compound with fluorine. Radon fluoride was first prepared in a metal vacuum line of the type shown in Figure 1. Gaseous radon-222, obtained from 2 mg. of radium chloride in aqueous solution, was passed through a magnesium perchlorate drying tube into a 100-cc. Pyrex bulb. The bulb was then attached to the prefluorinated nickel and Monel vacuum line, and part of the radon was condensed in *Trap B* at −195°. When the trap was warmed to −78°, it was demonstrated that the radon moved readily under vacuum into other parts of the line which were cooled to −195°. Since radon decays by emission of alpha particles, which cannot penetrate heavy-walled vessels, the radon position was determined by counting the 1.8-mev. gamma activity of a subsequent daughter, Bi^{214}. The radioactive decay scheme, from Ra^{226} to Pb^{206}, is shown in Figure 2. The Bi^{214} grew into equilibrium with radon wherever the latter appeared and decayed where it disappeared, within several hours. The gamma counting was done with sodium iodide scintillation detectors, shielded by lead bricks, and a 400-channel pulse-height analyzer.

Radon fluoride was prepared in several experiments by condensing 5 to 100 microcuries of radon in a 5-cc. nickel reaction vessel at −195°, adding fluorine to 300 mm. pressure, and heating the mixture to 400° for thirty minutes. When the vessel was cooled to −78° and the excess fluorine was pumped off, it was found that the radon remained fixed in the reaction vessel. The compound which had been formed did not distil in a vacuum of 10^{-5} mm. to 10^{-6} mm. until heated to a temperature of 230° or higher. In some instances, a reaction vessel with a capillary inlet tube was used, and the compound could then be concentrated in a small section of the metal capillary by distillation at 230° to 250°.

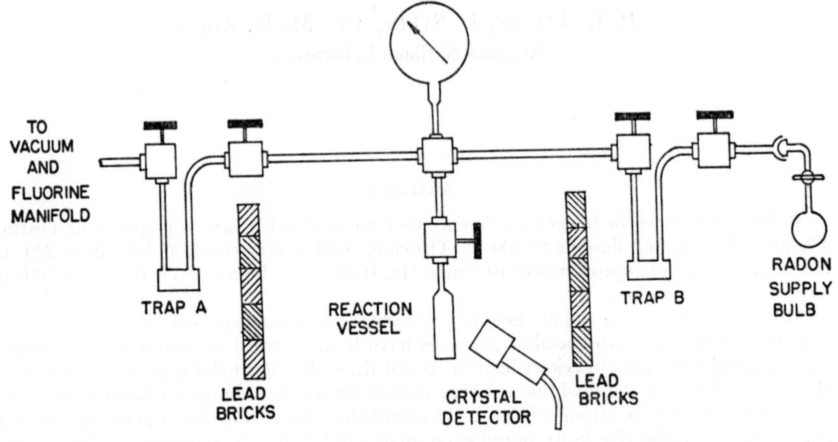

FIG. 1.—Vacuum line for the preparation of radon fluoride

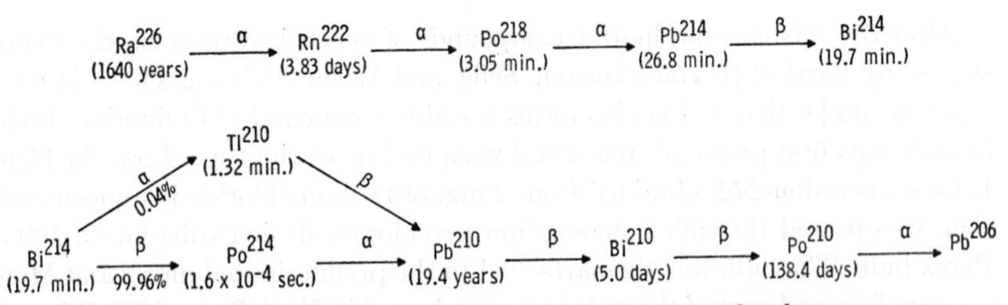

FIG. 2.—Radioactive decay of Ra^{226} and subsequent daughters

In one experiment, a mixture of xenon and radon was fluorinated; the xenon fluoride product was distilled at 50° into a cold trap at −195°, leaving the less volatile radon fluoride behind.

Radon fluoride can be reduced with hydrogen at 500° to recover elemental radon. At 200° the compound does not appear to react with hydrogen. The composition of the compound has not yet been determined, but it may be possible to vaporize samples inside a Time-of-Flight Mass Spectrometer. The tracer quantities of radon fluoride prepared thus far have shown no evidence of radiation decomposition from alpha-particle emission. However, the compound has been present in very dilute form on the inner surfaces of the container vessels, and most of the energy of the alpha particles has therefore been absorbed by the metal walls rather than by the compound.

Attempts have recently been made to prepare other radon compounds by thermal, photochemical, and electrical discharge methods, using 2 to 260 microcurie amounts of radon and milligram or gram amounts of other chemicals. Whereas elemental radon distils in vacuum from a cold trap at −78°, it is assumed that radon compounds in general will not distil at this temperature. Reduction in volatility of the radon has therefore been used as a simple criterion for compound formation, although it is recognized that very volatile compounds may not be detected by this method. In the first attempt to prepare a radon chloride, 16 microcuries of radon were mixed with chlorine at 150 mm. pressure in a quartz U-tube, and the mixture was heated to 400° for forty-five minutes. The U-tube was cooled to −195° to condense all the radon and chlorine, then warmed to −78°; at this temperature both the radon and chlorine distilled into a cold trap at −195°, indicating that no compound of low volatility was formed. When the same mixture was heated to 800° for ninety minutes and rapidly quenched in liquid nitrogen, similar results were obtained. A mixture of 2.8 microcuries of radon and approximately 0.3 ml. of iodine monochloride, heated to 500° for twenty minutes, also gave no evidence of reaction. The effect of ultraviolet light on 9.8 microcuries of radon and chlorine at approximately 150 mm. pressure was studied in a quartz cell of 4-in. path length. The mixture was irradiated with light from a high-pressure mercury arc for sixty-four hours, with the gas circulated through the cell by a convection loop operating between 0° and 80°. Again no compound appeared to be formed.

The effect of ozone on radon was first studied photochemically. The quartz cell was filled with 18 microcuries of radon and oxygen at 165 mm. pressure, then irradiated for forty hours at room temperature with intense light, predominantly of 2537 A. wave length. Although a small amount of ozone was produced, no evidence of reaction with radon was obtained. The effect of higher concentrations of ozone was then studied with the apparatus shown in Figure 3. Ozone was generated by circulating oxygen at 50 mm. to 500 mm. pressure through a Pyrex Berthelot tube [4], which was immersed in dilute copper sulfate solution with the electrodes connected to a 15,000-volt transformer. In some instances a heating tape was used above the cold trap to produce gas convection through the ozone generator; in others, liquid oxygen was condensed in the supply bulb and distilled slowly through the generator to the cold trap. A mixture of ozone and oxygen was collected in the trap at −195°, and most of the oxygen was then pumped off, leaving the less volatile ozone in the trap. With Halocarbon grease on the stopcocks, the ozone decomposed very slowly, even at room temperature. Several experiments were tried in which radon was mixed with the oxygen beforehand and circulated through the ozonizer to the cold trap, which was either at −195° or −78°. The radon in the trap appeared to be unchanged, since it remained volatile at −78°. In other experiments, radon was frozen initially in the trap at −195°, ozone was then generated and collected in the trap, and the mixture was allowed to warm to room temperature overnight. Again no evidence of compound formation was obtained.

It was observed, however, that whenever radon-oxygen mixtures were circulated through the ozone generator, approximately 5 to 20 per cent of the radon was fixed inside the generator within several hours. The tracer radon was bound to the glass of the Berthelot tube very strongly and could not be removed by vacuum distillation at 450° or by heating the tube with hydrogen at 500°.

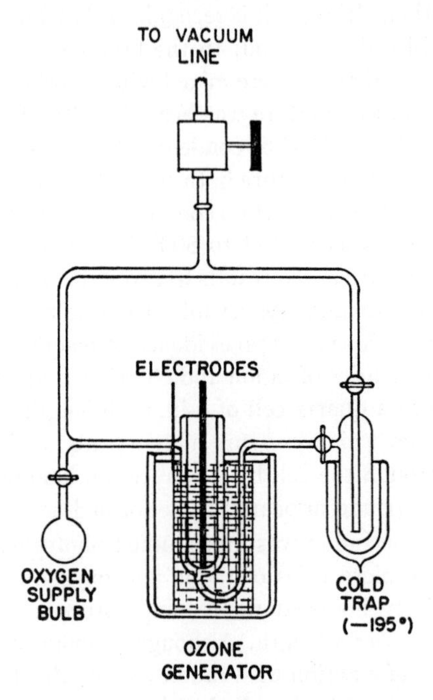

FIG. 3.—Apparatus for experiments with ozone and radon

When the ozone generator was cut from the line and the inside was washed with cold distilled water, 2 M nitric acid, and 6 M nitric acid, very little of the fixed radioactivity was removed. Hot 6 M nitric acid removed approximately 50 per cent of the radon daughters, with little effect on the fixed radon, since the gamma activity grew back within several hours. Cold 6 M sodium hydroxide removed both radon and daughters, probably by dissolving a thin layer of the glass.

The effect of microwave discharges was investigated with the apparatus shown in Figure 4. Approximately 260 microcuries of radon was first condensed in a small quartz bulb, oxygen was added to 5 mm. pressure, and the bulb was placed near the director of a Raytheon Model CMD4 Diathermy Unit, which generates microwaves of 12.3 cm wave length. A discharge was started with a

Tesla coil and allowed to continue for 145 minutes. When the oxygen was pumped out at −78°, all of the radon remained in the bulb. It was found to be distributed over the bottom half of the bulb in the region where the glow discharge had occurred. When the bottom section was strongly heated in vacuum, the radon daughters distilled out of the bottom at approximately 600° and condensed in a ring on the cooler upper section. It was clear that the radon did not move, however; the gamma activity of the ring decayed when the system was

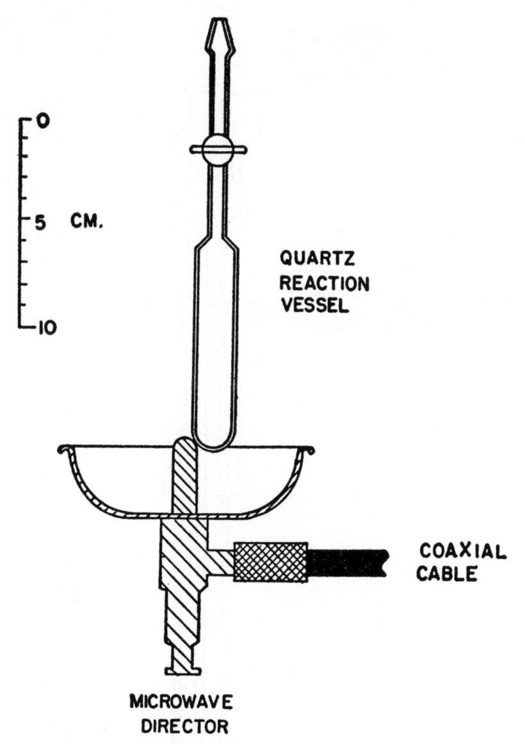

FIG. 4.—Apparatus for producing microwave discharges in quartz

left overnight at room temperature, and the activity grew back into the bottom section. The process was repeated at a temperature of 970° with similar results. The radon deposit was also heated with hydrogen at 970°, but very little radon was removed by this process.

Since the tracer quantity of radon had a negligible partial pressure in each experiment, it was necessary to add another gas to produce an ionizing discharge. The radon was fixed when the diluent gas was oxygen, nitrogen, or helium at about 5 mm. pressure. Quartz bulbs were generally used, but the results were the same when Pyrex bulbs were substituted. With discharges of 30 to 145 minutes' duration, the percentage of radon fixed varied from 10 to 100 per cent in different experiments. The effectiveness of water and solutions of nitric acid and sodium hydroxide in washing off the activity was the same as described for the ozone generator.

The microwave cell shown in Figure 5 was devised so that discharges could be produced in a metal system. A $\frac{3}{4}$-in. (outside diameter) nickel test tube was attached by a standard tapered joint to a brass chamber, which was connected to the vacuum line by a side arm. Another side arm, fitted with an insulated vacuum seal, transmitted energy from the microwave generator to a threaded brass

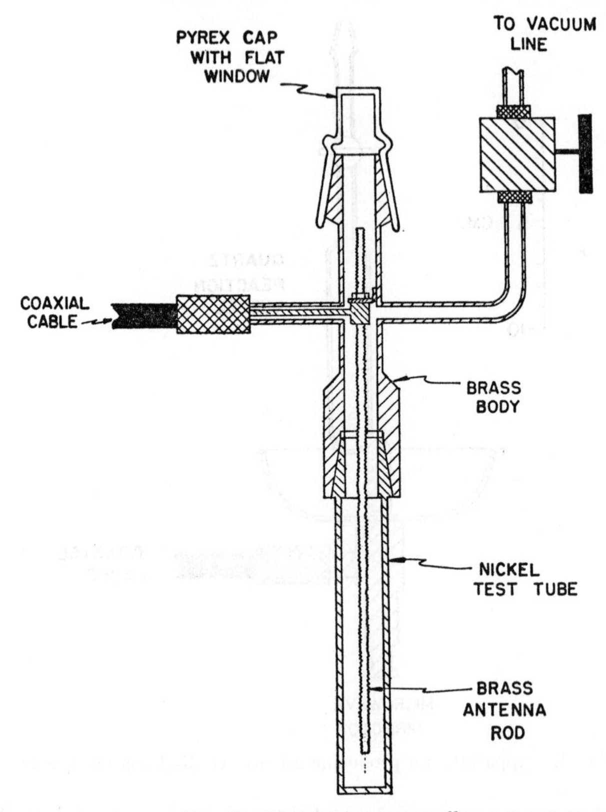

FIG. 5.—Metal microwave cell

rod, which served as an internal antenna. A mixture of radon and another gas could be added to the cell at low pressure, and a discharge could be produced in the region between the bottom of the brass rod and the nickel test tube. A Pyrex window at the top of the cell was used to watch the discharge, and the system was tuned by adjusting the position of the rod, which was approximately two wave lengths long. It was found that the efficiency was low, since the ionization was produced in only a small fraction of the total volume. In one experiment, a thirty-minute discharge with 23 microcuries of radon and oxygen at approximately 2 mm. pressure fixed 25 per cent of the radon. In a second trial, no radon was fixed, apparently as the result of too high a pressure of oxygen (> 10 mm.). Other results were: 42 microcuries of radon and 4 mm. oxygen pressure, 22 per cent fixed in one hour; 122 microcuries of radon and 2 mm. oxygen pressure, 37 per cent fixed in one hour.

Almost all of the fixed radon appeared on the bottom of the brass antenna rod; very little was found on the walls of the nickel test tube. The voltage gradients produced by the microwave field apparently caused radon ions to be preferentially driven into the antenna. The bound radon was present in a thin surface layer, since it could be removed by etching the brass with 6 M nitric acid. At the conclusion of the experiments with this apparatus, the bottom section of the brass antenna with a radon deposit was cut into small pieces and gradually heated in vacuum. At approximately 900°, near the softening point of the brass, the radon escaped and moved to a cold trap on the vacuum line.

From the foregoing experiments it can be concluded that only the reaction of radon and fluorine yielded a definite compound in the present investigation. Chlorine, iodine monochloride, oxygen, and ozone gave no evidence of reaction with radon under the conditions which were used. The results obtained with high-voltage and microwave discharges can probably be attributed to physical processes involving ion bombardment of the walls, since similar results with other elements have been reported by previous investigators [5–8]. However, the possibility of chemical stabilization of the radon by formation of an oxide or silicate cannot be entirely discounted. The stability of radon imbedded in quartz is quite remarkable, considering that radon daughters such as Bi^{214} or its precursors, Po^{218} and Pb^{214}, can be removed at 600°, leaving the radon firmly bound. No doubt practical applications of the electrical discharge methods can be found, such as the preparation of radon sources for medical and other purposes.

Comment.—

The lack of volatility of the radon fluoride would be explained if radon formed an ionic fluoride. The lesser volatility of ionic compounds compared to covalent compounds is well known. The behavior of the radon in the quartz microwave discharge experiment would also be explained in a similar fashion. In this case, apparently a radon silicate is formed.—*B. Weinstock.*

REFERENCES

1. N. Bartlett, *Proc. Chem. Soc.* 218 (1962).
2. H. H. Claassen, H. Selig, and J. G. Malm, *J. Am. Chem. Soc.* **84,** 3593 (1962).
3. P. R. Fields, L. Stein, and M. H. Zirin, *ibid.* 4164 (1962).
4. A. C. Jenkins, *Ozone Chemistry and Technology,* p. 13. Washington, D.C.: American Chemical Society, 1959.
5. K. B. Blodgett and T. A. Vanderslice, *J. Appl. Phys.* **31,** 1017 (1960).
6. C. Y. Bartholomew and A. R. La Padula, *ibid.* 445 (1960).
7. L. Holland, *J. Sci. Instruments* **36,** 105 (1959).
8. J. Koch, *Nature* **161,** 566 (1948).

PART 3 SOME PRACTICAL CONSIDERATIONS

The atomic energy industry has repeatedly come in contact with problems involving the noble gases. The high cross-section of the xenon isotope Xe^{135} has emphasized the importance of this element in reactor design, and the substantial fission-product yield of radioactive xenon and krypton nuclides has made their behavior in fuel-element reprocessing significant. It is interesting that the behavior of fused fluorides in corrosion-testing yielded data which, with hindsight, could only be interpreted in terms of the production of a non-volatile xenon fluoride under conditions where xenon and gaseous fluorine are reacted together. It is not surprising, therefore, that the Atomic Energy Commission exhibited an early and continuing interest in research on noble-gas compounds.

The contributions of Washington personnel in this section are based on continuing consultation with research workers at the various AEC laboratories throughout the country. As is obvious, most of the experimental work has been done by scientists who are primarily concerned with the effect of these observations on the general body of scientific knowledge. However, even at this stage it is appropriate to consider the potential practical results of these developments.

POTENTIAL APPLICATIONS OF THE NOBLE-GAS COMPOUNDS

J. H. Pomeroy
U.S. Atomic Energy Commission

I will try rubbing a clouded crystal ball to speculate on some of the possible future applications of the new compounds of the noble gases. I just received a copy of an interesting new book, *The Scientist Speculates* [1], which I took as an appropriate frame of reference, since the subtitle of the book is: *An Anthology of Partly-baked Ideas.*

I would like to offer you a few "partly-baked" ideas on the noble-gas compounds, with the hope that they may stimulate you to have a few "fully-baked" ideas of your own.

Because of its volatility and the large amounts which are produced, Xe^{133} has been one of the most troublesome of the fission products to keep under control in atomic power systems. If it were converted to a fluoride, would this be a useful way to store it safely or to use it? This isotope is available commercially today, potentially in megacurie amounts, if anyone wants that much. The notorious Xe^{135} has the highest known cross-section for the capture of thermal neutrons (over 10^6 barns). Would having either of these isotopes available in a solid, rather than a gaseous, form be of possible value to anyone in the nuclear field? From reported values of the density of XeF_4 [2, 3], you may calculate that this compound contains xenon atoms at a concentration equivalent to that of gaseous xenon at a pressure of almost 500 atmospheres. MacKenzie [see p. 81, this volume] has reported that XeF_4 breaks down under certain conditions of radiolysis. We will need additional measurements of the stability of these fluorides to various kinds of radiation before we can plan on using them in reactors, for instance. It is well known that large amounts of radioactive xenon and krypton are liberated during the processing of used reactor-fuel elements and are difficult to trap completely. It would be worthwhile to investigate the possibilities of trapping them as fluorides or as one of the other less volatile compounds. If there were serious problems of radiation stability, these might be overcome, perhaps by dilution with an inert salt.

Would xenon tetrafluoride be convenient for storing fluorine in a readily available form? Many people have already considered possible applications of these noble-gas fluorides as oxidants, for instance in rocket propellant systems.

Would there be advantages to using a suitable noble-gas fluoride as a fluorinating agent in a closed system where the noble gas could be recycled to prevent loss [4–6]? This could perhaps involve electrolytic generation of the reagents in anhydrous hydrogen fluoride. A solution of XeF_4 in anhydrous hydrogen fluoride is a moderately vigorous fluorinating agent at room temperature, attacking platinum, and has converted benzene to a fluorinated benzene [7].

There is a report from Argonne National Laboratory that a solution of XeF_4 in anhydrous hydrogen fluoride has caused benzene to form a dark, rubbery polymer of unknown structure. Will not chemists want to investigate this and similar reactions further? As a fluorinating agent, XeF_4 seems to resemble UF_6 more than it does ClF_3 [7].

Xenon oxides are sensitive to shock and highly explosive [11]. Someone might be able to find a useful specialized application for some of these materials. When they explode, they leave no solid residues.

Investigators at Argonne have suggested that metal hexafluorides might be used in the purification of the noble gases, since they would permit a clean separation of xenon from krypton, or that they might be used to trap radon. The reaction between hexafluoride and noble gases is reversible and so could be used in a recycling system.

Could we use the same kind of reaction to purify or separate some of the hexafluorides of uranium, neptunium, plutonium, or related materials? PuF_6 reacts; UF_6 does not. A study of the applications of such reactions to the processing of nuclear reactor fuels is already under way at several laboratories.

Radon is still being used in medicine today for radiation therapy. Usually, the gas is pumped out of a radium solution and is sealed in small glass ampules or metal tubes which are then inserted into or around a tumor. One of the problems with handling these "needles" or "seeds" is that sometimes the seal leaks, and the radon escapes to where it is not wanted. Would it be useful to convert radon to a non-volatile form, such as the fluoride, to prevent some of these difficulties [7, 8, and see Fields *et al.*, p. 113, this volume]?

Radon-beryllium mixtures have occasionally been used as neutron sources. Would having the radon in a non-volatile form be useful here?

There may be considerable interest in the xenates and perxenates because they are powerful oxidizing agents. Since the xenate ion decomposes to give only volatile products [9], would this be of possible use to someone who wants to leave a system neat and tidy at the end of a reaction?

Malm has reported that the solubility of sodium xenate is about 0.2 mg/ml in 0.5 *N* NaOH [see p. 172, this volume]. Would this be a possible reagent for the separation or gravimetric determination of the sodium ion? We would like to know the solubilities and general properties of many other salts, both inorganic and organic. Obviously, much electrochemistry, polarography, etc., of many of these materials still remains to be done, although Appelman and Kilpatrick have made a good start [see pp. 155 and 185, this volume].

How do these volatile fluorides behave under conditions of vapor-phase chromatography?

What other atoms or radicals can be hooked on to the noble-gas atoms? Allen and Horrocks [10] have suggested that possible BF_3 or other Lewis acids may also be incorporated into some of these molecules. How long will the list of known compounds of the noble gases be by 1970? There may be a few more surprises for us, possibly including molecular species with an unpaired electron or so.

There will have to be a critical re-examination of some of the existing body of already recorded information in this fast-growing field. Much of the early work on the xenon fluorides was carried out on mixtures of XeF_2, XeF_4, and XeF_6 and should be repeated to see how the individual pure materials behave.

I also suspect that the discovery of the rare-gas compounds has been greeted with particular enthusiasm by the producers of textbooks in chemistry, since it gives them a fine excuse to come out with new editions, as a sort of windfall of unplanned obsolescence.

In conclusion, it is my feeling that one of the greatest possible values of the discovery of these compounds will be the sharp lesson to all of us involved with the profession of science. We cannot afford to take our universe too much for granted; it must still be full of many other surprises.

REFERENCES

1. L. J. GOOD, *The Scientist Speculates*, pp. 413. New York: Basic Books, 1962.
2. S. SIEGEL and E. GEBERT, *J. Am. Chem. Soc.* **85**, 240 (1963).
3. J. A. IBERS and W. C. HAMILTON, *Science* **139**, 106 (1963).
4. J. L. WEEKS, C. L. CHERNICK, and M. S. MATHESON, *J. Am. Chem. Soc.* **84**, 4612 (1962).
5. D. H. TEMPLETON, A. ZALKIN, J. D. FORRESTER, and S. M. WILLIAMSON, *ibid.* **85**, 242 (1963).
6. P. A. AGRON *et al.*, *Science* **139**, 842 (1963).
7. C. L. CHERNICK *et al.*, *ibid.* **138**, 136 (1962).
8. P. R. FIELDS, L. STEIN, and M. ZIRIN, *J. Am. Chem. Soc.* **84**, 4164 (1962).
9. F. B. DUDLEY, G. GARD, and G. H. CADY, *Inorg. Chem.* **2**, 228 (1963).
10. L. C. ALLEN and W. DEW. HORROCKS, JR., *J. Am. Chem. Soc.* **84**, 4344 (1962).
11. D. F. SMITH, *J. Am Chem. Soc.* **85**, 816 (1963).

SAFETY PRECAUTIONS IN HANDLING NOBLE-GAS COMPOUNDS

R. B. SMITH
Atomic Energy Commission

Scores of noble-gas compound explosions and a few fires have occurred since their discovery less than a year ago. To date, most of the work and all of the reported incidents with noble-gas compounds have involved simple xenon fluorides, oxyfluorides, or other solid compounds derived from them by hydrolysis. Many explosions have occurred in liquid-nitrogen cold traps while being warmed to room temperature or while removing product. With but one exception, these accidents have not caused personal injury or serious property damage, a fact partially attributable to the very small (normally milligram) quantities handled at one time. None have taken place during transportation or during product preparation at elevated temperatures.

The various xenon fluorides and oxyfluorides have all been prepared at elevated temperatures without incident, thus suggesting that these compounds possess at least a fair degree of thermal stability. By contrast, all unforeseen explosions experienced with these compounds appear to have been initiated at temperatures below 40° C. by very mild thermal or mechanical exposure. This occasionally observed change of an apparently relatively thermally stable material into a highly thermally explosive material has been frequently attributed to contaminants, notably organic matter or moisture. The xenon fluorides, and especially XeF_6, are known to be reactive fluorinating agents, and the possibility of an explosion in organically contaminated xenon fluorides is certainly plausible. However, events of this type known to have been experienced to date have been limited to small spontaneous fires.

The interest in moisture as a xenon fluoride contaminant was considerably increased by the observation of several investigators that the xenon fluorides did not appear explosive until after exposure to some form of moisture, and by publicity given an experiment in which 0.39 gm. of XeF_4 was dissolved in 1.5 ml. of water, the resulting clear solution evaporated under vacuum at room temperature to yield a white solid, then suspected of being either $Xe(OH)_4$ or $XeO_2 \cdot 2H_2O$. While heating the silica container under vacuum with warm (30°–40° C.) air from a portable blower, a detonation took place, shattering the container and inflicting eye injuries to both persons in the immediate area [1].

Recent work at Berkeley demonstrated that XeO_3 is formed by carefully evaporating aqueous solutions containing dissolved XeF_4 [see Williamson and Koch, this volume, p. 164], while work at Oak Ridge showed XeO_3 could also be formed by hydrolysis of XeF_6 or $XeOF_4$ [2]. While solid XeO_3 appears to be a very sensitive high explosive, no explosions during handling of any xenon compound in solution are known to have occurred.

Investigators who have experienced noble-gas compound explosions attributed to XeO_3 report the following general impression. XeO_3 is comparable in detonation sensitivity to nitrogen tri-iodide and in explosive force to TNT. All explosions are classed as sharp reports. Detonation of 25 mg. will cause the ears to ring for one-half hour. Detonation may be induced by mild heating, by mild mechanical disturbance, or by pouring; or, they may take place spontaneously. Sensitivity of individual samples varies somewhat, and a few samples have been stored for several weeks. This explosive material may easily be inadvertently formed by exposing xenon compounds to moisture such as that in room air.

The only known test data on noble-gas compound explosion hazards other than XeO_3 were furnished by a firm who reported that XeF_4 could be detonated by shock from a 2 kg. weight falling 20 cm. in a drop test instrument. Thus the current available test data on noble-gas compound explosion hazards seem to add up as follows:

1. Solid XeO_3 is a very unstable high explosive readily detonated by mild heating, gentle rubbing or friction, and, on a few occasions, has exploded spontaneously while undisturbed.
2. XeO_3 is readily formed at room temperatures by hydrolysis of XeF_4, XeF_6, or $XeOF_4$ (e.g., by exposure to moisture in room air).
3. XeF_4 is susceptible to detonation by shock according to information received on March 13, 1963 (although no information was given as to the purity of the samples tested), and no explosions have been reported for purified XeF_4.

Solid xenon oxides other than XeO_3 would probably also be explosive if they could be produced. Such oxides might be formed by exposing any xenon compound to moisture (such as that in room air) or by evaporating any aqueous solution containing dissolved xenon compounds. The opinion that noble-gas fluorides may be susceptible to detonation by shock does not appear widespread. The xenon fluorides appear to be thermodynamically stable; XeO_3 and probably all other oxides are thermodynamically unstable [see Part IV, this volume].

Test data are certainly needed to clarify factors contributing to the variations observed in the sensitivity of XeO_3 to detonation. More information on the conditions under which explosive compounds may be formed by hydrolysis is also needed. Further evaluation of the degree of sensitivity of noble-gas compounds to detonation by shock would help clarify precautions that should be observed during their handling.

Several investigators have observed formation of a yellow surface coloration on XeF_4 in contact with water. In a few cases it was noted that such a yellow solid readily detonated. More work is needed to establish the composition and hazard properties of this apparently explosive material.

Alternatively, it has been reported that hydrolysis of xenon tetrafluoride or hexafluoride in alkaline solution yields an alkali salt which is relatively stable or at any event does not detonate under the usual laboratory manipulation. [See Malm, Bane, and Holt, this volume, p. 171.]

With these uncertainties in mind, the following suggested set of interim safeguards have been prepared.

INTERIM SUGGESTED SAFEGUARDS PENDING FURTHER EXPERIENCE

1. *Xenon trioxide* (XeO_3)—

a) Limit amounts handled manually to not over 25 mg. per batch.

b) Use safety glasses (preferably supplemented by a face shield).

c) Use ear plugs when handling over 10 mg.

d) Use gloves for hand protection. Avoid skin contact with XeO_3.

e) Use portable transparent shield between XeO_3 and operator when XeO_3 is handled in equipment (e.g., glass) subject to fragmentation. (Note: The same objective may also be achieved using non-transparent shields plus mirrors to permit indirect viewing of operations.)

f) When applicable, metal cold traps specially designed to incorporate adequate safety (e.g., such as that described in report KL-237 (Rev. 9/26/59) are recommended. When using such traps, keep trap opening pointed in direction such that explosion forces will be directed away from operator. Use mirrors for viewing inside of trap or when inserting or removing product.

g) Deactivate scrap before disposal.

h) Provide shields around stored product to minimize possible blast effects on personnel or adjacent property. Non-reactive non-fragile containers are preferable.

i) Label containers to denote explosion hazard. Place warning signs denoting explosion hazard in areas used for storing or handling XeO_3.

j) Do not ship offsite.

k) Anticipate unexpected explosions.

2. *Xenon fluorides and oxyfluorides*—

a) If known or suspected that any of these compounds contain moisture or easily fluorinated organic contaminants, handle as sensitive explosives [see 1 above].

b) Treat all solids derived by evaporation of solutions containing dissolved xenon compounds as sensitive explosives [see 1 above].

c) When any of the subject compounds are exposed to or in contact with water or aqueous solutions, treat as potentially explosive. Recommendations in 1 above are suggested as interim guides.

d) Take particular care to avoid inadvertent contact of subject products with moisture, e.g., by storing and handling under vacuum or dry inert-gas atmosphere. Storage of pure dry product in fluorination resistant or inert metal (e.g., passivated nickel, Monel, stainless) containers preferable.

e) Ship pure dry solids in accordance with ICC Regulations. As interim precaution (and supplementing information in [*d*] above) ship only in non-friable sealed containers cushioned against excessive external mechanical shock. Limit single shipments to one gram maximum. [Note: Precautions given in this section were not motivated by any adverse experience during shipments made to date but are intended to afford recognition (1) of possibility of forming explosive compounds if container leakage permits product exposure to moisture, (2) uncertainty as to shock sensitivity of the subject compounds and (3) of the general inadequacy of current knowledge and experience relative to hazards of these materials.]

f) Deactivate scrap prior to disposal.

g) Use safety glasses. Avoid skin contact with subject materials.

h) Vapors of the subject materials (particularly XeF_6) form sensitive explosives on exposure to moisture and may be formed on external surfaces in contact with air—e.g., near cracks, "weak spots" in plastic containers, around container closure, etc. Treat white solids resulting from such exposures as XeO_3 [see 1 above].

i) Work behind safety shields and limit batch size to minimum consistent with work objectives when using subject compounds as fluorinating agents (e.g., in conducting reactions with hydrogen). Anticipate possible explosions.

3. *Noble-gas compounds other than simple xenon fluorides, oxyfluorides, and* XeO_3—

a) Until such time as adequate hazard knowledge and experience are developed locally to warrant lesser precautions, use of safeguards detailed in 1 above is suggested.

REFERENCES

1. N. Bartlett and P. R. Rao, *Science* **139,** 506 (1963).
2. D. F. Smith, *J. Am. Chem. Soc.* **85,** 815 (1963).

PART 4 THERMOCHEMISTRY

The interest in the thermodynamic stability of noble-gas compounds is too obvious to be emphasized. Today calorimetry is done, for the most part, in laboratories designed to produce very precise results on highly purified samples. To some extent, the research reported here is more reminiscent of an earlier age in which preliminary observations on the best material available were all that could be expected. The qualitative conclusions are straightforward enough. The xenon fluorides are thermodynamically stable, and the bonds are chemical bonds with an energy content well into the range normally observed in chemical compounds. If the oxygen-containing compounds are less stable with respect to formation from the elements, this is primarily a result of the great stability of the oxygen molecule and not of the lack of strength in xenon-oxygen bonds.

THE HEAT OF FORMATION OF XENON TETRAFLUORIDE

STUART R. GUNN AND STANLEY M. WILLIAMSON
University of California, Livermore and Berkeley

ABSTRACT

The heat of reaction of XeF_4 with aqueous iodide solution has been measured. The standard heat of formation of the solid is -60 kcal. mole^{-1}, and the average thermochemical bond energy is 30 kcal.

INTRODUCTION

Among the important properties to be considered in the elucidation of the nature of the recently discovered rare-gas compounds are their heats of formation, from which bond energies can be derived. Accordingly, we have determined by solution-reaction calorimetry the heat of formation of xenon tetrafluoride.

Prior to these calorimetry experiments, Williamson and Koch [1] found in their studies on the reaction of xenon tetrafluoride with aqueous solutions that, in addition to the reaction with water,

$$XeF_4 + 2H_2O \rightarrow Xe + O_2 + 4HF\,, \qquad (1)$$

which Claassen, Selig, and Malm [2] and Chernick *et al.* [3] had reported earlier, a considerable part of the xenon in XeF_4 remained in the solution as a xenon (VI) species from which XeO_3 can be obtained [4]. When XeF_4 is reacted with an aqueous iodide solution, because of the vigor of the reaction a small part reacts by equation 1, and the rest is rapidly and quantitatively reduced by iodide:

$$XeF_4 + 4I^- \rightarrow 2I_2 + Xe + 4F^-\,. \qquad (2)$$

Holloway and Peacock [5] have used reaction 2 as a method of analysis for XeF_4 from their preparation.

This reaction was performed in a sealed calorimeter with a degassed solution to permit determination of both the O_2 and I_2 produced and hence the fraction of the XeF_4 reacting by the two different paths. Because the reaction is heterogeneous, the ratio of O_2 to I_2 is not a constant from run to run.

EXPERIMENTAL

Xenon tetrafluoride and preliminary experiments.—The XeF_4 samples used in these calorimetry experiments were taken from the material prepared by the flow-reaction method described by Templeton *et al.* [6], which is similar to that described by Holloway and Peacock. It was observed that this XeF_4 could be stored and handled at room temperature without apparent reaction or decomposition in Pyrex glass which had been heated well under vacuum before introduction of the XeF_4.

After preparation of the XeF_4, the inlets to the reactor were closed, and the XeF_4 was sublimed under vacuum to a trap downstream which was at $-30°$. HF and SiF_4, or any other volatile species, passed on through. By means of a break-seal, the XeF_4 from this trap could readily be sublimed through a manifold either into other traps, quartz X-ray capillaries, or preweighed, fragile, sample bulbs, etc. For the preliminary hydrolysis experiments, the sample bulbs, with a typical volume of 0.07 ml., were weighed to ± 5 μg. containing XeF_4 weighing from 3 to 20 mg. For the calorimetry experiments, the sample bulbs had a volume of about 3 ml., were weighed to ± 0.1 mg. containing XeF_4 weighing from 112 to 193 mg., and were stored at $-196°$ until assembled in the calorimeter about twenty hours before the run.

TABLE 1

XeF_4 HYDROLYSIS DATA

XeF_4 (μ MOLE)	H_2SO_4 (M)	BEFORE KI (MOLE %)		AFTER KI (MOLE %)		TOTAL (MOLE %)	
		O_2	Xe	I_2 as O_2	Xe	O_2	Xe
132.72....	3.0	51.6	71.5	44.3	29.8	95.9	101.3
87.24....	9.0	50.4	73.0	43.4	27.8	93.8	100.8

In addition to the X-ray analysis, hydrolysis experiments were done from which analytical data could be obtained. Table 1 presents the data and conditions for two such experiments. XeF_4 samples were reacted in an all glass system by crushing the sample bulb in degassed sulfuric acid solution at 0°. Oxygen and xenon were first liberated and measured separately by a Toepler pumping system. Then solid KI was added to the remaining solution still under vacuum, and Xe and I_3^- were formed. Xe gas was again measured, and the I_3^- was titrated with standard thiosulfate to the starch end-point.

Since the Xe is observed to be greater than 100 per cent of theoretical, the total O_2 (both as directly measured O_2 and as I_2) would be expected to be less than 100 per cent. If removal of the xenon from the aqueous solution were a problem, the percentage of Xe should be below 100 per cent. Our percentage of Xe is in close agreement with that which Kirshenbaum *et al.* [7] obtained from their hydrolysis of XeF_4 in dilute KOH solution; however, they do not report the corresponding percentage of O_2 for our comparison. Our data might suggest a slight contamination of XeF_2. Since the vapor pressures of XeF_2 and XeF_4 at room temperature are reported to be comparable [8, 9], it may be impossible to get very pure samples by sublimation.

Calorimetry.—Rocking-bomb calorimeter I [10] was used with bomb IE, which is fabricated entirely of tantalum (bored and turned from a 4-in. rod). The bulb-breaker assembly and spring were also entirely tantalum. The bulb was supported in a jig of platinum wire and sheet. A Pyrex valve was used; the valve and bomb gaskets were lubricated with Kel-F grease.

The internal volume of the bomb is 654 ml. A solution of 380 ml. water containing 0.0190 mole KI and 0.00038 mole HCl was used for the XeF_4 runs. The solution was thoroughly degassed by repeated stirring and pumping.

In order to minimize the time during which the solution was in the bomb after the reaction, duplicate electrical calibrations were performed before the reaction, over about the same temperature interval as that anticipated for the reaction. The heat of reaction is thus referred to the final temperature of the reaction period, which in all cases was $25.00 \pm 0.03°$ C.

Two measurements of the heat of solution of xenon in a degassed solution similar to the final XeF_4-reaction solutions were performed. Weighed samples of xenon (3.77 and 5.46 millimoles) in bulbs of about 68 ml. internal volume (similar to those used for B_2H_6 and HCl [11, 12]) were used.

To compensate for the endothermic evaporation of water, the observed heat values in all cases were corrected by addition of 0.0120V cal., where V is the internal volume of the bulb [13, 14].

Analyses.—Following the reaction, the bomb was attached to a vacuum line and the gas transferred by means of a Toepler pump through three traps at $-196°$ to a gas buret (calibrated by weighing mercury), the bomb being repeatedly removed and shaken. The oxygen was measured; then the traps were warmed to $-78°$ and the xenon pumped to the buret, the ice being sublimed from the first trap to the second to free any occluded gas. Mass-spectrometric analysis of both batches of gas was performed to establish their purity. Gas imperfection corrections [14] were applied to the xenon measurement.

Some of the iodine was transferred to the cold traps with the gas. This was washed out and combined with the bomb solution. Aliquots were then titrated with standard sodium thiosulfate to determine iodine and with standard thorium nitrate to determine fluoride. The thorium nitrate was standardized both by titration with acidimetrically standardized hydrofluoric acid and with standardized EDTA. These titration procedures were shown to give good results with both separate and mixed KF and $KI\text{-}I_2$ solutions.

RESULTS

From the solubility data of Seidell [15] and the ratio of vapor and liquid volumes in the bomb, it may be estimated that at equilibrium about 15 per cent of the xenon would be dissolved in the solution. Alexander [16] has given the heat of solution of xenon as -4.1 ± 0.2 kcal. $mole^{-1}$ from calorimetric measurements; a similar value is obtained from the temperature coefficient of the solubility. Thus the expected heat effect per mole of xenon in our experiments would be -0.6 kcal. for ΔH, or -0.5 kcal. for ΔE. Our two runs with xenon were consistent with this, but were uncertain due to the prolonged evolution of heat from thirty to fifty minutes. Under the stirring conditions in the calorimeter, the approach to equilibrium is evidently slow. The uncertainty about this heat effect and about whether equilibrium is attained during the reaction period of the XeF_4 runs contributes an uncertainty of perhaps ± 0.2 kcal. mole^{-1} to the XeF_4 reaction heats.

Results of the XeF_4 runs are given in Table 2. $\Delta H°$ is calculated solely from the sample weight; it includes a ΔnRT term to convert from ΔE to ΔH and the -0.5 kcal. term to convert the Xe(aq) to Xe(g).

The xenon found corresponds in all runs to 99.0 per cent of the sample. It is felt that the analysis should be better than this; however, the fairly high solubility of xenon in water may have resulted in a relatively constant and small amount being regularly left behind in the shaking and pumping procedure.

TABLE 2

HEAT OF REACTION OF XeF_4(C) WITH KI(AQ)

WEIGHT IN MILLIMOLES					q_{obs} (CAL.)	$-\Delta H°$ (KCAL. MOLE^{-1})
XeF_4	Xe	O_2	I_2	F^-		
0.5442	0.5387	0.0305	0.956	2.295	104.93	191.7
.8326	.8246	.0517	1.484	3.224	160.14	191.2
.8360	.8277	.0838	1.424	3.259	158.78	188.8
0.9388	0.9308	0.0484	1.690	3.707	182.26	193.0

The amount of O_2 found varies considerably, the percentage with respect to XeF_4 being 5.60, 6.21, 10.02, and 5.16 in the four runs. In the calculations below, it is assumed that these percentages represent the amount of XeF_4 reacting by equation 1 and that the rest reacted by equation 2. The combined amounts of O_2 and I_2 in the four runs are 93.4, 95.4, 95.2, and 95.2 per cent of the theoretical. Small amounts of iodine may have been lost in the handling procedure and in greased joints of the gas-analysis line. However, comparison experiments in which similar KF-KI-I_2-HCl solutions stood in the bomb indicated some loss of I_2 and perhaps also of F^- with time, presumably by reaction with the tantalum. The tantalum bomb was used because much larger I_2 losses were found with a coinage-gold bomb. The F^- analyses fluctuate considerably, amounting to 105.4, 96.8, 97.5, and 98.7 per cent of the theoretical.

An error would be introduced into our results if there were a significant heat effect due to reaction of I_2 or F^- with the bomb during the reaction period, or if there were a continuing reaction of O_2 with I^- after the reaction period, changing the apparent amounts of reactions 1 and 2. The initial reaction was almost instantaneous upon breaking the XeF_4 bulb; there then ensued a smaller heat effect—perhaps 1 to 3 per cent of the total—at a rate steadily decreasing for about fifteen minutes, at which time the expected drift rate of the calorimeter was reached, indicating any continuing reaction to be at a very low rate. It is possible that the slower heat effect is due to reaction of O_2 and I^-, the rate decreasing and becoming essentially zero as the acidity of the solution and the partial pressure of oxygen decrease. This would introduce no error provided that the O_2/I_2 ratio at the end of the calorimetric main period is the same as that later determined by the analyses.

Since the concentrations and the relative amounts of O_2 and I_2 differ in the four runs, $\Delta H_f°$ (XeF_4) must be calculated separately from each. For the equilibria

$$HF \rightleftharpoons H^+ + F^- \tag{3}$$

and

$$I^- + I_2 \rightleftharpoons I_3^-, \tag{4}$$

we take $K_3 = 6.7 \times 10^{-4}$ and $K_4 = 750$ (see Bjerrum [17] *et al.* and Latimer [18]). Neglecting activity coefficients, we then calculate the net reaction for each run; as an example, for the second,

$$XeF_4 + .1242\ H_2O + 5.5635\ I^- + .3971\ H^- \rightarrow$$

$$Xe + .0621\ O_2 + .6459\ HF + 3.3541\ F^- + 1.8119\ I_3^- + .0639\ I_2\ (aq)\ . \tag{5}$$

The F^-/HF ratio for the four runs ranges from 4.0 to 6.1, and the I_3^-/I_2 ratio ranges from 27.1 to 31.6. HF_2^- and higher iodine complexes are negligible. We then use the NBS values [19] for the heats of formation of the aqueous species: H_2O, -68.317; I^-, -13.37; I_3^-, -12.4; I_2, $+5.0$; F^-, -78.66; HF, -75.48 (the HF value is calculated from the NBS value for F^- and the heat of ionization of HF, $+3.18$, given by Hepler, Jolly, and Latimer [20]), and the $\Delta H°$ values of Table 1 to calculate the standard heat of formation of XeF_4(c) from each run; -60.2, -60.7, -59.6, and -60.0 kcal. mole^{-1}, an average of -60.1. The random deviation is small compared with possible systematic errors, including those in the tabulated heats of formation for the F and I species.

The analytical data do not rule out the possible presence of XeF_2 or XeF_6, the former being more probable. Estimating from Pitzer's analogy [21] with ClF_3 and BrF_3, -40 kcal. mole^{-1} for $\Delta H_f°$ of XeF_2(g), or -50 for XeF_2(c), and assuming its reaction to be similarly divided between oxidation of H_2O and I^-, its heat of reaction would be about half that of XeF_4 on a weight basis. Hence presence of 10 per cent by weight of XeF_2 in our samples would make our reaction heat values low by 5 per cent, or 10 kcal., and the correct heat of formation of XeF_4(c) would be -50 instead of -60.

Jortner, Wilson, and Rice [22] have reported 15.3 ± 0.2 kcal. mole^{-1} for the heat of sublimation of XeF_4, giving -45 kcal. mole^{-1} for $\Delta H_f°$(XeF_4 [g]). This value is in agreement with Pitzer's estimate that it should be smaller than the -105.6 for $\Delta H°$ of

$$BrF(g) + 2F_2(g) \rightarrow BrF_5(g)\ .$$

Using +37 for the heat of dissociation of F_2 [23], the heat of formation of gaseous XeF_4 from gaseous XeF_4 from the gaseous atoms is −119 and the average thermochemical bond energy is 30 kcal.

Thus XeF_4 is thermodynamically quite stable. It is of interest that KrF_4 [24] is unstable with respect to the elements at room temperature, implying a positive free energy of formation of the gas and a heat of formation positive or near zero. From this, one may guess that the heat of formation of RnF_4 might be around −80 to −100 kcal. mole^{-1}, since the ionization potential of radon is considerably lower than that of xenon.

Acknowledgment.—We thank Lewis J. Gregory for performance of the analyses of the calorimeter solutions.

REFERENCES

1. S. M. Williamson and C. W. Koch, *Science* **139**, 1046 (1963).
2. H. H. Claassen, H. Selig, and J. G. Malm, *ibid.* **84**, 3593 (1962).
3. C. L. Chernick *et al.*, *Science* **138**, 136 (1962).
4. D. H. Templeton, A. Zalkin, J. D. Forrester, and S. M. Williamson, *J. Am. Chem. Soc.* **85**, 817 (1963).
5. J. H. Holloway and R. D. Peacock, *Proc. Chem. Soc.* 389 (1962).
6. D. H. Templeton, A. Zalkin, J. D. Forrester, and S. M. Williamson, *J. Am. Chem. Soc.* **85**, 242 (1963).
7. A. D. Kirshenbaum, L. V. Streng, A. G. Streng, and A. V. Grosse, *ibid.* 360 (1963).
8. D. F. Smith, *J. Chem. Phys.* **38**, 270 (1963).
9. P. A. Agron *et al.*, *Science* **139**, 842 (1963).
10. S. R. Gunn, *Rev. Sci. Instruments* **29**, 377 (1958).
11. S. R. Gunn and L. G. Green, *J. Phys. Chem.* **64**, 61 (1960).
12. ———, *J. Chem. Eng. Data* **8**, 180 (1963).
13. ———, *J. Phys. Chem.* **64**, 1066 (1960).
14. J. A. Beattie, R. J. Bariault, and J. S. Brierley, *J. Chem. Phys.* **19**, 1222 (1951).
15. A. Seidell, *Solubilities of Inorganic and Metal Organic Compounds*, vol. 1, p. 1569. 3d ed.; New York: D. Van Nostrand Co., 1940.
16. D. M. Alexander, *J. Phys. Chem.* **63**, 994 (1959).
17. J. Bjerrum, G. Schwarzenbach, and L. G. Sillen, *Stability Constants, Part II*, Special Publication No. 7, The Chemical Society, London, 1958.
18. W. M. Latimer, *Oxidation Potentials.* 2d ed.; New York: Prentice-Hall, 1952.
19. F. D. Rossini *et al.*, *Circular of the National Bureau of Standards*, 500 (1952).
20. L. G. Hepler, W. L. Jolly, and W. M. Latimer, *J. Am. Chem. Soc.* **75**, 2809 (1953).
21. K. S. Pitzer, *Science* **139**, 414 (1963).
22. J. Jortner, E. G. Wilson, and S. A. Rice, *J. Am. Chem. Soc.* **85**, 814 (1963).
23. T. L. Cottrell, *The Strengths of Chemical Bonds.* London: Butterworth Scientific Publications, 1958.
24. A. V. Grosse, A. D. Kirshenbaum, A. G. Streng, and L. V. Streng, *Science* **139**, 1047 (1963).

THE HEAT CAPACITY AND RELATED THERMODYNAMIC FUNCTIONS OF XENON TETRAFLUORIDE

W. V. JOHNSTON, D. PILIPOVICH, AND D. E. SHEEHAN
North American Aviation Science Center

INTRODUCTION

Considerable interest has developed among chemists as to the reasons for the stability of the recently discovered inert-gas compounds, and much theoretical effort has been expended in this direction. There has been, however, a lack of data on the thermodynamic properties of these compounds with which to compare these theoretical predictions. The possible existence of several modifications of XeF_4 have also been reported [1, 2, 3]. For these reasons an early measurement of such properties as heat capacity and heat of formation would seem especially desirable. This paper presents our initial measurements of the heat capacity of XeF_4 from approximately 20° K. to room temperature, from which the standard entropy and entropy of formation of the solid have been determined. Since our own heat-of-formation measurements have not been completed, we have used the recently reported data of Gunn and Williamson [4] to compute the standard free energy of formation.

EXPERIMENTAL PROCEDURE

Sample preparation.—Two methods of making XeF_4 have been utilized. Originally the pressure-temperature bomb reaction reported in the original papers by Claassen, Selig, and Malm [5] was used. A stainless steel or Monel Hoke bomb was used. The Monel bomb cracked because of thermal stresses developed during the quench. Approximately stoichiometric amounts of xenon and fluorine were mixed to prepare XeF_4. High-purity xenon and fluorine (1 per cent HF, max.) were used. The mixture was heated to 450° C. for $2\frac{1}{2}$ hours and then quenched rapidly in a bucket of water. The fluffy white XeF_4 was sublimed into Kel-F ampules and stored at −80° C. The resulting product was analyzed by reacting it with excess mercury to form mercurous fluoride and measuring both the amount of xenon liberated and the gain in weight of the mercury. It was found necessary to both warm and shake the reaction vessel to complete the reaction. The yields of XeF_4 were found to be low when made by the bomb method. XeF_4 made in this way was used for the specific-heat determinations.

The electric-discharge method [6] was utilized to make the larger quantities of material needed for the combustion calorimetry. A stoichiometric mixture of xenon and fluorine was passed through an electric-discharge apparatus at 15–20 mm. back pressure and a flow rate of 15–20 cc/min. Impurities such as SiF_4 and some unreacted xenon were trapped downstream by a liquid-nitrogen trap. The yield was estimated to be above 90 per cent. The discharge apparatus was constructed of glass with the copper electrodes recessed out of the gas stream. The electrodes were 90 mm. apart. A neon-sign transformer was used to provide the voltage required.

Specific-heat calorimeter.—The newly constructed specific-heat calorimeter was similar in design to the low-temperature calorimeter described recently by F. J. Morin [7]. Significant differences are the use of a gold 2.1 per cent cobalt *vs.* copper thermocouple for temperature measurement and the enclosure of the sample holder. The silver cup was in the form of a cylinder $\frac{3}{8}$ in. in diameter and 1 in. long. A stainless steel re-entrant well extended from the bottom to the center of the calorimeter. The heater, made of $\sim$100 ohms of 0.001 in. diameter manganin wire wound on a tiny silicon thimble, was inserted into this well. After loading 1.8727 gm. of XeF_4 into the sample holder in a helium-filled dry box, the top was crimped and soldered with soft solder. The empty mass of the calorimeter was 2.274+ gm.

The sample holder was thermally isolated from its surroundings by imbedding it in 600-mesh alumina powder contained in a relatively massive copper container and evacuating the dead space. The calorimeter may be operated adiabatically by removing the Al_2O_3 powder and installing a differential thermocouple between the sample holder and adiabatic shield. A temperature drift downward was established by cooling the surroundings. Specific-heat measurements were made by measuring the temperature rise of the sample holder accompanying power input into the sample holder.

The gold-cobalt thermocouple was selected because it retains its temperature sensitivity to very low temperatures. This sensitivity coupled with the use of a very sensitive galvanometer-micro ammeter combination enabled temperature differences of less than 0.001° to be resolved. The temperature rises were recorded on a high-speed galvanometer recorder. The temperature rise was determined by graphically extrapolating to the midpoint of the heating curve from the time at which the drift ratio before and after the heating period were the same. The heating periods were of the order of thirty seconds and the temperature rise was typically 0.2 to 0.3° K. The electrical power input and timing measurements were made by the conventional methods of precise calorimetry.

EXPERIMENTAL RESULTS

Specific heat of XeF_4.—The smoothed data for the specific heat and heat capacity per mole of XeF_4 are given in Table 1. The plotted curve of the specific heat is shown in Figure 1. There were no apparent abnormalities in the heat

TABLE 1

HEAT CAPACITY AND SPECIFIC HEAT OF XeF_4

Temperature (° K.)	Specific Heat (cal/gram)	C_p Sat. (cal/mole)
10	0.0012	0.259
20	.0093	1.933
30	.0217	4.494
40	.0318	6.600
50	.0403	8.361
60	.0474	9.826
70	.0532	11.033
80	.0583	12.092
90	.0631	13.089
100	.0676	14.025
110	.0719	14.899
120	.0760	15.761
130	.0800	16.574
140	.0838	17.362
150	.0874	18.125
160	.0912	18.901
170	.0947	19.640
180	.0982	20.366
190	.1016	21.068
200	.1051	21.782
210	.1082	22.435
220	.1116	23.125
230	.1147	23.778
240	.1181	24.479
250	.1213	25.145
260	.1247	25.859
270	.1279	26.523
280	.1312	27.200
290	.1342	27.829
298.16	.1367	28.334
300	0.1373	28.457

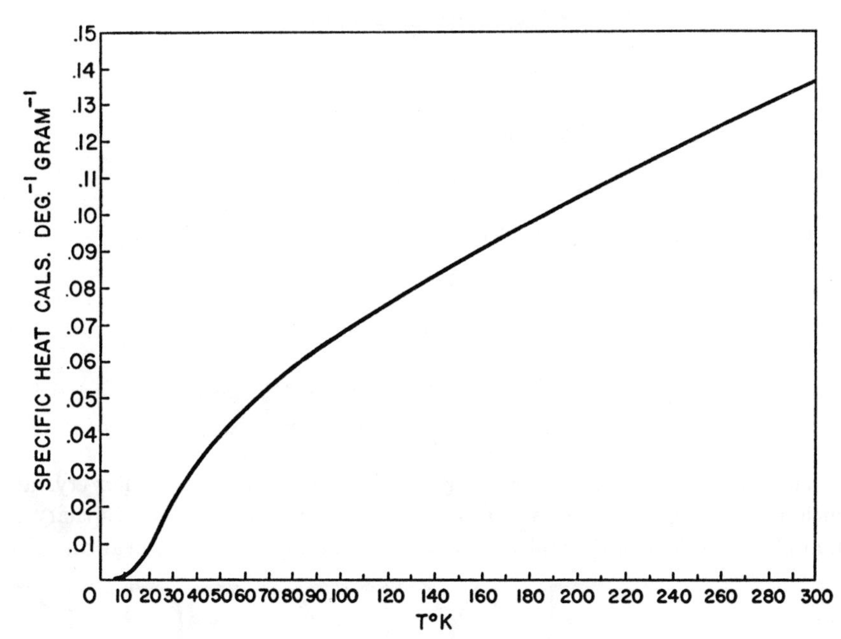

FIG. 1.—Specific heat of XeF_4

capacity from 20° K. to about 268° K. Thermal data taken between this temperature and 300° K. during both heating and cooling have exhibited small abnormalities that have not as yet been explained. The data have not been consistent in that of the three traverses of this temperature range that have been made, two have produced high values of heat capacity indicative of a second-order transformation, but the third, very careful determination—made under equilibrium conditions using the adiabatic modification of the calorimeter—produced a smooth curve with no points deviating more than 0.8 per cent from the curve.

Entropy of XeF_4 to 298.16° K.—The entropy of XeF_4 from 0° K. to 298.16° has been determined from the heat capacities. By plotting ^{C}pk *vs.* T.^{2}, the θ_D was computed and found to be about 122. The entropy from 0° K. to 20° K. was obtained from the computed heat capacities using this value of θ_D. The value of θ_D was observed to decrease with decreasing temperature. Results are summarized in Table 2.

TABLE 2

ENTROPY OF XeF_4

Temperature	eu
0°–20° K.	0.68
20°–290° K.	33.55
290°–298.16° K.	0.78
Entropy of XeF_4 at 298.16°	35.0 eu ± 1.0

DERIVED QUANTITIES

Entropy of formation of XeF_4.—For the reaction at 25° C. and assuming the third law of thermodynamics to hold for this compound:

$$Xe(g) + 2F_2(g) = XeF_4(s)$$

$$\Delta S^\circ = S^\circ_{XeF_4} - [S^\circ_{Xe} + 2S^\circ_{F_2}]$$

$$\Delta S^\circ = 35.0 - 137.5$$

$$\Delta S^\circ = -102.5 \text{ cal. mole}^{-1} \text{ deg.}^{-1}$$

Free energy of formation at 25° C.—

$$\Delta F^\circ = \Delta H^\circ - T\,\Delta S^\circ$$

$$\Delta F^\circ = -60 \text{ kcal/mole} - [298.16\ (-102.5)]$$

$$\Delta F^\circ = -29.4 \text{ kcal. mole}^{-1}$$

DISCUSSION

The shape of the heat-capacity curve is similar to that of molecular crystals such as benzene, CO_2, SO_2, and CH_4, in which the observed heat capacity is that of the motions of the molecules as a whole, acting as a lattice heat capacity plus the oscillations between the atoms of the compound as an internal heat capacity. Since the theoretical heat capacity for a crystal such as this with five atoms and six degrees of freedom is 30 cals/deg/mole and the measured heat capacity is 28 at 300° K., it is to be expected that the heat capacity is still rising. The heat capacity at the lowest temperatures is mostly due to the lattice vibrations since the heat capacity due to the lowest frequency vibrational modes of the molecule is less than 1 per cent of the observed heat capacity at 20° K. Since value of θ_D is observed to vary with temperature in this region it is probable that the Debye distribution of frequencies is changing or that the vibration frequencies of the molecules are not the same as their librational motion [8]. The obtaining of additional heat-capacity data at these very low temperatures would be desirable. In view of the variations of the heat-capacity data obtained near room temperature, the absence of any transition cannot be stated with finality. It is well known that too rapid cooling through a second-order transition can result in only partial transformation to the lower temperature state and that on reheating the temperature for the onset of the transformation is raised. There are usually prolonged downward drifts after a heating period in this region as thermally excited atoms or molecules achieve the lower temperature configuration. We have evidence of these latter effects but have not been able to measure above 300° K. because our thermocouple is not calibrated above this temperature.

The large decrease in entropy occasioned by the formation of the solid compound causes the free energy of formation to be considerably less than that of the heat of formation, thus providing additional evidence that equilibrium constants and over-all stability estimated from heats of formation are, at best, approximations.

Acknowledgment.—The authors wish to thank Mr. Gary Lindberg and Mr. Fred Lieu for their assistance in constructing the calorimeter and in helping with the calculations.

REFERENCES

1. J. A. Ibers and W. C. Hamilton, *Science* **139,** 106 (1963).
2. J. H. Burns, *J. Phys. Chem.* **67,** 536 (1963).
3. S. Siegel and E. Gebert, *J. Am. Chem. Soc.* **85,** 24 (1963).
4. S. R. Gunn and S. M. Williamson, *Science* **140,** 177 (1963).
5. H. H. Claassen, H. Selig, and J. G. Malm, *J. Am. Chem. Soc.* **84,** 3593 (1962).
6. A. D. Kirshenbaum, L. V. Streng, A. G. Streng, and A. V. Grosse, *ibid.* **85,** 360 (1963).
7. F. J. Morin and J. P. Maita, *Phys. Rev.* **129,** 1115 (1963).
8. R. C. Lord, Jr., *J. Chem. Phys.* **9,** 693 (1941).

THERMOCHEMICAL STUDIES OF XENON TETRAFLUORIDE AND XENON HEXAFLUORIDE

L. STEIN AND P. L. PLURIEN*
Argonne National Laboratory

ABSTRACT

The reduction of xenon tetrafluoride and xenon hexafluoride with hydrogen is being studied with an isothermal calorimeter at 120° to 130° C. Each compound reacts rapidly with hydrogen in this temperature region, forming xenon and hydrogen fluoride:

$$XeF_4(g) + 2H_2(g) \rightarrow Xe(g) + 4HF(g) \quad (1)$$

$$XeF_6(g) + 3H_2(g) \rightarrow Xe(g) + 6HF(g) . \quad (2)$$

Preliminary values of the heats of reaction obtained thus far are $\Delta H = -202$ kcal/mole XeF_4 for reaction 1 and $\Delta H = -306$ kcal/mole XeF_6 for reaction 2. The approximate heats of formation are calculated to be:

$$\Delta H_{f\,XeF_4(g)} = -55 \text{ kcal/mole}$$

and

$$\Delta H_{f\,XeF_6(g)} = -79 \text{ kcal/mole} .$$

From the present measurements the average xenon-fluorine bond energy is found to be 32.0 kcal. in the tetrafluoride and 31.5 kcal. in the hexafluoride.

Previous investigators [1–3] have shown that xenon fluorides can be reduced quantitatively with excess hydrogen at 400° to form xenon and hydrogen fluoride. The reactions of xenon tetrafluoride and hexafluoride with hydrogen are now being studied at lower temperatures with an isothermal calorimeter in order to obtain heats of formation of these compounds. The tetrafluoride does not react with hydrogen to any noticeable extent at 25° but reacts slowly at 70° and rapidly at 130°. The hexafluoride behaves erratically when mixed with hydrogen at 25°; in some instances a spontaneous reaction occurs; in other instances no reaction is observed. At 130°, the hexafluoride is rapidly reduced by hydrogen. The first experiments with the calorimeter have therefore been carried out in the interval 120° to 130°, with the reactions regulated by the rate of hydrogen addition, and some preliminary results are available.

APPARATUS

The calorimeter, shown in Figure 1, consists of a silvered Pyrex Dewar vessel mounted inside a cylindrical copper jacket. The top, side, and bottom of the

* Permanent address: Centre d'Études Nucléaires de Saclay, Seine-et-Oise, France.

jacket are heated by separate electric windings. The Dewar contains a nickel reaction vessel, 300 cc. of mineral oil, a twin-bladed stirrer, a 25-ohm platinum resistance thermometer (not shown), and a 400-ohm calibration heater (not shown). Both the Dewar cover, which is made of thin Transite, and the top of the jacket are split into halves for easy assembly.

The reaction vessel consists of a nickel can of approximately 130 cc. volume attached to a small bellows valve and a vacuum coupling by thin-walled Monel

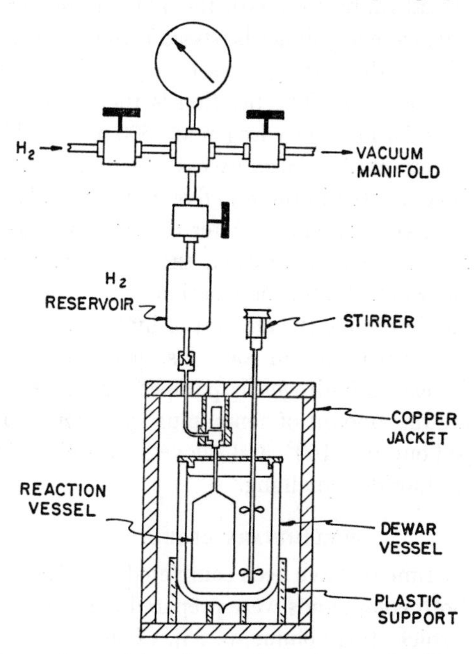

FIG. 1.—Calorimeter for studying the reactions of xenon fluorides with hydrogen

tubing. The volume of the tubing between the can and the valve point comprises less than 0.1 per cent of the total volume. When the calorimeter is assembled, the valve is inserted into an electrically heated copper sleeve, where it is clamped with a set screw. The vacuum coupling is then attached to the hydrogen reservoir, which is also electrically heated. Above the reservoir, the metal vacuum line contains a 0–4000 mm. bourdon gauge, several cold traps (not shown), and manifold connections to a hydrogen cylinder and oil diffusion pump.

Thermocouples indicate the temperature of the sleeve, reservoir, and jacket at several points. A double microvolt potentiometer, Rubicon Model No. 2773, is used for electrical measurements with the resistance thermometer and calibration heater. The current measurements are made with standard resistors manufactured by Leeds and Northrup Company and Gray Instrument Company. Time intervals are measured with a Beckman EPUT meter, Model No. 7350 C.

EXPERIMENTAL PROCEDURE

After the reaction vessel has been thoroughly prefluorinated at 500°, it is evacuated, weighed, filled on a vacuum line with XeF_4 or XeF_6, and again weighed to determine the amount of compound added. The calorimeter is assembled with 300 cc. of mineral oil in the Dewar and is then heated to 120° or slightly higher. When temperature equilibrium is established, the heat capacity is measured by electrical calibration with the 400-ohm heater. Prepurified hydrogen (99.97 per cent) from a cylinder is passed through a Deoxo unit, through a cold trap at −195°, into the hydrogen reservoir, which is initially at the same temperature as the calorimeter. The reaction is then carried out by cracking open the small valve and allowing hydrogen at 3500 mm. to 4000 mm. pressure to enter the calorimeter until a large excess has been added. Generally the hydrogen addition is completed in three to five minutes, and a temperature rise of the order of one degree is observed. After the reaction, the heat capacity is again determined by electrical calibration, the calorimeter is cooled to room temperature and disassembled, and the reaction vessel is washed with acetone, dried, and weighed to determine the amount of hydrogen added. As a check on the stoichiometry of the reaction, in some instances additional weighings are made as follows: the excess hydrogen is pumped out at −195°; the vessel is weighed to determine the amount of xenon plus hydrogen fluoride present; the xenon is then pumped out at −160°; and the vessel is weighed to determine the amount of hydrogen fluoride remaining.

PRELIMINARY RESULTS

Thus far three experiments have been completed with XeF_4 and three experiments with XeF_6. Both compounds were prepared elsewhere in this laboratory as described by Chernick [this volume, p. 35]. Chemical analysis and infrared examination suggest that the purity of the XeF_4 exceeds 99 per cent. The XeF_6 contains possibly XeF_4 and another unidentified substance, and its purity may be as low as 95 per cent. For further calorimetric experiments, attempts are being made to purify both compounds more extensively.

Preliminary results obtained from the first experiments are shown in Table 1. Under the conditions which are used at present, all reactants and products are in the gaseous state, and no corrections for vaporization are necessary. The reactions occurring in the calorimeter are as follows:

$$XeF_4(g) + 2H_2(g) \rightarrow Xe(g) + 4HF(g) \tag{1}$$

$$XeF_6(g) + 3H_2(g) \rightarrow Xe(g) + 6HF(g)\,. \tag{2}$$

The hydrogen fluoride is assumed to be entirely monomeric, since the concentration of polymers is very low at 120°.

The total energy change in each experiment is shown in column 2. Since hydrogen is added from a source outside of the calorimeter, the incoming gas does work on the contents of the calorimeter which must be subtracted from the total energy change to obtain $\Delta E_{reaction}$. The correction term, calculated as nRT, where n is the number of moles of hydrogen added, R is the gas constant, and T is the absolute temperature, is given in column 3. From $\Delta E_{reaction}$ and the increase in the total number of moles of gas (equation 1 or 2), $\Delta H_{reaction}$ is readily calculated. No correction of the enthalpy from the actual temperature to 25° is made at present, since the experimental error is larger than the correction and since the heat capacities of gaseous XeF_4 and XeF_6 have not yet been determined.

TABLE 1

A. REDUCTION OF XeF_4 WITH EXCESS H_2

Weight of XeF_4 (mg.)	Total Energy Change (cal.)	Correction for PV Work (cal.)	$\Delta H_{reaction}$ (kcal/mole XeF_4)
690.9	−685.8	7.4	−202.2
326.3	−325.6	9.9	−199.0
329.2	−337.1	9.5	−204.7
		average = −202.0	

B. REDUCTION OF XeF_6 WITH EXCESS H_2

Weight of XeF_6 (mg.)	Total Energy Change (cal.)	Correction for PV Work (cal.)	$\Delta H_{reaction}$ (kcal/mole XeF_6)
493.0	−626.0	7.8	−305.2
382.1	−489.9	9.4	−306.0
298.4	−388.1	10.2	−308.3
		average = −306.5	

Using the value, −64.2 kcal/mole, for the heat of formation of hydrogen fluoride [4], the approximate heats of formation of XeF_4 and XeF_6 are calculated as follows.

From equation 1:

$$\Delta H_{reaction} = 4\Delta H_{f_{HF(g)}} - \Delta H_{f_{XeF_4(g)}}$$

$$-202.0 = 4\,(-64.2) - \Delta H_{f_{XeF_4(g)}}$$

$$\Delta H_{f_{XeF_4(g)}} = -54.8 \text{ kcal/mole.}$$

From equation 2:

$$\Delta H_{reaction} = 6\Delta H_{f_{HF(g)}} - \Delta H_{f_{XeF_6(g)}}$$

$$-306.5 = 6\,(-64.2) - \Delta H_{f_{XeF_6(g)}}$$

$$\Delta H_{f_{XeF_6(g)}} = -78.7 \text{ kcal/mole}.$$

The present value for the heat of formation of gaseous XeF_4 can be compared with the value obtained by Gunn and Williamson [5] for crystalline XeF_4 (-60 kcal/mole).

Taking the heat of dissociation of fluorine to be 36.7 kcal/mole [6], the average bond energies in XeF_4 and XeF_6 are calculated as follows:

$$\text{average BE in } XeF_4 = \frac{54.8 + 2(36.7)}{4} = 32.0 \text{ kcal}.$$

$$\text{average BE in } XeF_6 = \frac{78.7 + 3(36.7)}{6} = 31.5 \text{ kcal}.$$

REFERENCES

1. H. H. Claassen, H. Selig, and J. G. Malm, *J. Am. Chem. Soc.* **84**, 3593 (1962).
2. J. L. Weeks, C. L. Chernick, and M. S. Matheson, *ibid.* 4612 (1962).
3. J. G. Malm, I. Sheft, and C. L. Chernick, *ibid.* **85**, 110 (1963).
4. T. L. Higgins and E. F. Westrum, Jr., *J. Phys. Chem.* **65**, 830 (1961).
5. S. R. Gunn and S. M. Williamson, *Science* **140**, 177 (1963).
6. W. H. Evans, T. R. Munson, and D. D. Wagman, *J. Res. N. B. S.* **55**, 147 (1955).

THE HEAT OF FORMATION OF XENON TRIOXIDE

STUART R. GUNN
University of California, Livermore

ABSTRACT

The heat of explosion of XeO_3 has been measured calorimetrically. The derived heat of formation is $+ 96 \pm 2$kcal. mole^{-1}.

Xenon trioxide has recently been identified and reported to be a sensitive and powerful explosive [1–3]. In this respect it contrasts with the xenon fluorides, which are stable. Determination of the heat of formation of the compound thus appeared to be of interest. This has been done by calorimetry of the explosive decomposition to the elements.

EXPERIMENTAL

Xenon tetrafluoride (obtained from Peninsular Chemical Research Co., Gainesville, Fla.) was sublimed into a U-trap of borosilicate glass at $-60°$, 0.3 to 0.4 gm. usually being taken. The trap was filled with helium and removed from the vacuum line. Then 3 ml. H_2O containing 1 to 2 μmoles H_2SO_4 were placed in the bottom of the trap with a pipette. The trap was agitated manually with occasional cooling to gradually wash down the deposit of XeF_4. The yellow deposit reported by Williamson and Koch [1] persisted for a few minutes. Then the clear solution was transferred to a platinum or Teflon boat and evaporated to dryness over barium oxide. One batch exploded in the dessicator soon after reaching dryness. The deposit was taken up in about 0.3 ml. water, placed in the sample cup of the calorimeter bomb, and evaporated nearly to dryness with a stream of dry oxygen.

The calorimeter bomb was of stainless steel, 0.625 in. inside diameter, 1.115 in. outside diameter, 7 in. long inside, closed by a heavy threaded plug seated against a gasket. The gasket was exposed to the blast of hot gas from the explosion; neoprene was used for runs I and II, Teflon for runs III and IV, and lead for runs V and VI. A stainless steel tube 0.040 in. inside diameter, 12 in. long, extended upward from the plug to a stopcock. Two electrodes passed through an insulated seal in the plug. Platinum leads were attached to the electrodes; a loop in one of these supported the Pyrex sample cup of about 0.5 ml. volume (a platinum cup was used for run III), and a fuse of platinum wire

0.002 in. in diameter and 2 in. long was attached to the leads and dipped into the cup. The cup was suspended about 3 in. below the plug (6 in. in run V). The bomb was lined with platinum sheet 5 in. long, centered near the cup, except for runs I and V.

After the XeO_3 in the cup was nearly dry, the bomb was assembled and pumped on the vacuum line overnight. Four batches exploded spontaneously during the pumping.

The bomb was then placed in the calorimeter and the heat determination performed. The calorimeter, laboratory designation XXIX, consists essentially of a copper cylinder 2 in. outside diameter and 9.50 in. long, with a central bore 1.125 in. diameter and 9.35 in. deep, having a copper resistance thermometer wound on the outer surface, suspended by a thin-walled stainless steel tube 1.15 in. inside diameter in an evacuated submarine surrounded by a thermostat. Four calibrations at 130 to 150 cal. were performed with a heater wrapped on the explosion bomb. The maximum deviation from the mean was 0.13 per cent. The fuse was fired with a 24-v. storage battery. Oscilloscopic time-current-voltage measurements on similar systems have shown the energy dissipated to be essentially that required to heat the mass of platinum in the fuse to a liquid at its melting point. Half of the temperature rise occurred in about 1.4 min., and the final drift rate was attained in about 18 min.

After the run, the oxygen was transferred by a Toepler pump through three traps at $-196°$ to a calibrated gas buret and measured. The traps were then warmed to $-78°$, and the remaining gas measured. Samples of both batches of gas were analyzed by mass spectrometry. Usually about 0.2 per cent of the xenon was found in the oxygen fraction. Some CO_2 was found in the xenon fraction, presumably from attack on the gaskets; Smith [2] has reported the same observation. The xenon fraction was also found to contain SiF_4 and BF_3, in amounts corresponding to 0.5 to 2.0 per cent and 0.1 to 0.5 per cent, respectively, on the xenon. These presumably came from HF attack on the glass during the hydrolysis of the XeF_4, although it seems surprising that they were not removed from the solid XeO_3 by the prolonged pumping. Water was not determined. However, on the basis of other observations [1, 4], it is expected that pumping overnight would reduce water in the solid to a rather low amount.

RESULTS

Results of the six runs successfully performed are given in Table 1. The sum of the O_2 and CO_2 ranges from 99.0 to 100.9 per cent of the theoretical, referred to xenon. The observed heats q are corrected for the firing energy and the heat of formation of the CO_2 found. ΔE is calculated from q and the amount of xenon found. The average is -97.0; converted to constant pressure, this gives $+95.5$ for $\Delta H_f°(XeO_3)$.

Some possible errors may be considered. The sample may detonate so quickly as to break the fuse before the usual amount of electrical energy has been dissipated. Any water bound in the sample would volatilize, absorbing heat. A small heat effect might be associated with release of SiF_4 and BF_3. There might be a small amount of oxidation of the bomb by the oxygen. In view of these uncertainties, it is felt that the best value for $\Delta H_f°$ is $+96 \pm 2$ kcal. mole^{-1}.

Exact calculation of the thermochemical bond energy E(Xe-O) would require knowledge of the heat of sublimation, ΔH_{sub}. However, the vapor pressure at room temperature is unobservable, and the compound explodes upon slight heating. The crystal and molecular structure have been determined, but calculations of the heat of sublimation for such an unusual substance would still appear to be rather difficult. Hence, using 119 kcal. mole^{-1} for the dissociation energy of oxygen [5], E(Xe-O) may only be expressed as $27.5 - .33\Delta H_{sub}$. Since ΔH_{sub}

TABLE 1

HEAT OF EXPLOSION OF XeO_3

Run	Xe (mmoles)	O_2/Xe	CO_2/Xe	q (cal.)	$-\Delta E$ (kcal. mole^{-1})
I.	0.3015	1.457	0.033	28.39	94.2
II.	.3111	1.453	.051	30.89	99.3
III.	.2414	1.476	.037	23.84	98.8
IV.	.2596	1.458	.027	25.25	97.3
V.	.1066	1.489	.009	10.24	96.1
VI.	0.2763	1.495	0.010	26.67	96.5

is expected to be fairly large, the bond energy is significantly less than the value of -30 for E(Xe-F) [Gunn and Williamson, this volume, p. 138].

Exact calculation of the oxidation potential in solution would require knowledge of the entropy of XeO_3 and its free energy of solution. However, neglecting these secondary contributions, the standard oxidation potential in acid solution may be estimated as about -1.9 v.

Acknowledgment.—Thanks are due to S. M. Williamson of the University of California, Berkeley, for consultation.

REFERENCES

1. S. M. WILLIAMSON and C. W. KOCH, *Science* **139,** 1046 (1963).
2. D. F. SMITH, *J. Am. Chem. Soc.* **85,** 816 (1963).
3. D. H. TEMPLETON, A. ZALKIN, J. D. FORRESTER, and S. M. WILLIAMSON, *ibid.* 817 (1963).
4. S. M. WILLIAMSON, private communication.
5. T. L. COTTRELL, *The Strengths of Chemical Bonds.* London: Butterworth Scientific Publications, 1958.

PART 5 AQUEOUS CHEMISTRY OF NOBLE-GAS COMPOUNDS

The first report that xenon formed simple fluorides also suggested that aqueous solutions containing relatively stable xenon species could be prepared. Subsequent investigations have proved the aqueous chemistry of xenon to be both complicated and exceptionally interesting. Both hexavalent and octavalent species seem to be well established. A variety of salts with xenon containing anions have been isolated, and there seems to be no doubt that this chemistry will proliferate and ultimately include krypton and radon compounds as well.

AQUEOUS SOLUTION CHEMISTRY OF XENON
AN INTRODUCTION

MARTIN KILPATRICK
Argonne National Laboratory

The chemistry of aqueous solutions of xenon to date is based on the products of hydrolysis of XeF_4 [1, 2, 3], discussed in the volume by Williamson and Koch, p. 158, and the alkaline hydrolysis of XeF_6 [4; see also this volume, Malm, Bane, and Holt, p. 167]. In the first case the solid product is the unstable XeO_3 [3, 5], and in the second case the sodium salt of octavalent xenon of the general formula $M_{2n}XeO_{4+n} \cdot yH_2O$. Heavy metal perxenates of barium, copper, silver, lead, and uranium have been prepared from the sodium salts [Gruen, this volume, p. 174].

The term perxenate is reserved for this type of salt which contains octavalent xenon and the corresponding acid is perxenic acid, $H_{2n}XeO_{4+n}$. Following the IUP report on nomenclature [6] the xenates would correspond to xenic acid, $H_{2n}XeO_{3+n}$, containing hexavalent xenon. To date, xenon has exhibited valences of II, IV, VI, and VIII in stable compounds, but only one oxide, three fluorides, two anion salts, and two oxyfluorides have been isolated and characterized. There is no problem with naming the oxides, and the acids can be named hypoxenous II, xenous IV, xenic VI, and perxenic VIII with the salts hypoxenites, xenites, xenates, and perxenates. The oxyfluoride $XeOF_4$ has been called xenonoxytetrafluoride [Chernick, Claassen, Malm, and Plurien, this volume, p. 106] but the nomenclature report recommends xenon oxide tetrafluoride. In either case, di-, tri-, and tetra- should be used, for example XeO_2F_2 would be xenon dioxydifluoride or xenon dioxide difluoride. In the case of doubt as to the valence exhibited, the Roman numerals (II), (IV), (VI), and (VIII) are recommended. If compounds exhibiting other valences are found in the future the Roman numeral will be necessary.

HYDROLYSIS OF XENON FLUORIDES

From the results of hydrolysis experiments it is evident that in addition to the oxidation of water to oxygen there is a disproportionation of the xenon compound to higher and lower valence states for the tetra- and hexafluorides. At Argonne National Laboratory these hydrolyses have been carried out in alkaline solution, while at California and elsewhere the reaction has been carried out in acid solution.

The usual technique is to add the aqueous solution to the solid fluoride at low temperature, measure and identify the gas evolved, and determine the fluoride after the solution has come to room temperature. For XeF_2 the surface of the solid becomes yellow in contact with the aqueous alkaline solution, but the yellow species is destroyed rapidly, and the heterogeneous reaction

$$XeF_2(c) \rightleftharpoons XeF_2(aq) + H_2O \xrightarrow{OH^-} Xe(g) + \tfrac{1}{2}O_2(g) + 2HF(aq) \quad (1)$$

is complete in ten to twenty minutes. The temperature of the reaction varies but can be considered to be in the neighborhood of 0° C. This is interpreted to mean that the yellow species has a short lifetime and there is no appreciable disproportionation.

When the starting material is XeF_4, alkaline hydrolysis yields a yellow solution which eventually yields a white solid of the composition $Na_{2n}XeO_{3+n} \cdot xH_2O$. During the early stages of the reaction of XeF_4 some two-thirds of the xenon and one-half of the oxygen is given off, but the solution can be evaporated to dryness without further loss of xenon. This result is not in accord with the results of Williamson [this volume, p. 158], who finds lower yields of Xe(VI) species. He also reports that the lifetime of the yellow-colored species at the surface of the solid XeF_4 is much less in sodium hydroxide than in water. However, the average stoichiometry indicates that two-thirds of the xenon and one-half the oxygen goes off as gas and one-third remains as $Xe(VI)O_3$ which is not in disagreement with the Argonne results [this volume, p. 167].

The hydrolysis of XeF_6 in water has been studied by Dudley, Gard, and Cady [7], who find that practically all of the xenon remains in solution, possibly as $Xe(OH)_6$. Titration indicates the acid is very weak, and this has been confirmed by Appelman [this volume, p. 185].

The alkaline hydrolysis of XeF_6 on better than gram samples yields a white precipitate of the general formula $Na_{2n}XeO_{4+n} \cdot yH_2O$. The yield varies from 27 to 58 per cent and depends on the time and the concentration of sodium hydroxide, as well as its excess. An aqueous solution of solid compounds [Xe(VIII)] can be converted to Xe(VI) by treatment with H_2SO_4 or $HClO_4$. Treatment of the aqueous solution Xe(VI) with ozone reconverts it to Xe(VIII). The oxidation equivalents are six and eight.

To summarize, XeF_2 in alkaline solution oxidizes water to oxygen and xenon is reduced from +2 to 0. XeF_4, in addition to oxidizing water, also disproportionates to Xe(VI) and Xe(II) which would give Xe(O). This happens in acid or alkaline solution, but the Xe(VI) is more stable in acid solution.

XeF_6 oxidizes water and in addition disproportionates to Xe(VIII) and lower valence states. The Xe(VIII) anion is fairly stable in alkaline solution but unstable in acid solution, so that the only stable acid in solution prepared to date is the Xe(VI) acid. The acid is polybasic and weak. Its dissociation constants could be determined by distribution measurements or spectrophotometrically. This has been done for OsO_4 [8, 9]. Neither XeO_3 dissolved in water nor the corresponding Xe(VIII), perxenic acid, show any acid properties relative to the base water but do show their acid properties in concentrated strong base by adding hydroxyl ions to form fairly stable anions. In this respect they resemble OsO_4, which is soluble in water but does not exhibit any acid reaction with litmus or release carbonate from a solution of bicarbonate but does form salts such as $K_2[OsO_4(OH)_2]$ by adding hydroxyl ion [8]. When anhydrides are dissolved in water, they may react with water to form acids which supply protons to base, while the alkaline-earth oxides dissolve to give hydroxyl ions in aqueous solution. In between we have oxides like OsO_4 or XeO_3, which form such weak or unstable acids that the salts can only be formed by fusion with caustic or concentrated solutions of alkali. We may regard these oxides as hydroxyl-ion acceptors, just as one regards fluoride acceptors in the solvent hydrogen fluoride [10].

$$\begin{array}{c} H_2O + H_2O \rightleftharpoons H_3O^+ + OH^- \\ + \\ XeO_3 \\ \uparrow\downarrow \\ \text{polymers} \rightarrow [\text{trimers}]^{3n-} \rightleftharpoons [\text{dimers}]^{2n-} \rightleftharpoons [XeO_3(OH)_n]^{n-}. \end{array} \tag{2}$$

The exhibition of acid properties will depend on the stability of the anion as well as the solubility of the oxide. When the solvent is a hydroxide solution rather than water, the possibility of an appreciable formation of the xenate is increased. The Xe(VI) acid is stable in acid solution, but both the Xe(VI) and Xe(VIII) species slowly decompose in alkaline solution. The instability of the perxenate in acid solution is best illustrated by the potentiometric titration curves of Appelman [this volume, p. 189], who shows that there is a break in the titration curve between pH's 8 and 5. The explanation is that the perxenic [Xe(VIII)] acid species in solution loses oxygen fairly rapidly at hydrogen-ion concentrations above 10^{-8}, leaving xenic acids [Xe(VI)]. The resulting solution can be titrated back and forth without appreciable decomposition. The solid sodium perxenate appears to be stable and can be used as a starting material to prepare XeO_3. The spectrophotometric measurement with perxenates indicates two species over the whole pH region. There is no evidence of a peroxide or ozonide.

REFERENCES

1. N. Bartlett and P. R. Rao, *Science* **139,** 506 (1963).
2. S. M. Williamson and C. W. Koch, *ibid.* 1046 (1963).
3. D. F. Smith, *J. Am. Chem. Soc.* **85,** 816 (1963).
4. C. Chernick *et al., Science* **138,** 136 (1963).
5. D. H. Templeton, A. Zalkin, S. D. Forrester, and S. M. Williamson, *ibid.* 817 (1963).
6. *J. Am. Chem. Soc.* **82,** 5523 (1960).
7. F. B. Dudley, G. Gard, and G. H. Cady, *Inorg. Chem.* **2,** 228 (1963).
8. D. M. Yost and R. J. White, *J. Am. Chem. Soc.* **50,** 81 (1928).
9. R. D. Sauerbrunn and E. B. Sandell, *ibid.* **75,** 4170 (1953).
10. M. Kilpatrick, *J. Chem. Ed.* **37,** 403 (1960).

THE REACTION OF XENON TETRAFLUORIDE WITH AQUEOUS SOLUTION: CHEMISTRY OF XENON TRIOXIDE

STANLEY M. WILLIAMSON AND CHARLES W. KOCH
University of California, Berkeley

The reaction of xenon tetrafluoride with water or aqueous solutions was first mentioned by Claassen, Selig, and Malm [1] and has been studied in more detail by us [2]. The detailed studies showed that, in addition to the direct oxidation of water by XeF_4 to give oxygen, xenon, and hydrofluoric acid, the reaction gave a very soluble xenon(VI) species. Evaporation of this resulting solution in Teflon ware at room temperature gave the white, crystalline xenon(VI) oxide, XeO_3 [3].

The nature of the hydrolysis reaction.—The reaction is rapid and vigorous. The system used to study the reaction quantitatively consisted of the reaction vessel, two U-traps, a 600-ml. automatic Toepler pump, and a 3-ml. gas buret. The reaction vessel could hold 3 ml. of solution and was equipped with a container bulb-breaking device and a rotating side arm so that solid reagents could be added at any time.

The procedure for the hydrolysis experiments was to introduce a weighed amount of XeF_4 into a degassed solution, measure independently the oxygen and xenon, then add solid potassium iodide to the resulting acidic solution or to the solution which was made acidic by addition of H_2SO_4—in order to recover any remaining xenon. The liberated iodine, when compared to the remaining xenon which was liberated and measured, gave the oxidation state of the soluble xenon species.

For the hydrolysis experiments, 3- to 20-mg. samples of XeF_4 were sublimed into weighed, thin-walled Pyrex glass bulbs. The volume of a typical bulb was 0.07 ml. When the empty weights of the bulbs were determined, they were not evacuated. Consequently, the volumes were determined in order to correct for the weight of air prior to subliming the XeF_4 into the bulb. The XeF_4 was prepared by the method described by Templeton, Zalkin, Forrester, and Williamson [4]. The thin-walled bulbs could easily be broken in the reaction vessel which contained the degassed reaction medium. For the accuracy required in these experiments it was necessary to use a trap at $-210°$ to collect the xenon quantitatively while the oxygen was being transferred into the measuring buret by the Toepler pump. By pumping on the liquid-N_2 trap with a mechanical pump, this lower temperature was attained; another trap at −95° held back any water. It was found that if a trap at −196° was used, a sample of 100 μl. to 200 μl. of Xe (STP) would lose 5 μl. over a period of fifteen minutes. For large samples this loss was not important.

Since the reaction of XeF_4 with an aqueous solution is heterogeneous, reproducible results were not obtained from identical samples because of the concomitant reactions. The reaction of XeF_4 with basic solutions has been observed to be quite different from the reaction with acidic or neutral solutions. The time

TABLE 1

XeF_4 HYDROLYSIS DATA

RUN	XeF_4* (μmole)	REACTION MEDIUM	BEFORE KI-ACID QUENCH		AFTER KI-ACID QUENCH†		
			O_2 (μmole)	Xe (μmole)	I°÷6 (μmole)	Xe(VI) (%)	Xe° measured (μmole)
I......	26.72	H_2O	13.59	30.88	2.1	7.8	.......‡
II.....	16.71	H_2O§	5.98	13.77	4.7	28.2	4.3
III....	46.77	0.1 M NaOH	30.29	....	4.8	10.3	....
IV....	72.02	1.0 M NaOH	....	....	3.9	5.4	....
V.....	132.72	3 M H_2SO_4	68.46	94.95	39.2	29.6	39.6
VI....	81.09	3 M H_2SO_4	38.02	71.30	17.0	21.0	17.1
VII....	87.24	9 M H_2SO_4	43.99	63.71	25.2	28.9	24.2

* The sample weights are expressed in terms of μmoles of XeF_4.

† No non-condensible gas at −210° in amount larger than the blank of the pumping system was liberated when KI(s) was added.

‡ Ellipses indicate that a measurement for that particular quantity was not made.

§ Water was distilled onto the XeF_4 instead of reacting the XeF_4 with bulk water.

lag between the initial reaction of XeF_4 with basic solutions and the second reaction of quenching with iodide and acid is important in the determination of the yield of the soluble xenon(VI) species. The yield of the xenon(VI) species is greater when the hydrolysis is carried out in acidic or neutral solution than it is in basic solution, and the species decomposes at a measurable rate in basic solution. It is stable in acidic solution. Table 1 gives data for several hydrolyses in different aqueous media. Even though the hydrolyses were started below 0°, the actual temperature was difficult to control and estimate once the reaction started. In all cases the sample bulb was broken directly on or in the frozen solution, and the reaction was moderated by holding the reaction vessel at a temperature such that a slush was maintained inside.

The possibility of any gases other than oxygen and xenon being present as hydrolysis products was ruled out by mass analysis of the gas mixture from a run similar to those in Table 1. A sample was taken before the gas contacted any mercury or other reducing materials. It can be seen in runs V and VI that the iodine atom/xenon atom ratio is 5.9 and 6.0, respectively; whereas, in runs II and VII the ratio is 6.6 and 6.3, respectively. It is difficult to assign limits of experimental error to this measurement, but the oxidation state of 6 seems real. However, these measurements do not rule out the possibility of other oxidation states of less stability.

For the runs in which all of the xenon was recovered and measured, the total percentage is somewhat greater than 100. Likewise, the total per cent of oxidizing power is less than 100. This can be explained by the presence of a small amount of XeF_2 as a contaminate in the XeF_4 sample. It has been reported by Smith [5] and Jortner, Wilson, and Rice [6] that the vapor pressures of XeF_2 and XeF_4 are not too different, so simple sublimation would not be expected to separate the two compounds efficiently.

TABLE 2

LIFETIMES OF YELLOW HYDROLYSIS COMPOUND

Reaction Medium	Time (minutes)
H_2O	8.0
0.1 M NaOH	0.2
6 M NaOH	0.2
3 M H_2SO_4	20.0
9 M H_2SO_4	1.0

An interesting observation during the hydrolysis reaction was that the surfaces of the white XeF_4 crystals become a bright, canary-yellow in contact with water. This yellow solid dissolves by reaction with the aqueous solution, liberating gas bubbles at the solid-liquid interface. A sample of the yellow solid, the composition of which is unknown to us, frozen in an ice matrix did not show an electron spin–resonance spectrum.

The lifetime of this yellow solid species was measured qualitatively in the neighborhood of from $-5°$ to $+5°$ as a function of the reaction medium (Table 2). On no occasion did the solution in contact with the yellow solid appear colored.

Examination of the data in Table 1 indicates that in the over-all reaction of XeF_4 with aqueous solutions, about 50 per cent of the XeF_4 directly oxidizes water, while the other 50 per cent undergoes a disproportionation reaction to produce aqueous XeO_3 and Xe by a mechanism as yet unknown. The following stoichiometry represents the average of the reactions studied:

$$XeF_4 + 2H_2O \rightarrow .67Xe\uparrow + .5O_2\uparrow + .33XeO_3 + 4HF\,. \qquad (1)$$

The nature of the aqueous xenon(VI) species.—Data from Table 1 indicate the presence of a species in a $+6$ oxidation state. All of the measurements in Table 1 were made on solutions which contained approximately 12 moles of HF per mole of xenon(VI) species. It was found that the hydrofluoric acid in a solution of hydrolyzed XeF_4 can be removed by passing the solution through a chromatographic column packed with a slurry of MgO. A solution treated in this manner gave an I/Xe ratio of 6.06. The aqueous species with or without the HF is stable in sulfuric acid solutions (other acids have not been examined). The concentration of a 3.85×10^{-3} M xenon(VI) solution in 3 M H_2SO_4 has remained unchanged for one hundred days. The concentration was checked both by iodine-thiosulfate titration and by iodine colorimetry. The concentration also remained unchanged when the solution was refluxed for five minutes. The concentration of this same solution also remained unchanged when the solution was stored in platinum instead of glass. This aqueous species seems without doubt to be the same one that Dudley, Gard, and Cady [7] and Smith [8] described from the hydrolysis of XeF_6.

Since it was known that non-basic solutions yield XeO_3(s) on evaporation [2, 3], the nature of the aqueous species was investigated via four routes. The

TABLE 3

OXYGEN ISOTOPIC DATA

Length of Equilibration (minutes)	O^{18} (%)	O/Xe
15.	51.4	2.76
60*.	54.0, 53.9†	2.95

* Two determinations were made under different instrument conditions.

† For complete exchange, O^{18} = 57.0%.

degree of hydration of the XeO_3 is not known by us, but the oxygen atoms can be exchanged by equilibrating the solid with O^{18}-enriched water. If the aqueous species is fully hydroxylated, i.e., xenic acid, as has been suggested by Dudley, Gard, and Cady [7], then an equilibrium such as

$$XeO_3 + 3H_2O \rightleftharpoons Xe(OH)_6 \tag{2}$$

must be operative. Pauling [9] has named H_4XeO_6 "xenic acid." If and when H_4XeO_6 is discovered, it perhaps should be called "paraperxenic acid" to be consistent with H_5IO_6.

For the experiment, XeO_3(s) made from XeF_4 and normal isotopic water was dissolved in 57.0 per cent O^{18}-enriched water and left standing in a quartz tube in one experiment for fifteen minutes and in another for one hour before the excess water was pumped off. An approximate 7000:1 mole ratio of water to compound was used so that the O^{16} brought in with the XeO_3 would be negligible. No attempt was made to weigh the solid, but about 0.5 mg. was used. The water (0.4 ml.) was vacuum-distilled to free it of any non-volatile material. After evaporation of the water, the residual XeO_3 was decomposed, and the gases were analyzed. Table 3 gives the data for the two experiments. The mass analysis of known mixtures of pure xenon and oxygen is complicated because of the reaction

$$O_2^+ + Xe \rightarrow Xe^+ + O_2 . \quad (3)$$

This resonance charge–exchange reaction increases the xenon-ion yield, and it has been observed on known mixtures of xenon and oxygen. One hour still may not have been enough to allow exchange to go to completion. This experiment still does not give the true nature of the aqueous species, because the major equilibrium could just as well be

$$XeO_3 + H_2O \rightleftharpoons XeO_2(OH)_2 \quad (4)$$

or

$$XeO_3 + 2H_2O \rightleftharpoons XeO(OH)_4 . \quad (5)$$

For a second approach, the NMR spectrum for hydrogen was obtained on an acidified aqueous solution. Small intensity resonances were observed at both lower and higher field strength than the water resonance. The water resonance sharpened and became more intense as the acidity of the solution was increased, but the small resonances remained nearly unchanged. It is not apparent that this experiment will elucidate the nature of the species.

The third approach was a direct titration of the solution with strong base. Dudley, Gard, and Cady [7] reported that no inflection in the curve of pH versus volume of 0.1 M NaOH was obtained in the pH range of 8 to 11. Likewise the weak acidity of any proton in xenic acid has been observed in this laboratory. Titration of a weighed sample of XeO_3 dissolved in H_2O with 1 M NaOH produced a linear increase in pH until pH = 10. Between pH 10 and pH 11 there was a slight change in slope, but it did not have the appearance of an end-point. If this change in slope is related to an end-point, the corresponding equivalent weight would be 254. This weight might suggest the titration of one hydrogen from $Xe(OH)_6$. (Formula weight = 233.)

As an alternative explanation to the behavior of the species in basic pH ranges, its stability with respect to decomposition, i.e., loss of oxidizing power, was determined. Since the great stability of this species in acid solution had been observed, the experiment involved the preparation of basic solutions of various pH values and taking aliquots at known times. The concentration of the xenon-oxidizing species in these aliquots could be determined by quenching the aliquot in acidified potassium iodide solution. The tri-iodide produced could then be accurately measured colorimetrically at 2870 A. on a Beckman DU Spectrophotometer. In such acidified solutions so dilute in I_3^-, air oxidation of iodide had to be taken into account. A blank was prepared each time a new aliquot was quenched, so that the iodide, when added to the same amount of acid into the blank and sample, would be exposed to the air for an equal time. The blank was exactly like the sample, excluding the xenon species. No attempt was made to hold the ionic strength of the different pH solutions a constant, and no investigation was carried out to see if there were any reactions with the buffers.

Table 4 presents data at four different pH values and for two different xenon-containing solutions. The optical density represents, in terms of tri-iodide, the remaining oxidizing power at some elapsed time. Zero time represents the total oxidizing power. The quenching reaction can be represented by

$$XeO_3 + 6H^+ + 9I^- \rightarrow 3I_3^- + 3H_2O + Xe\,. \tag{6}$$

The quench solution for $[OH^-]$ ranging from 10^{-6} to 10^{-2} was 5×10^{-3} M H_2SO_4 in 0.1 M KI. Of the basic solution containing the xenon compound, 0.250 ml. was added to 4.75 ml. of the quench. For $[OH^-] = 6$ M, the quench was 1.8 M H_2SO_4 in 0.1 M KI. Constant $[I^-]$ was used so that no needless error would be present in the colorimetry. $[I^-] = .1$ reduced the iodine discrepancy [10] to 0.7 per cent with

$$\frac{[I_3^-]}{[I_2][I^-]} = 714\,. \tag{7}$$

TABLE 4

DATA ON THE DECOMPOSITION OF XeO_3 (AQ)

$[OH^-]=10^{-6}$ M*		$[OH^-]=10^{-4}$ M*		$[OH^-]=10^{-2}$ M*	
Elapsed Time	Optical Density‡	Elapsed Time	Optical Density‡	Elapsed Time	Optical Density‡
0	0.439	0	0.439	0	0.439
24	.393	174	0.095	148	.340
74	.288	....	....	208	.305
99	.253	....	....	370	.283
137	.207	....	....	642	0.237
192	0.159	....	....	....	....
$t_{1/2}\sim125$ min. phosphate buffer		$t_{1/2}\sim90$ min. borate buffer		$t_{1/2}>800$ min. 0.01 M NaOH	

$[OH^-]=6$ M*		$[OH^-]=10^{-6}$ M†		$[OH^-]=6$ M†	
Elapsed Time	Optical Density‡	Elapsed Time	Optical Density‡	Elapsed Time	Optical Density‡
0	0.398	0	0.521	0	0.521
2	.376	8	.471	25	0.162
9	.345	56	.316	....	....
36	.280	116	0.253	....	....
62	.222	....	....	....	....
114	0.168	....	....	....	....
$t_{1/2}\sim90$ min. 6.0 M NaOH		$t_{1/2}\sim106$ min. phosphate buffer		$t_{1/2}\sim18$ min. 6.0 M NaOH	

* These solutions contained fluoride anion and the concentration of the xenon species in the basic medium was initially 3.82×10^{-6} M.

† These solutions were made by dissolving XeO_3(s) in distilled water (fluoride free) and the concentrations of the xenon species in the basic medium was initially 9.07×10^{-5} M.

‡ The Beer's Law constant for these measurements is $C/A = 2.612 \times 10^{-5}$.

The enhanced stability of the species at $[OH^-] = 10^{-2}$ M and in the presence of fluoride is not explained at this stage in the investigation. The effect of ionic strength may explain the difference in $t_{1/2}$ for solutions of different $[OH^-]$, but the difference in $t_{1/2}$ for the $[OH^-] = 6.0$ M appears real. For the $[OH^-]$ range of 10^{-2} to 10^{-6} M, these data seem to indicate zero-order dependence of the rate on hydroxide. But in the region of 10^{-2} to 6 M, the dependency on hydroxide may be greater than zero. However, at $[OH^-] > 10^{-7}$ M the aqueous xenon species is definitely unstable.

The fourth route to the nature of the aqueous species was the determination of its ultraviolet absorption spectrum. The absorptions at pH = 1.2 and at pH = 1.8 were broad beginning at 2700 A. and extending without peaking to 1850 A., the lower limit of the Cary 14 Spectrophotometer. At pH = 5.5 the absorption sharpened and peaked at 1980 A. and began at 2500 A. The absorption of the species at the same concentration in 5×10^{-4} M hydroxide ion had its maximum at 2020 A., and in 6×10^{-3} M hydroxide the maximum was at 2100 A. The shift appeared to be linear between these two hydroxide-ion concentrations. At pH = 5.5 the molar extinction coefficient was 3600, and it increased approximately 8 per cent in going from 1×10^{-3} to 6×10^{-3} M hydroxide ion. An exact measurement in the change of the coefficient was not possible because of the slow decomposition of the species. This decrease in absorbency at a given hydroxide concentration over a time checked with the decomposition study described using iodine colorimetry. The UV absorption, titration, and decomposition data suggest the presence of one or more charged species in basic solutions.

Xenon(VI) oxide.—Regardless of the nature of the aqueous species, it can readily be converted to the white, crystalline, non-volatile anhydrous $XeO_3(s)$ [2, 3, 8, 11]. This anhydrous material explodes violently without provocation. Investigators have had samples last up to six weeks before the last of it was used and have had samples explode almost immediately after going to dryness. In our laboratory the $XeO_3(s)$ was prepared from XeF_4 hydrolysis in Teflon ware open to the atmosphere. Gunn [12] finds platinum ware just as satisfactory. The resulting solutions are evaporated at room temperature in a desiccator over Drierite or, even better, over barium oxide. Reduced pressure in the desiccator does not decrease the probability of a successful preparation.

Gunn has observed, as we have, that for some preparations of XeO_3 a concentrated hydrolysis solution is noticeably pink. The visible or ultraviolet absorption of this has not been measured. We are also in agreement with Gunn in believing that the stability of the xenon(VI) oxide seems to be greatly increased with respect to detonation if the oxide is originally prepared from distilled water with a trace of added sulfuric acid, for example, 10 ml. H_2O + 4 μl. of 1 M H_2SO_4 per gram of XeF_4.

The solid XeO_3 can be detonated by rubbing it between a nickel spatula and the Teflon cup or by touching it with anything containing cellulose. Several hundred milligrams is more than sufficient to blow a hole through Teflon ¾ in. thick or to completely shatter a desiccator. This tendency to detonate violently made direct determination of the ratio of xenon to oxygen difficult. The procedure would be to weigh a sample, place it in a quartz tube, evacuate, seal, heat the tube to decompose the XeO_3 to Xe and O_2, and measure the amounts of Xe and O_2 with the Toepler pumping system described.

Because of the extreme hygroscopic nature of the oxide, the sample weights were not useful for obtaining data on percentage composition, but this procedure

TABLE 5

CHEMICAL ANALYSIS OF XeO_3(s)

	μMole of			O/Xe	H_2O/Xe
	O_2*	Xe	H_2†		
Sample 1......	11.500	7.692	1.243	2.990	0.16
Sample 2......	11.228	7.493	0.544	2.997	0.07

* To confirm the quantitative recovery of O_2 from the Toepler pumping system, weighed $KClO_3$ was used as a standard. It gave 2.994 oxygen atoms per mole of $KClO_3$.

† Forty-five minutes of pumping did not remove all the water, which was measured as H_2 gas after the H_2O had been passed over uranium turnings at 750° C.

TABLE 6

CHEMICAL REACTIONS OF XeO_3 (AQ)

Reactant	pH	Product	Comments
NH_3...........	6 M NH_4OH	presumably N_2	very vigorous
Excess I^-......	7 or less	I_3^-	fast
3% H_2O_2.......	neutral	O_2	very vigorous
Excess Br^-.....	1 or less	Br_2	slow
Fe^{+2}...........	neutral	Fe^{+3}	fast
Hg............	1 M H_2SO_4	Hg_2SO_4	fast

would still give the ratio of xenon to oxygen for an arbitrary weight of sample, if the sample tube could withstand the shock of the decomposition. A sample as small as 0.7 mg. ruptured the quartz tubing used as the decomposition vessel. Quartz wool was found to cushion the shock sufficiently well to save a couple of tubes. Before the sample was added, the quartz tube and wool were heated to 950° C. in an oxygen stream to insure the removal of any organic material. The XeO_3 sample was then placed directly in the 10 mm. quartz tube on quartz wool and covered with more quartz wool. The tube was constricted, evacuated under high-vacuum conditions for forty-five minutes, sealed, and heated in an iron container at 500° C. for fifteen minutes. Sample tubes heated at lower temperatures had no greater chance of surviving the detonation. A visual melting point on a hot stage was attempted, and a temperature of 150° was reached before the crystal detonated. Many tubes were broken by the shock of the detonation, but the data in Table 5 were obtained from two that remained intact.

Some chemical reactions of $XeO_3(aq)$.—Since pure $XeO_3(s)$ can be readily obtained, solutions of known composition and concentration were made. It has been found that a solution where $[XeO_3] = 3 \times 10^{-2}$ acts as a moderately good oxidizing agent. Table 6 lists some qualitative results.

Acknowledgment.—The authors would like to thank A. S. Newton of the Lawrence Radiation Laboratory, Berkeley, for the mass-spectrometric analyses.

REFERENCES

1. H. H. Claassen, H. Selig, and J. C. Malm, *J. Am. Chem. Soc.* **84**, 3593 (1962).
2. S. M. Williamson and C. W. Koch, *Science* **139**, 1046 (1963).
3. D. H. Templeton, A. Zalkin, J. D. Forrester, and S. M. Williamson, *J. Am. Chem. Soc.* **85**, 817 (1963).
4. *Ibid.* **85**, 242 (1963).
5. D. F. Smith, *J. Chem. Phys.* **38**, 270 (1963).
6. J. Jortner, E. G. Wilson, and S. A. Rice, *J. Am. Chem. Soc.* **85**, 814 (1963).
7. F. B. Dudley, G. Gard, and G. H. Cady, *Inorg. Chem.* **2**, 228 (1963).
8. D. F. Smith, *J. Am. Chem. Soc.* **85**, 816 (1963).
9. L. Pauling, *ibid.* **55**, 1895 (1933).
10. W. M. Latimer, *Oxidation Potentials*, p. 64. 2d ed.; Englewood Cliffs, N.J.: Prentice-Hall, 1952.
11. N. Bartlett and P. R. Rao, *Science* **139**, 506 (1963).
12. S. R. Gunn, Lawrence Radiation Laboratory, Livermore, private communication.

REACTIONS OF XENON FLUORIDES WITH AQUEOUS SOLUTIONS AND THE ISOLATION OF STABLE PERXENATES

JOHN G. MALM, BEN D. HOLT, AND RALPH W. BANE
Argonne National Laboratory

It has been shown previously [1, 2, 3] that upon hydrolysis of XeF_4, only a fraction of the xenon was recovered as gas. This suggested the existence of a xenon species in aqueous solution. Studies [4] of the hydrolysis of XeF_6 have shown that no xenon was liberated. These solutions are highly oxidizing and upon evaporation yield XeO_3. This chapter will describe some further studies of these reactions with aqueous solutions (in particular with NaOH), which led to the isolation of stable sodium perxenates containing octavalent xenon.

EXPERIMENTAL

Hydrolysis of XeF_2, XeF_4, and XeF_6.—The fluorides were prepared and purified in the manner described previously [1, 5, 6; see also C. L. Chernick, this volume, p. 35]. The hydrolysis was carried out in the following manner: A weighed sample of the fluoride was sublimed, *in vacuo*, into a Pyrex glass bulb and sealed off. The bulb was attached to a vacuum system and Toepler pump. A bulb containing a known excess of 1 *N* NaOH solution was also attached, and this solution was thoroughly degassed by freezing and thawing under low pressure. When a good vacuum was achieved, the system was isolated from the vacuum pump. With the fluoride sample at liquid-nitrogen temperature, the break seal leading to the sample bulb was broken, and the NaOH solution slowly admitted. As the sample and the caustic-soda solution slowly warmed to room temperature, reaction occurred, and the gases given off were collected in a known volume by means of a Toepler pump. The amount of gas evolved was calculated from gas law and pressure and temperature measurement. The proportions of xenon and oxygen were determined with a mass spectrometer. By means of hydrolyses carried out with excess dilute caustic, all the fluorine was retained as fluoride ion. The fluoride was determined by acid titration of the excess base, and by $Th(NO_3)_4$ titration with sodium alizarin sulfonate indicator.

The xenon species present in the solutions was found to react with the indicator used and had to be destroyed prior to the fluoride titration, either by heating or by the addition of H_2O_2.

Preparation of sodium perxenate.—For further study we undertook the preparation of sizable batches of the sodium perxenate. Gram quantities of XeF_6 were sublimed into a Pyrex glass bulb and maintained close to liquid-nitrogen temperature while a large excess of approximately 2 M NaOH solution was slowly added. The mixture was warmed to room temperature, diluted with an equal volume of water, and allowed to stand for about a day. At this stage the product of the sodium- and fluoride-ion concentrations does not exceed the solubility product of sodium fluoride. The less soluble sodium perxenate slowly precipitates as a white solid, while the initial yellow color fades away. The phenomena observed during the precipitation appear to depend somewhat on the final sodium hydroxide concentration. Below 1 M, the length of time required for complete precipitation increased, and the precipitate was more crystalline and had a pale yellow color. Above this concentration some precipitation occurred almost immediately on warming to room temperature. The product was then in the form of a finely divided white solid. The precipitate was separated by centrifuging, washed two or three times with water, and dried at room temperature in a stream of nitrogen. The dried material contained some water of crystallization, which was easily removed by further drying over $CaSO_4$ or silica gel in a vacuum dessicator.

Analysis of sodium perxenate.—The analysis of this white solid has posed some formidable problems. A number of procedures have been tried to determine the principal constituents and suspected impurities. A formula is suggested for the bulk phase but some uncertainty remains about its purity and detailed composition.

The water content is not easily determined. If the starting material is taken as the air-dried sample, the loss in weight on drying at room temperature in a vacuum dessicator is taken as water of crystallization. Additional water may be recovered by igniting in oxygen at 800° C., collecting on magnesium perchlorate, and weighing.

For sodium analysis, the sample was dissolved in dilute acid, and the sodium was determined by flame photometry. In one instance the sodium was determined by the gravimetric method using $NaMg(UO_2)_3(C_2H_3O_2)_9 \cdot 6H_2O$, and the agreement with the flame photometer method was better than 1 per cent.

The solid was decomposed at 700° C., the mixture of xenon and oxygen collected in a known volume, and the Xe/O_2 ratio determined with a mass spectrometer. The oxygen, of course, is not the total oxygen in the compound, but it was hoped that the xenon recovery would be quantitative. Observations on the oxidizing power of this compound suggest this may not be the case.

Determination of the oxidizing equivalents of the sodium perxenate is discussed in more detail by Appelman [this volume, p. 186], but the procedure is briefly described here. An excess of iodide was added to an aliquot of the normally alkaline perxenate solution, the solution acidified with perchloric acid, and the resulting tri-iodide was titrated with thiosulfate to give the total oxidizing power of the solution. A second aliquot was acidified and allowed to stand for a few minutes; excess iodide was added to it and again titrated. If this is interpreted as titration of Xe(VI) to Xe(0) and used to evaluate the total xenon concentration, these concentrations were always 2–5 per cent higher than those obtained by the thermal-decomposition method.

Fluoride was determined by standard colorimetric procedures using 3-aminomethyl alizarin-N, N-diacetic acid indicator. However, it was necessary to destroy the oxidizing xenate species before adding the indicator. An excess of H_2O_2 in the normally alkaline solution obtained on solution of the original compound appeared to be satisfactory, and the H_2O_2 itself could be destroyed by heating. Contamination by fluoride proved less troublesome than was feared initially, since no more than 0.2 per cent fluoride was found in any preparation.

When the sodium perxenate is ignited in air, sodium carbonate is an important constituent of the residue. [See D. Gruen, this volume, p. 175]. Carbonate contamination in the preparation was assayed by acidifying a sample measuring the volume of the evolved gases and determining the CO_2 content with the aid of the mass spectrometer.

For experiments requiring a determination of combined xenon, the radioactive isotope Xe^{133} was used. The original fluoride was prepared with xenon containing the radioactive nuclide, and the sodium salt prepared in the usual way. The 80 kev. X-ray was measured with a scintillation detector and γ-ray pulse–analysis apparatus, and the activity per gram of dry sample determined.

It was then possible to measure the solubility and stability of the combined compound in any solution by determining the activity in the supernatant after centrifuging and degassing the solution.

The combined oxygen was not determined directly. Total oxygen is reported as the sum of oxygen associated with the sodium (as Na_2O) and that found accompanying the xenon on thermal decomposition. It does not include oxygen found as water.

RESULTS AND DISCUSSION

Hydrolysis of XeF_2, XeF_4, and XeF_6.—The analytical results obtained in these hydrolyses are shown in Table 1. The hydrolysis of XeF_2 produces a bright yellow color which rapidly disappears as the ice mixture melts; the solution is colorless. The xenon and oxygen are quantitatively liberated upon reaction with water:

$$XeF_2 + H_2O \rightarrow Xe + \tfrac{1}{2}O_2 + 2HF \,. \tag{1}$$

The alkaline hydrolyses of both XeF_4 and XeF_6 produce a yellow color but also yield pale yellow solutions. The color in these solutions persists for days, which suggests that this color arises from a different aqueous xenon species than that associated with the hydrolysis of XeF_2.

The values determined for xenon and oxygen liberated on reaction of XeF_4 with NaOH solution are in good agreement with the results reported by Williamson and Koch [3] for the hydrolysis of this compound with acid or neutral solution, where they recovered 0.72 mole of xenon and 0.52 mole of oxygen per mole of XeF_4. The amounts of xenon and oxygen liberated can be accounted for by the following equations, assuming the first step in the hydrolysis is the disproportionation of Xe(IV) to give Xe(II) and Xe(VIII):

$$3Xe(IV) \xrightarrow{H_2O} 2Xe(II) + Xe(VIII)\,, \tag{2}$$

then

$$2Xe(II) \xrightarrow{H_2O} 2Xe + O_2\,. \tag{3}$$

TABLE 1

RESULTS OF HYDROLYSIS OF XENON FLUORIDES IN 1 M NAOH

COMPOUND	WEIGHT OF COMPOUND (GM.)	TIME OF PUMPING (HRS.)	MOLES OF GAS EVOLVED PER MOLE OF COMPOUND		MOLE HF/ MOLE COMPOUND
			Xe	O_2	
XeF_2......	0.212	4	1.01	0.49	1.94
XeF_4......	.379	4	0.61	.43	4.07
XeF_4......	.400	4	0.75	.54	3.94
XeF_4......	.292	2	0.67	.46	3.95
Av. XeF_4..	....	....	0.677	.477	....
XeF_6......	0.277	18	0.15	0.10	5.66

As will be shown later, Xe(VIII) decomposes rapidly in acid solution:

$$Xe(VIII) \xrightarrow{H^+} Xe(VI) + \tfrac{1}{2}O_2\,. \tag{4}$$

Reaction 4 also occurs in dilute alkali, but much more slowly. The over-all reaction of XeF_4 with H_2O in either acid or dilute base is given by equation 5,

$$3XeF_4 + 6H_2O \rightarrow 2Xe + \tfrac{3}{2}O_2 + Xe(VI) + 6HF\,, \tag{5}$$

which is in good agreement with the results given here.

Cady [4] has found that no xenon or oxygen was evolved when XeF_6 was hydrolyzed in acid or neutral aqueous solutions. The results given in Table 1 show this is true for hydrolysis in 1 *N* NaOH as well. The small amounts of xenon and oxygen evolved are probably from XeF_4 present as an impurity, as is indicated by the low fluoride analysis. If no xenon or oxygen is evolved, then the immediate product of the hydrolysis either in acid or in dilute base must be an aqueous (VI) species. However, there is considerable evidence that disproportionation products are formed either at higher concentrations of sodium hydroxide or on prolonged standing in the dilute base.

It appears that the xenon species produced and retained in solution is the same for the hydrolysis of XeF_4 as for the hydrolysis of XeF_6.

Isolation of stable perxenates.—The Xe(VI) solutions in dilute sodium hydroxide slowly lose the yellow color, liberate gas, and a salt precipitates in rather good yield based on the original xenon content; some of the yields are shown in Table 2. In general, higher NaOH concentrations precipitate the salt more rapidly. Upon completion of the reaction, the colorless supernatant has lost essentially all of the xenon. Assuming the insoluble Xe(VIII) salt arises by disproportionation of Xe(VI) to give gaseous Xe(0) and Xe(VIII), then the maximum yield of solid would be slightly less than 75 per cent, depending on its solubility in the supernatant.

In two experiments on a 100-mg. scale a 90 per cent yield of sodium perxenate was obtained by passing ozone into the 1 M NaOH solution of Xe(VI) rather than waiting for the internal rearrangement. The products appeared to be identical.

TABLE 2

YIELDS OF SODIUM PERXENATE FROM XeF_6 HYDROLYSIS

Run No.	XeF_6 (Gm.)	Initial Xe Concentration (M)	Final OH^- Concentration (M)	Time (Hrs.)	Yield Perxenate (%)	Oxidation Number meq/gm		High/Low
						High	Low	
33......	2.96	0.1	0.6	40	54	19.78	14.91	1.327
35......	5.7	0.15	0.8	40	31	20.93	15.6	1.342
43......	8.1	0.1	0.8	80	43	20.9	15.7	1.33
47......	4.8	0.2	0.6	1	27	21.5	15.95	1.332
49......	6.3	0.1	1.2	15	48	....	....	....

Generic sodium perxenate is a very stable salt, can be stored unchanged at room temperature, and decomposes to yield xenon and oxygen only above 300° C. It was not observed to melt up to the temperature of decomposition.

Characterization of Xe(VIII) salts.—The results of the analysis of two preparations of sodium perxenate are shown in Table 3. Spectrographic analysis of these salts showed sodium to be the only cation present. It can be seen from the table that fluoride and carbonate content account for a negligible part of the sodium as impurity. In all preparations analyzed to date, sodium to xenon ratios have varied from 4.5 to 5.6, the higher ratios being obtained with batches prepared by hydrolysis at higher NaOH concentrations. X-ray powder photographs have been taken for many preparations of "sodium perxenate." At least two phases and their hydrates have been distinguished. All preparations, however, gave the same ratio of the high oxidation number to low oxidation number (1.33), showing that solid products are mixtures containing octavalent xenon only.

Since in all cases the material balance is only about 95 per cent, it is hard to arrive at a formula for the product. The difficulty of removing water from a sodium salt by the methods used may be the reason for this incomplete analysis.

If one assumes that the missing material is water, then the results of the analysis lie between the formula $Na_4XeO_6 \cdot 2H_2O$ and the formula $Na_6XeO_7 \cdot H_2O$. It is convenient to treat Xe(VIII) compounds as derivatives of perxenic acid, $Xe(OH)_8$. The hydrate of the salt with the lesser amount of sodium contains four additional waters of crystallization.

Solubility of sodium perxenate.—At 25° C. in H_2O the dry salt showed a solubility of 7 gm/1; in 0.1 M NaOH, 1.2 gm/1; in 0.5 M NaOH, 0.2 gm/1.

Stability of Xe(VIII) solutions.—The solid salt stored under 1 M NaOH or 5.6 M NaOH showed no loss of xenon over a fourteen-day period. When dissolved in water at a concentration of 0.02 M, xenon was lost at the rate of about 3.5 per cent per day. Xe(VIII) salts decompose rapidly in acid, liberating O_2 and Xe(VI).

TABLE 3

ANALYSIS OF SODIUM PERXENATE

Analysis for:	Weight Per Cent Found	
	Batch No. 33	Batch No. 47
Sodium	32.4	28.1
Xenon	34.3	35.7
Oxygen	15.3	16.0
Water	3.2	6.5
Fluoride	......	0.2
CO_2	1	0.1
Na/Xe	5.4	4.5
High oxidation meq/gm	19.78	21.50
Low oxidation meq/gm	14.91	15.95
High/low	1.327	1.332

Stability of Xe(VI) solutions.—Sodium perxenate was dissolved in acid to a concentration of about 0.02 M. This gives rise to the Xe(VI) species. The stability was studied over a fourteen-day interval. With 1 *N* H_2SO_4 and 1 *N* $HClO_4$ no loss of xenon was observed. With 1.2 *N* HCl, however, xenon was lost at the rate of 4 per cent per day, presumably by oxidation of chloride.

The same concentration of Xe(VI) in 1 M OH^- loses xenon rapidly at first, disproportionating to give rise to Xe(VIII), which is then stable in basic solution and precipitates as sodium perxenate in NaOH.

Other Xe(VIII) compounds.—Solid salts of Pb^{++}, UO_2^{++}, Ag^+, and Cu^{++} are precipitated from solutions of sodium xenate in H_2O. No attempts have been made to characterize these products. (Compare D. Gruen, this volume, p. 174.)

Acknowledgment.—It is a pleasure to acknowledge the assistance of Miss Geraldine Knapp in all phases of these preparations. The many X-ray studies of Dr. Stanley Siegel and Miss Elizabeth Gebert aided in elucidating the nature of the solid compounds. Dr. Evan Appelman's solution-chemistry studies aided in establishing the existence of the Xe(VIII) species. We appreciate the many discussions with Drs. Martin Kilpatrick and Dieter Gruen.

REFERENCES

1. H. H. Claassen, H. Selig, and J. G. Malm, *J. Am. Chem. Soc.* **84**, 3593 (1962).
2. C. L. Chernick *et al.*, *Science* **138**, 136 (1962).
3. S. M. Williamson and C. W. Koch, *ibid.* **139**, 1046 (1963).
4. F. B. Dudley, G. Gard, and G. H. Cady, *Inorg. Chem.* **2**, 228 (1963).
5. J. G. Malm, I. Sheft, and C. L. Chernick, *J. Am. Chem. Soc.* **85**, 110 (1963).
6. J. L. Weeks, C. L. Chernick, and M. S. Matheson, *ibid.* **84**, 4612 (1962).

INFRARED ABSORPTION SPECTRA OF SOME HEAVY METAL PERXENATES

DIETER M. GRUEN
Argonne National Laboratory

The recent preparation of sodium perxenate from aqueous solution and the demonstration of its remarkable stability [Malm, Bane, and Holt, this volume, p. 167] will undoubtedly prove to be of crucial importance to studies of xenon-oxygen compounds.

Although perxenic acid itself is apparently not stable, a number of its metal salts have already been prepared, to be followed undoubtedly by many others. Those which are known at the present time are the copper, barium, silver, lead, and uranyl salts. The thermal stabilities of these substances vary, depending on the metal cation constituent, with silver forming the least, and sodium the most, stable of the salts so far prepared.

In order to understand the xenon-oxygen bonding properties in these compounds, a study of their infrared spectra in the region 4000–400 cm.$^{-1}$ was undertaken. The perxenates of uranium, lead, and silver were prepared by reacting solid sodium perxenate with aqueous solutions of uranyl nitrate, lead nitrate, and silver nitrate. The yellow uranyl and lead perxenates and the black silver perxenate were washed with distilled water and were vacuum dried for one hour prior to their incorporation in KBr discs. One to three milligrams of the perxenates in about 200 milligrams of KBr sufficed to give intense IR absorption bands. To show that the spectra were not influenced by compound or mixed crystal formation with the KBr matrix, a few of the spectra were also measured by depositing the powdered perxenates as thin films on AgCl sheets. Only minor spectral changes were observed in comparison with the measurements on KBr discs. All measurements were carried out on Perkin-Elmer 421 or 21 spectrophotometers.

RESULTS

The results of the study are presented chiefly in the form of the spectra themselves, plotted as transmittance versus wave number.

The spectrum of a sample of vacuum-dried sodium perxenate is shown in Figure 1 (upper curve). It is characterized by broad absorption bands in the regions 3600–2900 cm.$^{-1}$, 1700–1400 cm.$^{-1}$, and 900–500 cm.$^{-1}$. The intense

3600–2900 cm.$^{-1}$ band is associated with absorption by water or by hydroxyl groups. Some water was removed by heating the sample in air at 200° C. for three hours, as evidenced by a substantial decrease in the intensity of the water band. (Fig. 1, middle curve.) Pronounced spectral changes also occurred in the 900–500 cm.$^{-1}$ region. A subsidiary maximum at 870 cm.$^{-1}$ present in the sample dried at room temperature (Fig. 1, upper curve) becomes a well-resolved narrow absorption band on heating to 200° C. (Fig. 1, middle curve). The 870 cm.$^{-1}$ band was shown to be associated with carbonate-ion absorption, when it was found that the infrared spectrum of the product obtained by heating sodium perxenate at 500° C. in air is identical with the spectrum of pure sodium carbonate.

The spectrum of silver perxenate is shown in Figure 1, lower curve. It will be observed that this spectrum gives no indication of the 870 cm.$^{-1}$ carbonate band. It does, however, have a very intense band with maximum at 650 cm.$^{-1}$. Furthermore, the salt appears to be essentially anhydrous as evidenced by the low intensity of absorption in the 3600–2900 cm.$^{-1}$ region. On heating to ~150° C., silver perxenate was found to decompose violently!

The spectra of sodium and silver perxenate in the 1000–500 cm.$^{-1}$ region are shown in Figure 2, curves *A* and *B*. This particular sample of sodium perxenate was prepared by treating an acid decomposed sample of sodium perxenate with ozone in alkaline solution. This procedure yields a crystalline salt containing very little water of hydration. It can be seen that the 650 cm.$^{-1}$ band is present in both salts.

The spectrum of lead perxenate in the region 1900–600 cm.$^{-1}$ (Fig. 3, curve *A*) and in the region 1000–500 cm.$^{-1}$ (Fig. 4, curve *A*) has a band with maximum at 660 cm.$^{-1}$. When lead perxenate is heated to 150° C. in air, it decomposes to give a gaseous product and a yellow residue. The spectral features of the solid decomposition product (Fig. 3, curve *B;* Fig. 4, curve *B*) are virtually identical with those of lead perxenate itself, with the exception of the 660 cm.$^{-1}$, which is present in only very low intensity in the decomposed material.

A very similar situation to that just described for lead perxenate obtains in the case of uranyl perxenate. Again, the spectral features of the decomposed perxenate (Fig. 5, curve *B;* Fig. 6, curve *B*) are virtually identical with those of uranyl perxenate itself (Fig. 5, curve *A;* Fig. 6, curve *A*) except that the 670 cm.$^{-1}$ band present in the perxenate is absent in the decomposition product. The 880 cm.$^{-1}$ band in the uranyl compounds is due to the well-known uranium-oxygen stretching vibration.

DISCUSSION

The salient experimental result of this study is the finding in all of the perxenate salts of an intense band in the 650–680 cm.$^{-1}$ region whose position is insensitive to the metal-cation constituents. This band may be identified with the $\nu_3(f_{1u})$ vibration of the octahedral (XeO_6) grouping.

It is interesting to observe [1] that the ν_3 frequency in paraperiodic acid,

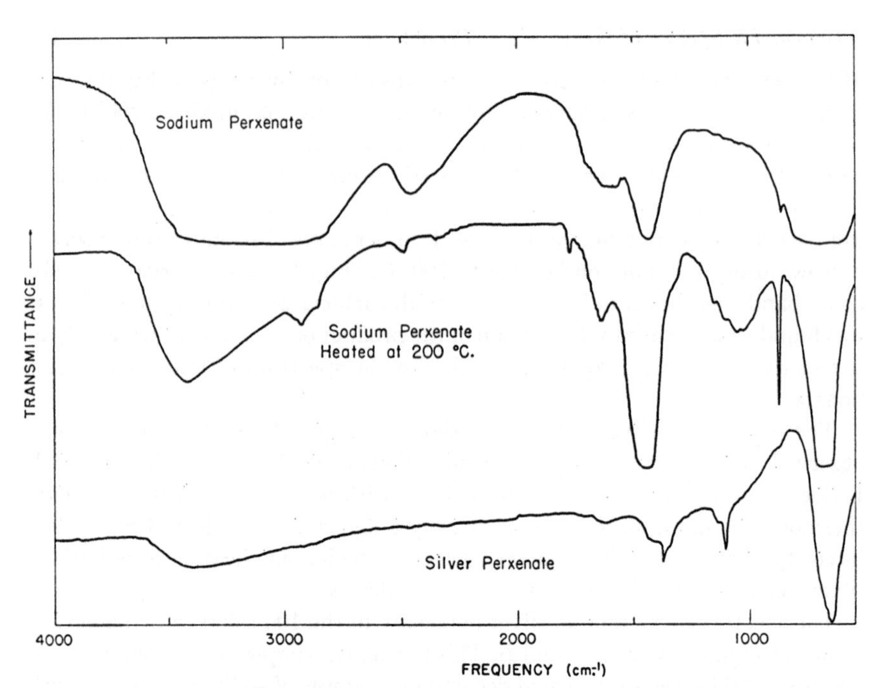

FIG. 1.—Spectra of sodium and silver perxenate from 4000 to 600 cm.$^{-1}$

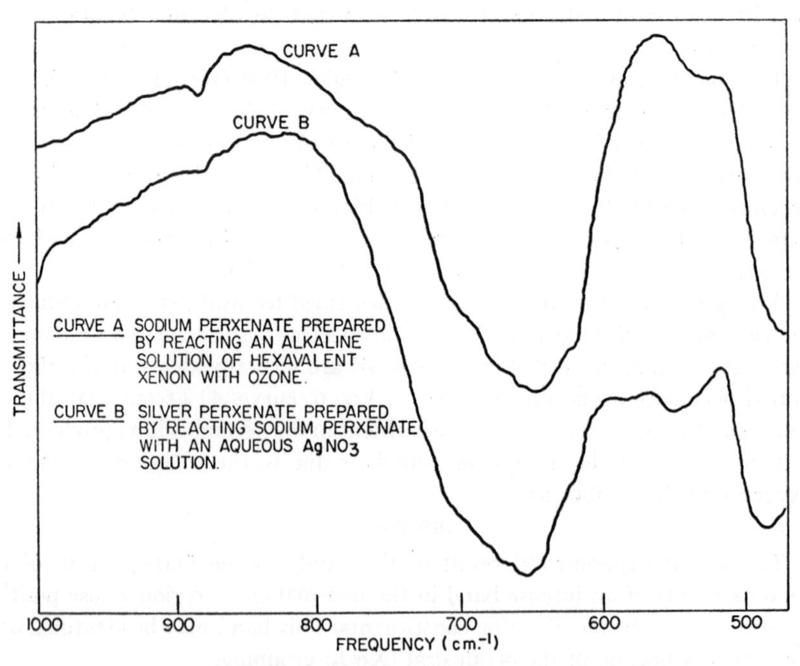

FIG. 2.—Spectra of sodium and silver perxenate from 1000 to 450 cm.$^{-1}$

H_5IO_6, and its salts occurs at ~700 cm.$^{-1}$. In tellurates and antimonates ν_3 is at ~650 cm.$^{-1}$ [2]. It would appear, therefore, that the Xe-O bond strength in the (XeO_6) complex is very similar to the Sb-O, Te-O, and I-O bond strengths in the octahedral groupings (SbO_6), (TeO_6), and (IO_6). It is worth noting that ν_3 in the tetrahedral metaperiodate ion IO_4^- is at 850 cm.$^{-1}$ indicating that the ratio of bond orders IO_4^-/IO_6^{5-} is about 1.5, a not unreasonable value assuming that the same number of electrons participate in bonding in the two ions [3].

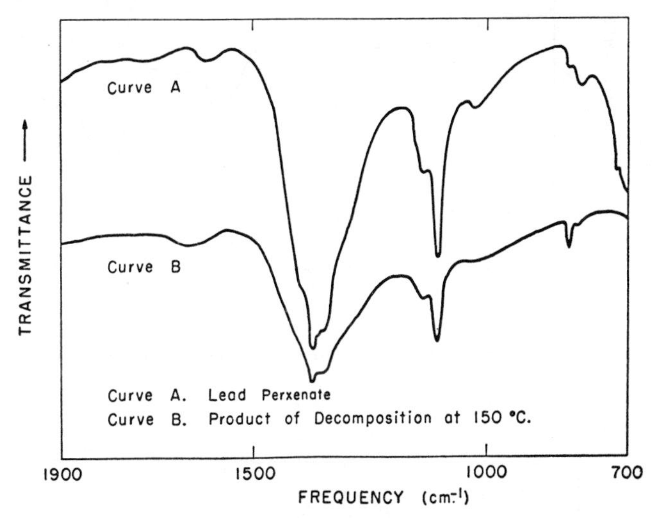

FIG. 3.—Spectra of lead perxenate and its decomposition product from 1900 to 700 cm.$^{-1}$

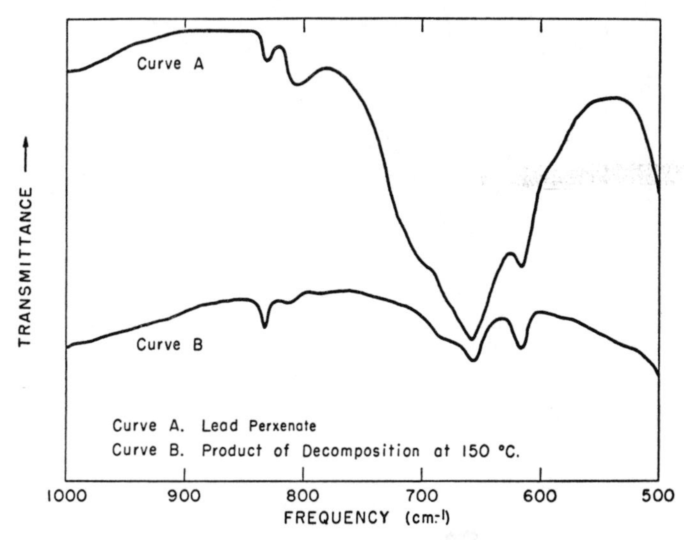

FIG. 4.—Spectra of lead perxenate and its decomposition product from 1000 to 500 cm.$^{-1}$

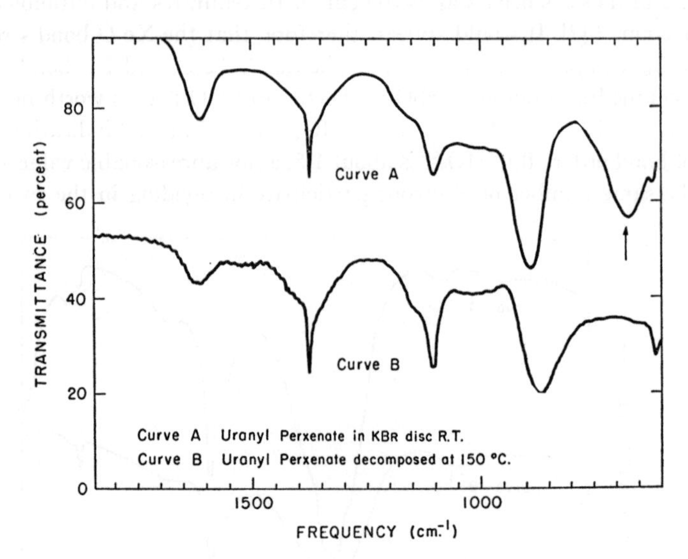

FIG. 5.—Spectra of uranyl perxenate and its decomposition product from 1850 to 600 cm.$^{-1}$.

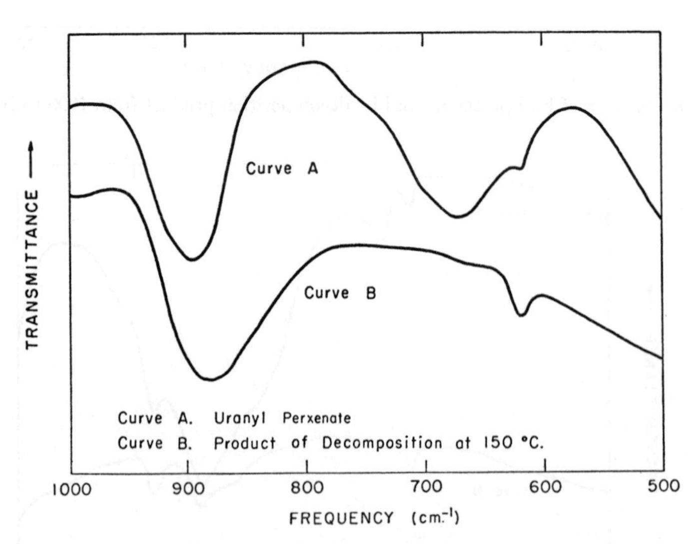

FIG. 6.—Spectra of uranyl perxenate and its decomposition product from 1000 to 500 cm.$^{-1}$.

Neither the exact stoichiometries of perxenate salts nor their crystal structures have as yet been determined. The following remarks concerning the structure of xenate and perxenate ions and compounds are therefore highly speculative.

The preparation of perxenates by ozonization of an alkaline xenate solution suggests an equilibrium of the following sort:

$$\left[XeO_4(OH)_2 \right]^{4-} + O_3 \rightarrow \left[XeO_6 \right]^{4-} + H_2O + O_2 \qquad (1)$$

The xenate and perxenate ions on this picture are both octahedral and closely related structurally.

The structures of (hypothetical) $M_nH_2XeO_6$ and M_nXeO_6 compounds may bear similarities to the structure of $(NH_4)_2H_3IO_6$ determined by Helmholz [4]. In this compound IO_6^{5-} octahedra are linked together by hydrogen bonding which tend to pull the octahedra closer together along the *c*-axis and to rotate them. A similar effect may occur in xenates of composition $M_nH_2XeO_6$. Hydrogen bonding would of course be absent in the perxenates of composition M_nXeO_6.

Helmholz has pointed out that in the absence of hydrogen bonds, the structure of $(NH_4)_2H_3IO_6$ would be determined by the NH_4^+ and IO_6^{5-} ions. The crystal could then have the structure of $(NH_4)_2PtCl_6$, and in fact the structure of $(NH_4)_2H_3IO_6$ is closely similar to the structure of $(NH_4)_2PtCl_6$.

Although the perxenate ion as formulated above leads to a formal oxidation state of eight for xenon, present evidence does not appear to the author to rule out a formulation involving peroxo bonds [5] which would leave the xenon in a hexavalent state:

$$\left[O_4Xe(O{-}O)_2XeO_4 \right]^{8-} \qquad (2)$$

The infrared spectra would not be expected to yield information on this point as peroxo linkages are known to be weak absorbers. However, the presence of peroxo linkages would be revealed by an X-ray structure determination as close oxygen distances in neighboring (XeO_6) groupings. In that event, the xenate-perxenate equilibrium could be looked on as one involving a change from a hydrogen-bonded structure to one involving peroxo-bonds.

Acknowledgment.—It is a pleasure to acknowledge the generosity of J. Malm in making a sample of sodium perxenate available for this work.

REFERENCES

1. H. Siebert, *Z. Anorg. allgem. Chem.* **303,** 162 (1960).
2. *Ibid.* **301,** 101 (1959).
3. *Ibid.* **273,** 21 (1953).
4. L. Helmholz, *J. Am. Chem. Soc.* **59,** 2036 (1937).
5. A similar suggestion has also been made by G. Goodman (private communication to the author).

THE REACTION OF AQUEOUS XENON TRIOXIDE WITH BROMIDE AND IODIDE ANION: A KINETIC STUDY

CHARLES W. KOCH AND STANLEY M. WILLIAMSON
University of California, Berkeley

The nature of the aqueous species that forms when $XeO_3(s)$ is dissolved in water is discussed by us in this volume [p. 161]. Since the heat of formation of $XeO_3(s)$ is about +96 kcal/mole [Gunn, p. 151, this volume], it would be quite significant if the electrical potential of this apparently highly irreversible half-reaction were known:

$$Xe + 3H_2O \rightarrow XeO_3(aq) + 6H^+ + 6e^- . \qquad (1)$$

The current study arose from an investigation of chemical reactions with $XeO_3(aq)$. It was observed that the oxidation of bromide to bromine was acid dependent and that the rate in the pH range of 1 to 0 was readily measurable. Iodide is oxidized to iodine extremely rapidly at pH values <6, but in the region 6 to 7, rate measurements have been made. Rates at 0.2°, 24.8°, and 39.9° C. have been observed to measure the energy of activation for the two reactions.

The appearance of Br_3^- and I_3^- as a function of time was observed colorimetrically on a Beckman DU Spectrophotometer at 2650 A. and 2870 A., respectively. Consideration of the following stoichiometric equation then directly gave the changes in concentrations of all species:

$$XeO_3(aq) + 6H^+ + 9X^- \rightarrow Xe + 3H_2O + 3X_3^- . \qquad (2)$$

The stock $XeO_3(aq)$ solution was prepared by dissolving $XeO_3(s)$ in distilled water. Its concentration of 0.03024 M was determined by iodimetry and by iodine colorimetry. The concentration was found to remain unchanged within experimental error during the investigation. The solid bromide and iodide salts were of analytical purity. The stock acid solution for the bromide runs was 1.065 M $HClO_4$; appropriate buffers were used for the iodide runs. To adjust the final volume of the bromide runs, 1.000 M $NaClO_4$ was used, so that the change in ionic strength was limited to the range from 1.01 to 1.05. The blanks for the spectrophotometry were identical to the samples except for the exclusion of the $XeO_3(aq)$.

The reaction of $XeO_3(aq)$ with Br^-.—Optical density versus time curves were obtained at different concentrations of the three species concerned in order to determine the dependence of the rate on each species. To determine the total Br_2 concentration, it was necessary to prepare standard curves at 2650 A. for each bromide-ion concentration employed. They were found to be consistent with the equilibrium constant reported by Latimer [1]:

$$Br_2 + Br^- \rightleftharpoons Br_3^- ; \qquad K = 17 . \tag{3}$$

Table 1 lists the concentrations, initial slopes, and experimental order of the rate law.

To further establish the order of the rate law, the hydrogen-ion concentrations were varied from 0.1148 to 0.959 M; and the bromide-ion, from 0.100 to 0.900 M. In addition, it seemed necessary to employ the fully integrated rate law if direct correlation of the rate constants to concentration changes were to be made.

The rate law,

$$\frac{d[Br_3^- + Br_2]}{dt} = -\frac{3dXeO_3}{dt} = k_5[XeO_3][H^+]^2[Br^-]^2 , \tag{4}$$

was integrated considering all three species as variables. The result is:

$$\begin{aligned} k_5 t = {} & \frac{8.530\cdot 10^{-2}}{\left\{\frac{[Br^-]_i}{3} - 3[XeO_3]_i\right\}^2 \left\{\frac{[H^+]_i}{3} - 2[XeO_3]_i\right\}^2} \cdot \log \frac{[XeO_3]_i}{[XeO_3]_f} \\ & \times \frac{-81}{[Br^-]_i \left\{\frac{[Br^-]_i}{3} - 3[XeO_3]_i\right\} \{3[H^+]_i - 2[Br^-]_i\}^2} \cdot \frac{[Br_3^- + Br_2]}{[Br^-]_f} \\ & \times \frac{-20.73\{[H^+]_i + 12[XeO_3]_i - 2[Br^-]_i\}}{\left\{\frac{[Br^-]_i}{3} - 3[XeO_3]_i\right\}^2 \{3[H^+]_i - 2[Br^-]_i\}^3} \cdot \log \frac{[Br^-]_i}{[Br^-]_f} \\ & \times \frac{-16}{[H^+]_i \left\{\frac{[H^+]_i}{3} - 2[XeO_3]_i\right\} \{3[H^+]_i - 2[Br^-]_i\}^2} \cdot \frac{[Br_3^- + Br_2]}{[H^+]_f} \\ & \times \frac{-9.212\{3[H^+]_i - 12[XeO_3]_i - \frac{2}{3}[Br^-]_i\}}{\left\{\frac{[H^+]_i}{3} - 2[XeO_3]_i\right\}^2 \{3[H^+]_i - 2[Br^-]_i\}^3} \cdot \log \frac{[H^+]_i}{[H^+]_f} . \end{aligned} \tag{5}$$

The subscripts i and f signify initial and final concentrations in the integrated expression.

Table 2 lists the values of the individual terms in the above rate law and the fifth-order rate constant as a function of varying concentrations of the three species at 24.8°.

An Arrhenius plot of the appropriate rate constants at 0.2°, 24.8°, and 39.9° gives an activation energy equal to 15.5 kcal/mole.

The reaction of $XeO_3(aq)$ with I^-.—Optical density versus time curves were obtained at different concentrations of the three species concerned. The total iodine concentration was determined by measuring the I_3^- band at 2870 A. These measurements at two different iodide concentrations also were in agreement with the equilibrium constant listed by Latimer [1]:

$$I_2 + I^- \rightleftharpoons I_3^- ; \qquad K = 714 . \tag{6}$$

TABLE 1

DETERMINATION OF ORDER OF RATE LAW

T = 24.8°

$[XeO_{3(aq)}]$ (M)	$[H^+]$ (M)	$[Br^-]$ (M)	Initial Slope $[Br_2+Br_3^-]/t$* (moles liters^{-1} min.$^{-1}$)
$15.08 \cdot 10^{-5}$	0.923	0.1000	$4.86 \cdot 10^{-7}$
$15.08 \cdot 10^{-5}$	.462	.1000	$1.12 \cdot 10^{-7}$
$15.08 \cdot 10^{-5}$	.462	.2000	$4.56 \cdot 10^{-7}$
$60.32 \cdot 10^{-5}$	0.959	0.1000	$25.0 \cdot 10^{-7}$

* $d[Br_2 + Br_3^-]/dt = k[XeO_3]aq^{1.19} [H^+]^{2.17} [Br^-]^{2.04}$

TABLE 2

DETERMINATIONS OF FIFTH-ORDER RATE CONSTANT: IONIC STRENGTH = 1.01–1.05

T = 24.8°

INITIAL CONCENTRATIONS OF 3 SPECIES								
$[XeO_3](aq)$ (M)	$[H^+]$ (M)	$[Br^-]$ (M)	*A* LOG ($[XeO_3]_i$*/$[XeO_3]_f$)	(*B*$[Br_3^- + Br_2]$†/$[Br^-]_f$)	*C* LOG ($[Br^-]_i$*/$[Br^-]_f$)	(*D*$[Br_3^- + Br_2]$*/$[H^+]_f$)	*E* LOG ($[H^+]_i$*/$[H^+]_f$)	k_5† LITERS^4 MOLES^{-4} MIN^{-1}
$6.032 \cdot 10^{-4}$	0.959	0.1000	0.512	0.090	0.010	....	....	0.412
"	"	.0400	0.785	0.289	0.112	....	....	0.384
"	"	.0200	1.54	0.91	0.42	....	....	0.21
"	"	.0100	5.96	4.42	1.08	....	....	0.46
"	0.480	.1000	0.422	0.086	0.032	....	....	0.304
"	0.240	"	0.400	0.122	0.036	....	....	0.252
"	0.120	"	0.578	0.422	−0.203	0.047	0.127	0.185
$1.508 \cdot 10^{-4}$	0.923	"	0.396	0.018	0.005	....	....	0.373
"	0.462	"	0.389	0.020	0.004	....	....	0.365
"	"	.2000	0.377	0.012	0.001	....	....	0.364
"	0.230	.500	0.787	0.030	0.068	.051	.121	0.517
"	0.344	.300	0.715	0.053	−0.033	.007	.018	0.670
"	0.115	.900	1.035	....	....	0.024	0.006	1.005
$1.508 \cdot 10^{-4}$	0.923 T=0.2°	0.100	0.0406	0.0020	0.0006	....	....	0.0380
"	0.923 T=39.9°	"	1.52	0.07	0.02	....	....	1.43

* The capital letters *A* through *E* correspond to the initial concentration terms listed in the integrated expression.
† Activity coefficient corrections have not been made.

The initial slope data show a relatively rapid deviation from a straight line relationship, and so suggest that there is more than one reaction occurring in the pH range investigated. This observation may be explained in part by the alkaline decomposition described in our previous paper. As a consequence of this uncertainty only initial slope calculations are listed in Table 3.

TABLE 3

DETERMINATION OF ORDER OF RATE LAW: IONIC-STRENGTH VARIABLE DEPENDING UPON pH OF PHOSPHATE BUFFER

T=24.8°

$[XeO_3]_{aq}$ (M)	pH	$[I^-]$ (M)	Initial Slope $[I_3^- + I_2]/t$* (moles liters^{-1} min.$^{-1}$)
$15.08 \cdot 10^{-5}$	7.01	0.0100	$7.15 \cdot 10^{-7}$
$9.05 \cdot 10^{-5}$	7.01	0.0100	$4.57 \cdot 10^{-7}$
$15.08 \cdot 10^{-5}$	7.01	0.00500	$3.76 \cdot 10^{-7}$
$9.05 \cdot 10^{-5}$	6.74	"	$2.15 \cdot 10^{-7}$
"	6.43	"	$2.40 \cdot 10^{-7}$
"	6.01	"	$3.55 \cdot 10^{-7}$
"	6.74	" at 39.9°	$5.81 \cdot 10^{-7}$

* $d[I_3^- + I_2]/dt = -3d[XeO_3]_{aq}/dt = k[XeO_3]^{0.94}[I^-]^{0.95}[H^+]^0$

An Arrhenius plot of the initial slopes at 24.8° and 39.9° gives an activation energy equal to 12.2 kcal.

REFERENCE

1. W. M. LATIMER, *Oxidation Potentials.* 2d ed.; Englewood Cliffs, N.J.: Prentice-Hall, 1952.

CHARACTERIZATION OF OCTAVALENT XENON IN AQUEOUS SOLUTION

EVAN H. APPELMAN
Argonne National Laboratory

ABSTRACT

Aqueous solutions of Xe(VIII) have been prepared from the solid sodium perxenate obtained by alkaline hydrolysis of XeF_6. At high pH these solutions decompose slowly to Xe(VI) and oxygen, the rate increasing with decreasing pH. The standard potential in basic solution for the Xe(VI)-Xe(VIII) couple is estimated to lie between 0.7 and 1.24 v. The ultraviolet spectrum of Xe(VIII) is given as a function of pH, and potentiometric acid-base titrations of both Xe(VIII) and Xe(VI) are presented. A series of protonation equilibria is postulated to explain the observations.

The formation of a solid sodium perxenate by alkaline hydrolysis of XeF_6 or by passing ozone through an alkaline solution of hexavalent xenon has been discussed elsewhere [see J. G. Malm *et al.*, p. 167, this volume]. This solid perxenate dissolves in water to form a highly alkaline solution which retains its oxidizing power for a moderate time. In this paper we will examine the species present in such solutions.

EXPERIMENTAL

Materials.—The xenon compounds used were samples of several batches of sodium perxenate prepared by Malm. All showed an oxidizing power of 8 equivalents per mole of xenon, but they had Na^+/Xe ratios ranging from 4.6 to 5.6 and contained varying amounts of water. Carbonate was also present in some of the samples to the extent of about 4 mole per cent of the xenon.

All other reagents used were commercial products of analytic grade. Distilled water was redistilled first from alkaline permanganate, then from dilute sulfuric acid, and distilled again without any additive before being used to prepare solutions.

Potentiometric titrations and pH measurements.—Beckman glass and calomel electrodes were used for all such measurements. Beckman buffers of pH's 4, 7, and 10 were used as pH standards. Solutions were stirred with Teflon-covered magnetic stirring bars during measurements. Measurements at 5° were made with a Beckman Model H-2 continuous-reading pH meter to ± 0.05 pH unit.

Measurements at room temperature were made to ±0.01 pH unit with either a Beckman Model G pH Meter or a Cary Model 31 Vibrating Reed Electrometer.

Spectrophotometry.—Spectrophotometric measurements were made at 25° with a Cary Model 14 Recording Spectrophotometer. Spectra were run in silica cells of 1.000 mm. light path, with xenon concentrations around 0.003 M. Borate, phosphate, and carbonate buffers were used to control the pH of solutions studied, and care was taken to run blanks with solutions nearly identical to those containing the xenon. Corrections were made for decomposition of the perxenate whenever necessary. Extinction coefficients were measured with a precision of about ±2 per cent, the principal sources of error being the decomposition and the weighing of small samples.

Analytical procedures.—It has been fairly well established that perxenate decomposes in acid to give hexavalent xenon. An iodide solution reduces both Xe(VIII) and Xe(VI) to elemental xenon, but the reaction of Xe(VI) with iodide is slow unless the solution is at least slightly acid. Therefore, the following iodometric method was used to analyze perxenate solutions.

An excess of iodide was added to an aliquot of the solution. After a few minutes, the solution was acidified with perchloric acid, and the tri-iodide was titrated with thiosulfate that had been standardized against KIO_3. A second aliquot was acidified and allowed to stand for a few minutes. An excess of iodide was added to it, and the tri-iodide was again titrated.

The first titration was taken to measure the total oxidizing power of the solution. The second was interpreted as titration of the xenon from the +6 state to Xe(0) and was used to evaluate the total xenon concentration. Xenon concentrations determined in this way were always 2 to 5 per cent lower than those determined by heating the dry solid, measuring the volume of gas evolved, and analyzing for xenon with a mass spectrometer. The reason for this discrepancy is not yet known.

Sodium was determined by flame photometry, water as the loss on drying in high vacuum at room temperature, and carbonate (as CO_2) by the mass spectrometer after acid decomposition.

RESULTS AND DISCUSSION

General observations.—The most pronounced characteristic of these solutions is their instability. Between pH 12 and pH 13 the half life of an 0.003 M perxenate solution is about thirty-six hours. At pH 8.5 it is only one hour, while at still lower pH the decomposition becomes almost instantaneous. More concentrated perxenate solutions seem to decompose still more rapidly. Mass-spectrometric analysis has shown that the gas evolved on decomposition is almost exclusively oxygen, and the amount corresponds closely to one-fourth of the total oxidizing titer of the original solution, which had been nearly 8 equivalents per mole of xenon. Testing with starch-iodide paper has shown that no appreciable quantities of ozone are released during decomposition.

We have found that hydrogen peroxide in either acid or basic solution reacts with any xenon compound to liberate xenon gas. Thus, the absence of appreciable xenon in the gas evolved on decomposition of perxenate means that little or no hydrogen peroxide is formed in the process. As further proof of this, addition of a titanium sulfate solution fails to produce the characteristic peroxide color during the perxenate decomposition. Hence, we conclude that we are likely dealing with a true Xe(VIII) and not with a peroxy compound.

Little information is yet available with which to bracket the oxidation potential of Xe(VIII) or, for that matter, of Xe(VI). Xenon(VIII) will oxidize alkaline iodate to periodate, hence its standard oxidation potential in basic solution, $E_B°$, must lie between 0.7 v. and the oxygen-ozone potential of 1.24 v. We suspect it is probably close to the latter, since, although ozone will precipitate the solid sodium perxenate from an Xe(VI) solution in molar base, in 0.1 M NaOH ozone will not form Xe(VIII) in solution within a short time. It is quite possible that the lattice energy of solid sodium perxenate is an important factor in the formation of Xe(VIII).

Spectrophotometric measurements.—Figure 1 shows the ultraviolet spectrum of perxenate as a function of pH. Since no effort was made to control the ionic strength, only qualitative conclusions can be drawn from the results. However, it is apparent that there is a change in the perxenate species below pH 11. Within the limits of experimental uncertainty there are two isobestic points, indicating that over the whole pH range only two principal absorbing species are present.

The broken curve in the figure is the spectrum of a sample that was acidified to decompose the perxenate and then returned to pH 13. It is therefore the spectrum of hexavalent xenon, which has no absorption maximum above 210 mμ, although it does have one at a slightly shorter wave length [see Williamson and Koch, p. 164, this volume].

The very large extinction coefficients of the perxenate in this spectral region are comparable to those of periodate solutions, which also show marked pH dependence [1, 2].

Potentiometric titrations.—Figure 2 shows a potentiometric titration of 50 ml. of 0.003 M perxenate with 0.1 M perchloric acid at 5°. In the region between 2 and 4 moles of H^+ per mole of xenon a small addition of acid would cause the pH to drop sharply to around 5.5 or 6, then rise asymptotically toward its original value. This would indicate that the addition of the third proton leads to decomposition of the perxenate and this decomposition is completed with the addition of the fourth proton. The broken curve is a similar titration of an NaOH solution containing 2 mole per cent sodium carbonate. This curve has been split to facilitate comparison with the two wings of the perxenate curve. We see that there is distinct buffering of the perxenate in the alkaline region. The first proton reacts with something close to a free hydroxide ion; the second, however, forms an acid of pK $\sim$ 10.5, while the acid formed by addition of the third proton, were it stable, would have a pK around 5.5. There is also some suggestion of a buffering region in the acid Xe(VI) solution remaining at the end of the titration.

When the titration was carried out at room temperature, the break at $H^+/Xe = 2$ became somewhat less sharp, but the general features of the curve were unchanged. Furthermore, the same curve, with breaks at $H^+/Xe = 2$ and 4, was obtained regardless of the Na^+/Xe ratio in the solid sample used. This strongly suggests that the extra sodium in the solid is associated with an impurity anion.

In order to look at any protonation reactions of the Xe(VI), a titration was carried out at room temperature and at higher concentration. Six ml. of 0.03 M perxenate were titrated with 1 M $HClO_4$. Decomposition began soon after the titration was started. The pH drifted severely until 4 moles of H^+ had been added per mole of xenon, and there was no sign of the break at $H^+/Xe = 2$.

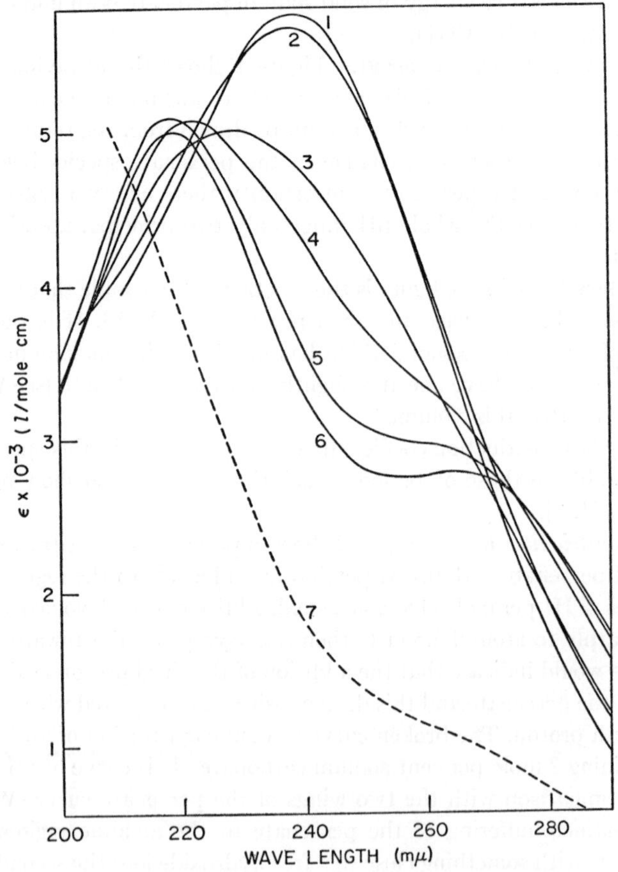

FIG. 1.—Ultraviolet absorption spectra of perxenate solutions at pH values: *1*—12.8 (NaOH solution), *2*—11.5 (unbuffered), *3*—10.4 (phosphate buffer), *4*—10.1 (carbonate buffer), *5*—9.4 (borate buffer), *6*—8.5 (borate buffer), *7*—Xe(VI) in NaOH solution at pH 13.

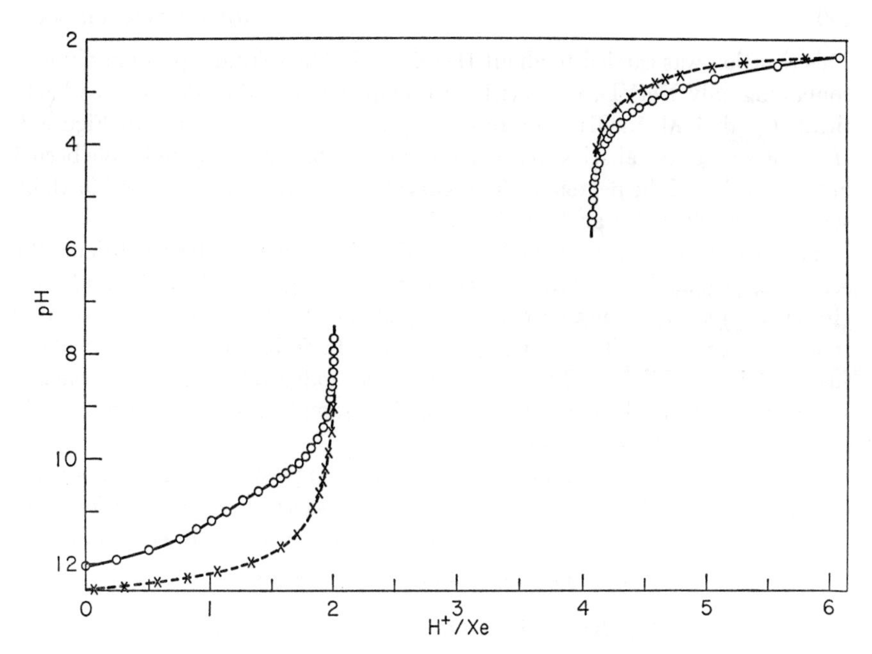

FIG. 2.—Acid-titration curve of a perxenate solution at 5°

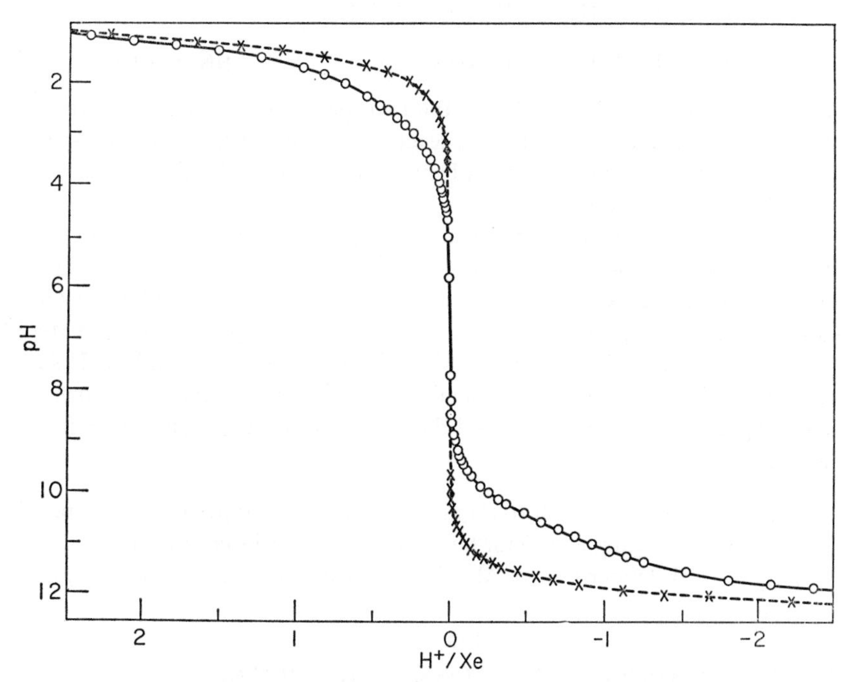

FIG. 3.—Acid-base titration curve of a xenon(VI) solution

This titration was carried to about $H^+/Xe = 7$. The solution, presumably now containing only Xe(VI), was swirled to remove carbon dioxide and was back-titrated with 1 M NaOH. The result appears as the solid curve in Figure 3. Here the zero of the abscissa has been set at the point which would have been 4 in the titration of the perxenate. This curve could be retraced back and forth by successive additions of acid and base.

The broken curve is an NaOH-$HClO_4$ titration under similar conditions. We see buffer regions in the xenate curve corresponding to pK's of 2.5 and 10.5. However, in view of the uncertain purity of the starting material, these pK's should be regarded with skepticism. The acid-buffering region is particularly suspect, since, although the alkaline pK was independent of the perxenate sample used, the acid pK varied from 2.5 in a sample of low Na^+/Xe to 3.5 in one with an unusually high sodium content.

The following reaction scheme is one of the simplest which can satisfy the experimental observations. We have chosen our species on the assumptions that no polymers are formed and that the xenon coordination is always octahedral.

$$Na_4XeO_6(s) + H_2O \rightarrow 4Na^+ + OH^- + HXeO_6^{-3} \qquad (1)$$

$$HXeO_6^{-3} + H^+ \rightleftharpoons H_2XeO_6^{-2} \qquad pK \sim 10.5 \qquad (2)$$

$$H_2XeO_6^{-2} + H^+ \rightleftharpoons H_3XeO_6^- \qquad pK \approx 5.5 \qquad (3)$$

$$H_3XeO_6^- + H_2O \rightarrow H_5XeO_6^- + \tfrac{1}{2}O_2 \qquad (4)$$

$$H_5XeO_6^- + H^+ \rightleftharpoons H_6XeO_6 \qquad pK \sim 10.5 \qquad (5)$$

We see that reaction 2 can provide the two species observed in the ultraviolet spectra of the perxenate. Similarly, reaction 5 is compatible with the spectrophotometric measurements of Xe(VI) made by Koch and Williamson, as well as with their titration results.

If the buffer region we observe in acid solutions of Xe(VI) is really significant, it must represent the formation of a positive ion such as $H_7XeO_6^+$, since an aqueous solution of XeO_3 is not acidic. We choose to reserve judgment on the existence of such a species until work can be done with a purer starting material.

The solution chemistry of xenon(VIII) is by no means outlandish. It shows obvious similarities to periodate chemistry, and it is suggestive of the chemistry of OsO_4 [3]. Further work will reveal the extent to which any such analogies will be of use.

Acknowledgments.—I wish to thank John G. Malm for preparing and providing the perxenate samples and for his continued active interest in this work, and I want to thank Dieter Gruen, Martin Kilpatrick, and Herbert H. Hyman for many helpful discussions.

REFERENCES

1. C. E. Crouthamel *et al.*, *J. Am. Chem. Soc.* **71,** 3031 (1949).
2. *Ibid.* **73,** 82 (1951).
3. D. M. Yost and R. J. White, *ibid.* **50,** 81 (1928).

PART 6 DIFFRACTION STUDIES AND THE STRUCTURE OF XENON COMPOUNDS

The importance of diffraction techniques in determining the structure of new compounds hardly needs emphasis at this time. Not only have these techniques been employed to yield very precise pictures of the simple fluorides, but a rather interesting mixed crystal, an equimolar mixture of XeF_2 and XeF_4, has been identified, a substance not easy to characterize in the absence of such detailed study.

The electron-diffraction analysis of the structure of gaseous xenon tetrafluoride originally led to the suggestion that the xenon was not co-planar with the square formed by the four fluorine atoms. This conclusion was strongly challenged at the conference on noble-gas compounds, and the authors were able to re-examine their data and modify their conclusions for their contribution included in this volume.

CRYSTALLOGRAPHIC STUDIES OF XENON DIFLUORIDE, XENON TETRAFLUORIDE, AND SOME SODIUM XENATE HYDRATES

S. Siegel and E. Gebert
Argonne National Laboratory

XeF_2 is tetragonal [1] with $a = 4.315 \pm .003$ A. and $c = 6.990 \pm 0.004$ A. The space group is $I4/mmm$ with xenon atoms in 0, 0, 0; $\frac{1}{2}, \frac{1}{2}, \frac{1}{2}$, and F atoms in $00z$, $00\bar{z}$ + *b.c.* The two molecules in the cell lead to a density of 4.32 gm/cc. A value $z = 0.306 \pm 0.02$ determined by X-rays leads to linear F-Xe-F molecules with Xe-F = 2.14 ± 0.14 A. However, neutron-diffraction results [2] give $z =$ 0.2837 leading to Xe-F distances of 2.000 A. [Levy and Agron, this volume, p. 221].

XeF_4 crystals are monoclinic with $a = 5.03$ A., $b = 5.92$ A., $c = 5.79$ A., and $\beta = 99°27'$. The space group is $P2_1/n$ with Xe atoms in 0, 0, 0; $\frac{1}{2}, \frac{1}{2}, \frac{1}{2}$, and F atoms in general positions. The computed density, based on the two molecules in the cell, is 4.04 gm/cc. A planar character for the molecule is indicated by the symmetry. The determination of the fluorine coordinates [3–5] shows the XeF_4 molecule to be square also, with Xe-F distances of 1.93 to 1.94 A. [see next three chapters, this volume].

A second phase has been found growing with the XeF_4 crystals. It is also monoclinic [6, 7] with $a = 6.61$ A., $b = 7.33$ A., $c = 6.40$ A., and $\beta = 92°35'$ and may be a new xenon compound or a second modification of XeF_4. [Compare Burns, Ellison, and Levy, this volume, p. 226.] The space group is $P2_1/c$ with Xe atoms in face-centered positions and F atoms in general positions. The calculated density is 4.44 gm/cc, a value substantially larger than observed for the first phase.

The hydrolysis product, $Na_4XeO_6 \cdot 5H_2O$,* is found to be orthorhombic with $a = 10.36$ A., $b = 10.45$ A., and $c = 11.87$ A. A product of composition, $Na_4XeO_6 \cdot 2H_2O$, is also orthorhombic with $a = 6.25$ A., $b = 5.77$ A., and $c =$ 10.28 A. The latter dimensions are based on observations of very disordered crystals. A third composition, Na_5HXeO_7, can be referred to cubic symmetry

* This has now been established as $Na_4XeO_6 \cdot 8H_2O$. [Cf. W. C. Hamilton, J. A. Ibers, and D. R. MacKenzie, *Science* **141,** 532 (1963).

with $a = 9.391$ A. With the exception of a few faint maxima, the indexing indicates that the cell is face centered.

Assignment of formulas for these compounds is tentative. [See Malm, Bane, and Holt, this volume, p. 167.]

REFERENCES

1. S. Siegel and E. Gebert, *J. Am. Chem. Soc.* **85,** 240 (1963).
2. H. Levy and P. Agron, *ibid.* 241 (1963).
3. J. Ibers and W. Hamilton, *Science* **139,** 106 (1963).
4. D. A. Templeton *et al., J. Am. Chem. Soc.* **85,** 242 (1963).
5. J. Burns *et al., Abstracts,* Annual Meeting, Am. Cryst. Assn., Cambridge, Mass., March 28, 1963.
6. S. Siegel, *Abstracts,* Sponsor Conference, University of Chicago, February 26, 1963.
7. J. Burns, *J. Phys. Chem.* **67,** 536 (1963).

X-RAY INVESTIGATION OF THE CRYSTAL STRUCTURE OF XENON TETRAFLUORIDE

WALTER C. HAMILTON AND JAMES A. IBERS
Chemistry Department, Brookhaven National Laboratory

ABSTRACT

The crystal structure of xenon tetrafluoride, reported earlier by us, has been refined further. The average Xe-F bond length, uncorrected for thermal motion, is 1.94 ± 0.02 A. The F-Xe-F bond angle is 90.8 ± 1.1°. As the molecule is required by the space-group symmetry to be planar and centrosymmetric, these results indicate that any departures from a square-planar configuration for the molecule are insignificant. The thermal vibrations of the molecule, as derived from the anisotropic thermal parameters, are readily interpreted in terms of rigid-body translations and rotations with root-mean-square amplitudes of approximately 0.12 A. and 6°, respectively. The latter correspond to torsional oscillations with frequencies of about 50 to 70 cm.$^{-1}$.

INTRODUCTION

Previously [1] we found from a preliminary refinement of X-ray data for XeF_4 that the Xe-F distances were 1.92 ± 0.03 A. and that the molecule showed insignificant deviations from a square-planar configuration. We present here the results of a more complete refinement of those data; the conclusions reached are not altered, but the limits of error on the results are now smaller.

XeF_4 is monoclinic, space group $C_{2h}^5 - P2_1/n$, with cell parameters $a = 5.03$, $b = 5.90$, $c = 5.75$ (all ±0.03) A., $\beta = 100 \pm 1°$. These values do not differ significantly from those reported subsequently by Siegel and Gebert [2] (5.03, 5.92, 5.79, 99°27′) and by Templeton, Zalkin, Forrester, and Williamson [3] (5.050, 5.922, 5.771 [all ±0.003], 99.6 ± 0.1°). With two molecules in the unit cell, the two xenon atoms may be placed at the centers of symmetry (0, 0, 0) and ($\frac{1}{2}, \frac{1}{2}, \frac{1}{2}$). The two crystallographically independent fluorine atoms are placed in two sets of fourfold general positions (4*e*) $\pm(x, y, z), \pm(\frac{1}{2} + x, \frac{1}{2} - y, \frac{1}{2} + z)$.

SOME EXPERIMENTAL DETAILS

The crystals of XeF_4 were contained in quartz capillaries and were photographed by the precession method with Mo $K\alpha$ radiation. Photographs were taken of the reciprocal lattice nets $h, h - l + n, l$ with n ranging from 0 to 4 (these nets are parallel to reciprocal lattice planes $[1\bar{1}\bar{1}]$), of a platelike crystal of linear dimensions 1 by 1 by 0.3 mm. The large face of the crystal was approximately parallel to the film for each exposure so that the absorption correction is nearly constant for each net. Since separate scale factors were carried in the least-squares refinements, errors introduced by absorption should be small. The crystal was stable during the photography and did not grow in size, although there were other small crystals in the capillary; in fact, it was the same size some three months later. A second crystal was used to obtain the $h, h + l, l$ net. Experimental details are described more fully in our earlier note [1].

DETERMINATION OF THE STRUCTURE

The structure was determined by calculation of a three-dimensional difference Fourier, with coefficients $(F_{obs} - F_{Xe})$ for those reflections for which $(h + k + l)$ is even. Such a difference Fourier necessarily exhibits a false mirror plane. The correct fluorine atom positions were chosen by comparison of observed and calculated intensities of several of the weak, fluorine-only $(h + k + l$ odd) reflections. The resultant parameters are listed in Table 1 together with the derived Xe-F bond distances and F_1-Xe-F_2 angle.

REFINEMENT OF THE STRUCTURE

A series of least-squares refinements was carried out, starting with the parameters determined earlier (Table 1). The atomic scattering factors for neutral fluorine tabulated by Ibers [4] and the values of Watson and Freeman [5]

TABLE 1

PARAMETERS AND MOLECULAR GEOMETRY FOR XENON TETRAFLUORIDE

Parameter	IH	Visual	Counter	Neutron
F_1 x.....	0.225±0.006	0.242±0.004	0.229±0.003	0.2356±0.0002
y.....	0.027±0.011	0.033±0.005	0.033±0.002	0.0297±0.0002
z.....	0.306±0.005	0.305±0.004	0.297±0.002	0.3002±0.0001
F_2 x.....	0.242±0.009	0.260±0.004	0.260±0.003	0.2643±0.0002
y.....	0.165±0.007	0.144±0.004	0.146±0.002	0.1481±0.0002
z.....	−0.162±0.007	−0.159±0.004	−0.153±0.002	−0.1536±0.0002
R(Xe-F_1)....	1.92 ±0.03	1.96 ±0.03	1.91 ±0.02	1.939 ±0.002
R(Xe-F_2)....	1.92 ±0.04	1.92 ±0.02	1.91 ±0.02	1.932 ±0.002
Angle(F_1-Xe-F_2)....	94 ±3°	90.8 ±1.1°	90.4 ±0.9°	90.0 ±0.1°

THERMAL PARAMETERS

	Xe	F_1	F_2
β_{11}.........	0.0123±0.0005	0.040±0.008	0.034±0.008
β_{22}.........	.0089± .0003	.034± .011	.024± .005
β_{33}.........	.0110± .0005	.026± .006	.033± .007
β_{12}.........	− .0003± .0017	.009± .007	− .015± .005
β_{13}.........	.0053± .0003	− .006± .005	.017± .005
β_{23}.........	0.0006±0.0011	−0.005±0.005	0.004±0.004

IH = preliminary results of Ibers and Hamilton [1]; Visual = our present results; Counter = results of Templeton *et al.* [3]; Neutron = results of Burns, Agron, and Levy [8]. The thermal parameters are those from the present refinement. The IH and visual distances are computed for our cell, the counter and neutron for that of Templeton *et al.*

The notation F_1 and F_2 agrees with that in our original paper, but is opposite to that of Templeton *et al.* [3] and Burns *et al.* [8].

for neutral xenon were used. A dispersion correction was applied in accordance with the tabulation of Templeton [6]. Refinements were carried out on F.

Of the 268 reflections used in the earlier refinement (where all reflections were weighted equally), 54 were given zero weight in the present calculations, either because they were partially obscured by the beam stop, occurred on portions of the film where the Lorentz-polarization factor was changing extremely rapidly, or were so weak on the particular films that it was difficult to estimate their intensities. These reflections are marked by a $ sign in Table 2. The remaining 214 reflections were assigned weights inversely proportional to their estimated variances $(0.10\ F)^2$. In addition, the 18 reflections observed on the $n = 1$ and $n = 3$ layers ($h + k + l$ odd) were included in the refinement with estimated variances $(0.4\ F)^2$.

The data were first refined with separate isotropic thermal parameters on each atom. The resultant weighted R factor,

$$R = \left[\frac{\Sigma w(\,|F_{\text{obs}}| - |F_{\text{calc}}|\,)^2}{\Sigma w\,|F_{\text{obs}}|^2}\right]^{1/2},$$

was 0.114, and there were some possibly significant shifts from our earlier parameters for the fluorine atoms. Next a series of refinements was carried out in which the xenon was allowed to vibrate anisotropically, but the fluorines were constrained to vibrate isotropically. The weighted R factor was reduced to 0.100, which, according to the R-factor ratio test [7], is a highly significant improvement. A final series of calculations was carried out with anisotropic thermal parameters on all the atoms; the final weighted R factor is 0.097. This differs little enough from the preceding R factor so that one cannot reject at the 0.10 level of significance the hypothesis that the vibrations of the fluorine atoms are isotropic. However, the physical reasonableness of the thermal motions derived from these parameters and their agreement with the other diffraction results [3, 8] supports the supposition that this refinement is meaningful. The final parameters from this anisotropic refinement are listed in Table 1 with the Xe-F distances and F_1-Xe-F_2 angle. Although there are some possibly significant shifts in the parameter values between our earlier refinement and this more reliable study, the derived bond distances and angles do not differ significantly. Table 2 lists the observed structure amplitudes and the structure factors (in electrons) derived from this refinement. The fluorine-only ($h + k + l$ odd) reflections are listed at the end of Table 2. Note that some reflections are listed twice; these were observed separately on different nets. Their internal consistency, as judged by a value of 0.12 for $R^{12} = \Sigma|F_{\text{obs}}^{(1)} - F_{\text{obs}}^{(2)}|/\frac{1}{2}\Sigma(F_{\text{obs}}^{(1)} + F_{\text{obs}}^{(2)})$ is in line with the generalized R factor of 0.097 and with the conventional R factor of 0.099 for all reflections. The conventional R factor for the reflections assigned zero weight ($) is 0.22. Structure amplitudes calculated for other reflections accessible on our films do not exceed our estimate of the minimum observable F.

TABLE 2

OBSERVED AND CALCULATED STRUCTURE FACTORS (IN ELECTRONS) FOR XeF_4

K	L	FØ	FC	$	K	L	FØ	FC	$	K	L	FØ	FC	$	K	L	FØ	FC	$
**** H = 0 ****					**** H = 1 ****					**** H = 3 ****					**** H = 4 ****				
0	0	179	179	$	6	-5	28	27		2	-1	49	51		7	1	15	23	$
0	4	53	53		7	-2	29	28		2	3	25	38	$	7	-1	24	24	
1	1	71	94	$	7	4	22	22		2	3	42	38		8	0	11	19	$
1	3	59	72		7	-4	25	27		2	-3	63	55						
1	5	40	40		8	-3	22	23		2	5	30	28		**** H = 5 ****				
2	2	63	62							2	5	26	28		0	1	37	36	
2	2	67	62		**** H = 2 ****					2	-5	46	46	$	0	3	29	31	
2	4	57	56		0	0	42	35	$	2	7	8	18	$	0	-5	35	39	
2	6	37	35		0	2	82	83		3	0	52	51		1	0	45	40	
3	-1	49	57		0	-2	86	99		3	2	40	47	$	1	2	36	36	
3	3	74	72		0	4	53	47		3	2	52	47		1	2	36	36	
3	3	75	72		0	6	24	25		3	-2	50	50		1	4	19	20	
3	5	38	35		1	1	68	69		3	4	30	33		1	4	20	20	
3	7	26	25		1	1	58	69	$	3	4	37	33		1	-4	43	44	
4	0	49	57		1	-1	95	82	$	3	-4	43	47		1	-6	25	30	
4	-2	50	55		1	3	54	50		3	6	19	19	$	2	1	31	31	
4	4	52	44		1	3	50	50		3	-6	36	35		2	-1	50	44	
4	4	41	44		1	-3	72	65		4	1	49	50		2	3	29	28	
4	6	26	27		1	-3	79	65	$	4	-1	39	38		2	3	28	28	
5	-1	44	47		1	5	49	40		4	3	30	33		2	-3	39	41	
5	-3	45	44		1	5	43	40		4	3	40	33		2	5	10	17	$
5	5	30	28		2	0	65	61		4	-3	52	46		2	5	16	17	$
5	5	30	28		2	2	65	67		4	5	21	22		3	0	45	37	
6	-2	38	36		2	2	60	67		4	-5	40	38		3	2	27	27	
6	-4	28	28		2	-2	86	76		5	0	37	36		3	-2	40	40	
7	-3	26	26		2	4	41	39		5	2	37	34		3	4	14	20	$
					2	4	39	39		5	-2	45	42		3	4	11	20	$
**** H = 1 ****					2	-4	56	50		5	4	30	26		4	1	27	27	$
0	1	71	84	$	2	-4	61	50		5	-4	34	35		4	-1	42	37	
0	3	64	75		2	6	26	26		5	-6	22	25		4	3	15	22	$
0	5	37	38		3	1	57	67		6	1	27	30		4	-3	31	31	
1	0	82	93	$	3	1	54	67		6	-1	34	35		5	0	29	30	
1	2	66	97	$	3	-1	88	74		6	3	26	24		5	2	21	22	
1	2	77	97		3	3	32	36		6	-3	30	33		5	-2	28	27	
1	-2	64	54		3	3	42	36		6	-5	19	24		5	-4	26	26	
1	-2	64	54	$	3	-3	47	40		7	0	29	27		6	1	19	23	
1	4	39	39		3	5	34	34		7	-2	30	30		6	-3	26	23	
1	4	38	39		3	-5	56	45		8	-1	26	24		6	-5	15	23	$
1	-4	82	74		3	-5	48	45											
1	6	33	35		3	7	14	19	$	**** H = 4 ****					**** H = 6 ****				
2	1	46	54	$	4	0	50	57		0	0	69	69		0	2	24	27	
2	1	49	54	$	4	2	47	49		0	2	41	36		1	1	27	27	
2	-1	98	102		4	2	54	49		0	4	29	30		1	3	19	20	$
2	3	57	64		4	-2	53	49		0	-4	46	44		1	3	18	20	$
2	3	63	64		4	4	34	31		0	6	13	20	$	1	-5	22	29	$
2	-3	59	61		4	-4	45	41		1	1	47	50		2	0	30	29	
2	-3	65	61	$	4	6	20	22		1	1	48	50		2	2	27	25	
2	5	49	40		4	-6	37	32		1	-1	56	51		2	4	11	16	$
2	-5	52	42		4	-6	29	32		1	3	38	38		2	-4	30	30	
2	7	15	22	$	5	1	42	45		1	3	40	38		3	1	25	26	
3	0	64	71		5	-1	50	46		1	-3	49	50		3	-1	37	31	
3	2	56	56	$	5	3	32	33		1	5	16	21	$	3	-3	27	26	
3	2	61	56		5	3	37	33		1	5	20	21		4	0	27	26	
3	-2	76	69		5	-3	39	38		1	-5	39	41		4	-2	16	25	$
3	4	40	42		5	5	25	25		2	0	53	52		5	-3	15	24	$
3	-4	54	49		5	-5	37	34		2	2	41	36		6	-4	11	22	$
3	-4	53	49		6	0	36	33		2	2	35	36	$					
3	6	32	31		6	-2	46	36		2	-2	52	47		**** H = 0 ****				
3	-6	35	31		6	4	28	25		2	4	32	29	$	1	2	31	19	
4	1	40	43	$	6	4	22	25		2	4	34	29		2	3	7	5	
4	1	36	43		6	-4	35	33		2	6	11	18	$	3	4	7	-7	
4	-1	63	73		7	-1	32	29		2	-6	30	33		4	-1	11	12	
4	3	45	46		7	-3	30	30		3	1	40	38	$					
4	3	55	46		8	-2	26	24		3	-1	45	41						
4	-3	50	48							3	3	36	35		**** H = 1 ****				
4	5	34	34		**** H = 3 ****					3	3	43	35		1	3	5	7	
4	-5	35	33		0	1	59	65		3	-3	55	52		1	-3	11	11	
4	-5	33	33		0	-1	86	70	$	3	5	10	19	$	2	0	29	-32	
4	7	17	19	$	0	3	34	38		3	5	19	19	$	2	2	9	10	
4	-7	27	26		0	-3	68	63		4	0	41	37		2	-2	22	15	
5	0	48	51	$	0	5	34	32		4	2	27	30		5	-3	6	-7	
5	2	43	44		0	7	11	19	$	4	-2	44	44						
5	-2	47	44		1	0	55	65		4	4	20	25		**** H = 2 ****				
5	4	33	32		1	0	57	65		4	4	30	25		1	2	16	-12	
5	4	35	32		1	2	43	42		4	-4	39	36		2	-1	16	11	
5	-4	40	37		1	2	45	42		5	1	29	30						
5	6	24	23	$	1	-2	91	78		5	-1	35	36						
5	-6	24	26		1	4	43	43		5	3	14	25	$	**** H = 3 ****				
5	-6	25	26		1	4	43	43		5	-3	32	35		1	1	10	5	
6	-1	40	38		1	-4	46	42		5	-5	21	25		1	1	6	5	
6	3	29	32		1	6	18	21	$	6	0	31	31		1	-1	5	5	
6	-3	39	34		1	6	14	21	$	6	2	22	22		2	0	22	16	
6	5	23	23	$	2	1	77	67		6	-2	31	29		2	-2	9	-10	
6	5	22	23		2	1	62	67		6	-4	23	23		5	-3	6	8	

DESCRIPTION OF THE STRUCTURE

The structure described by the parameters of Table 1 and the space group is one in which discrete, square-planar XeF_4 molecules are packed as shown in Figure 1. The dihedral angle between the planar molecule at the origin and at the body center is $54.2 \pm 1.5°$.

The bond distances derived from the present refinement are 1.961 ± 0.026 and 1.921 ± 0.021 A. The difference between these distances, 0.040 ± 0.037 A., is not significant, and thus it is appropriate to give a value of 1.94 ± 0.02 A. These distances are uncorrected for thermal motion. In the absence of a complete description of the vibrations of the solid an exact correction for motion is

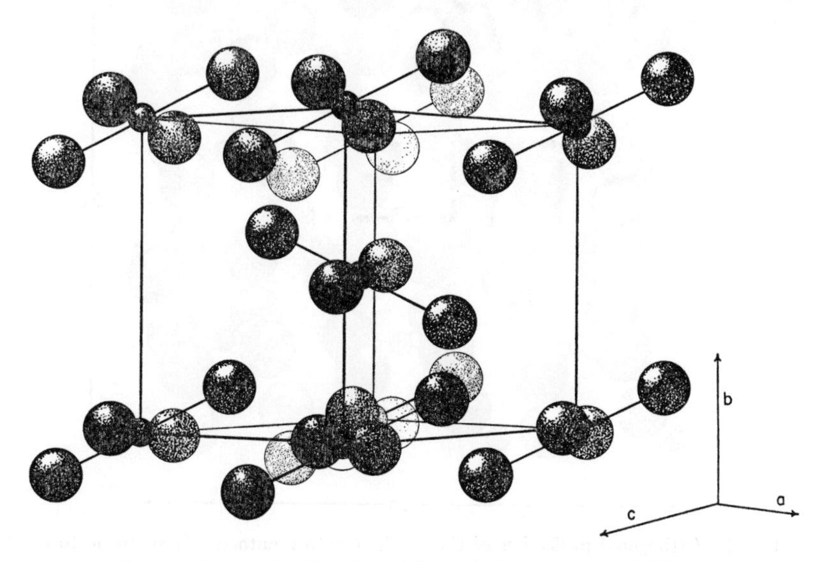

FIG. 1.—Perspective drawing of the crystal structure of XeF_4

not possible. Some idea of the effects of motion on the derived distances can be obtained by making the reasonable assumption that the fluorine atoms ride on the xenon atom [9]. With such an assumption the mean Xe-F distance is found to be 1.965 ± 0.022 A. The F_1-Xe-F_2 bond angle is $90.8 \pm 1.1°$. Thus, although the site symmetry in the crystal is only $\bar{1}$, it is clear that the molecule has a square-planar configuration (D_{4h} symmetry).

Packing.—The fluorine atoms lie approximately in the planes $x = \frac{1}{4}$ and $x = \frac{3}{4}$; the distance between these planes is 2.5 A. The xenon atoms lie in planes half-way in between. The packing is such that each fluorine atom has close contacts with five other fluorine atoms in the same plane, three in the plane on one side, and two in the plane on the other. Two of these contacts are to fluorines in the same molecule; these are slightly shorter than those between molecules. In addition to the short, bonded Xe-F distance, there is also a short, non-bonded distance of about 3.2 A. from each fluorine to a single xenon. Thus, each fluorine atom has a coordination number of 12. Figure 2 depicts this packing, and the distances are listed in Table 3.

If we exclude the intramolecular distances, the average value of the F···F contact is 3.13 (range 2.97 to 3.24). The van der Waals radius of fluorine may

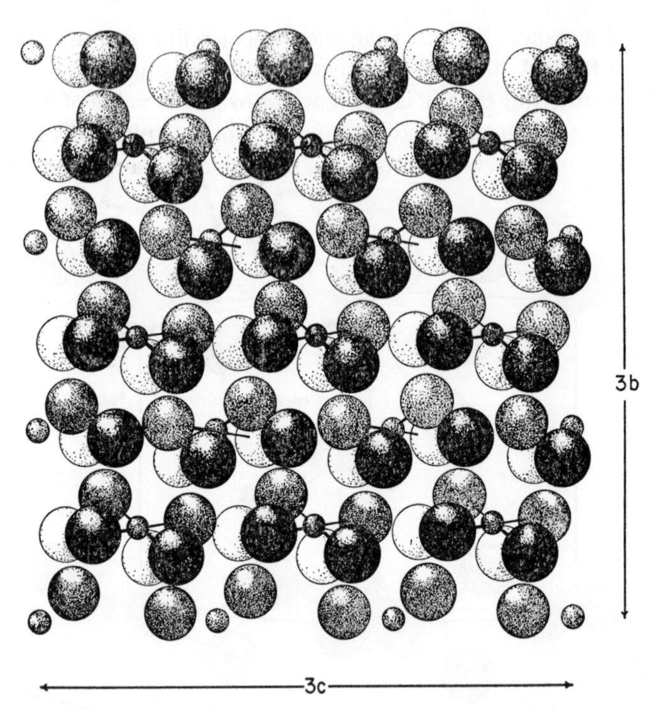

FIG. 2.—Orthogonal projection of the XeF_4 structure onto the (100) plane to illustrate the packing. The more heavily shaded fluorine atoms are at $x = \frac{3}{4}$, those of intermediate weight at $x = \frac{1}{4}$, and the lightest at $x = -\frac{1}{4}$. The xenon atoms are indicated by the small circles, and some of the bonds are shown. Each fluorine atom has five neighbors in the plane with two or three neighbors in the plane above and three or two neighbors in the plane below. The next closest distance in the plane is 4.14 A.

thus be assigned the value 1.56 A., and, if the two short non-bonded Xe···F contacts may be regarded as van der Waals contacts, the van der Waals radius of xenon is 1.65 A.

Description of the thermal motion.—The principal axes of the ellipsoids of vibration of all atoms lie very nearly along natural molecular axes defined by

$$n_1 = r_{F_1} - r_{Xe}$$

$$n_2 = r_{F_2} - r_{Xe}$$

$$n_3 = n_1 \times n_2 \, .$$

The root-mean-square components of motion in these directions are given in Table 4. These motions are adequately (as determined by a χ^2 test) described in terms of translation and libration of a rigid body (the molecule), and the components of these rigid-body motions were determined by the least-squares procedure formulated by Cruickshank [10, 11], weighted with the errors derived

TABLE 3

NON-BONDED DISTANCES LESS THAN 3.5 A. IN THE CRYSTAL OF XeF_4

Neighbors of F_1 at 24, 03, 30					Neighbors of F_2 at 26, 14, −16				
Atom	x	y	z	Distance	Atom	x	y	z	Distance
F_2..........	26	14	−16	2.74*	F_1..........	24	03	30	2.74*
F_1..........	26	−47	20	3.02	F_1..........	26	53	20	3.07
F_1..........	26	53	20	3.02	F_2..........	24	−36	−34	3.13
F_2..........	24	−36	66	3.07	F_2..........	24	64	−34	3.13
F_2..........	26	14	84	3.14	F_1..........	24	03	−70	3.14
F_2..........	74	−14	16	2.97	F_1..........	76	−03	−30	2.97
F_1..........	76	−03	70	3.16	F_1..........	74	47	−20	3.12
F_2..........	76	36	34	3.21	F_2..........	74	−14	16	3.24
F_2..........	−26	−14	16	2.71*	F_1..........	−24	−03	−30	2.71*
F_2..........	−24	36	34	3.12	F_1..........	−26	47	−20	3.21
Xe..........	50	50	50	3.17	Xe..........	50	50	−50	3.24
Xe..........	50	−50	50	3.51†					

The standard deviations range from 0.02 to 0.04 A. The approximate coordinates of the atoms in units of 0.01 are given for identification.

* Distances are intramolecular.

† Distance is not considered to be a van der Waals contact.

TABLE 4

ROOT-MEAN-SQUARE COMPONENTS OF THERMAL MOTION ALONG MOLECULAR AXES

	Xe	F_1	F_2
$U_{11}^{1/2}$ (A.)......	0.144±0.003	0.18±0.02	0.25±0.02
$U_{22}^{1/2}$ (A.)......	.105± .007	.28± .03	.10± .03
$U_{33}^{1/2}$ (A.)......	0.124±0.005	0.22±0.03	0.24±0.03

from the least-squares refinement of the structure factors as suggested by Hamilton [12]. The components of the translational motions are essentially the motions of the xenon atom as given in Table 4. The librational motions are approximately isotropic with a root-mean-square amplitude of 6 ± 1.4°. These librational motions correspond to torsional vibration frequencies of 70, 70, and 50 cm.$^{-1}$ if one adopts the approximation of Cruickshank [11] that

$$\langle \phi^2 \rangle = \frac{kT}{4\pi^2 I \nu^2}.$$

COMPARISONS WITH OTHER STUDIES

In Table 1 we also list the parameters derived by Templeton *et al.* [3] from an X-ray study in which counter methods were employed for the determination of the intensities of 329 independent reflections (including 96 fluorine-only reflections). We also list the parameters derived by Burns, Agron, and Levy [8] from a neutron-diffraction study in which approximately 600 reflections were recorded. Table 1 thus provides an interesting comparison between, in order of decreasing accuracy as judged by the estimated standard deviations, a neutron-diffraction study, an X-ray study of this heavy-atom–light-atom structure by counter methods, and the present X-ray study involving visual estimates of intensities. It can be seen from Table 1 that counter methods have not led to a structure that is significantly better (when compared with the neutron-diffraction study). Since counter methods should lead to more reliable estimates of intensities than visual methods, presumably other factors, such as absorption and crystal growth, have introduced (into one or both of the X-ray studies) systematic errors that outweigh the effects of purely random errors.

CONCLUSION

It is perhaps sobering yet amusing to realize that solid XeF_4, at present the object of two independent X-ray studies and one neutron-diffraction study, is better characterized and more thoroughly studied than are most simple molecular crystals. Certainly the accuracy with which the dimensions of the XeF_4 molecule are known far exceeds our ability to predict or calculate them, even semi-empirically.

REFERENCES

1. J. A. Ibers and W. C. Hamilton, *Science* **139,** 106 (1963).
2. S. Siegel and E. Gebert, *J. Am. Chem. Soc.* **85,** 240 (1963).
3. D. H. Templeton, A. Zalkin, J. D. Forrester, and S. M. Williamson, *ibid.* 242 (1963). [See also this volume, p. 203.]
4. J. A. Ibers in *International Tables for X-ray Crystallography*, vol. 3. Birmingham, England: Kynoch Press, 1962. Table 3.3.1 A.
5. R. E. Watson and A. J. Freeman, private communication (1963).
6. D. H. Templeton (see reference 4 above). Table 3.3.2 A.
7. W. C. Hamilton, Paper H-5, Am. Cryst. Assn. Meeting, Boulder, Colorado, 1961.
8. J. H. Burns, P. A. Agron, and H. A. Levy, *Science* **139,** 1209 (1963). [See also this volume, p. 211.]
9. W. R. Busing and H. A. Levy, unpublished work (1963).
10. D. W. J. Cruickshank, *Acta Cryst.* **9,** 754 (1956).
11. *Ibid.* 1005 (1956).
12. W. C. Hamilton, *Acta Cryst.* **15,** 353 (1962).

A DETERMINATION OF THE CRYSTAL STRUCTURE OF XENON TETRAFLUORIDE

DAVID H. TEMPLETON, ALLAN ZALKIN, J. D. FORRESTER, AND STANLEY M. WILLIAMSON
University of California, Berkeley

ABSTRACT

The crystal and molecular structure of XeF_4 has been determined by single-crystal X-ray diffraction techniques. The intensities of Mo $K\alpha$ X-rays diffracted by the crystal were measured with a scintillation counter. The monoclinic unit-cell dimensions are a = 5.050 A., b = 5.922 A., c = 5.771 A. (each ± 0.003 A.), and β = 99.6° ± 0.1°. The space group is $P2_1/n$ with two molecules per unit cell. The xenon atoms occupy the corners and body centers so that the molecular packing is pseudo-body-centered cubic. The molecule has a square-planar configuration. The Xe-F bond distance is 1.93 ± 0.02 A., after a correction of +0.02 A. for thermal vibration effects; the F-Xe-F bond angle is a right angle (90.4 ± 0.9°) within the accuracy of the determination.

INTRODUCTION

This paper is an extended and slightly modified version of our earlier report [1] which described our determination of the crystal and molecular structure of XeF_4.

The earliest X-ray study of this compound was by Siegel and Gebert [2], who determined the cell dimensions and space group. The atomic coordinates were determined simultaneously by Ibers and Hamilton [3] and ourselves [1], by X-ray diffraction. Ibers and Hamilton used photographic data from precession films, while we used stationary scintillation counter data. This work was soon followed by a neutron-diffraction study by Burns, Agron, and Levy [4] that gives somewhat higher precision for the fluorine coordinates than is feasible with the X-ray data.

EXPERIMENTAL

Xenon tetrafluoride was prepared by heating the elements to 300° in a flow system. Subsequently a slightly modified procedure was adopted. A 4:1 molar ratio mixture of fluorine and xenon was mixed well in 0.5-liter copper chamber which contained baffles with twice as much helium as a carrier gas. The gas mixture flowed through a copper U-trap at −120° into a 12-in. length of $\frac{3}{4}$-in. nickel tubing. The nickel and copper were joined by a silver-soldered connection. The last 6 in. of the reactor tube was heated to 350° by an electric furnace.

The reactor ended with 4 in. of $\frac{1}{2}$-in. copper tubing so that there was a thermal gradient before the copper-to-glass seal. A glass U-trap was then either sealed to the glass of the copper-to-glass seal or connected through an ungreased ground joint. The joint was used if the XeF_4 sample was to be transferred to other containers in a dry-box, and the seal was used if the trap were equipped with a break-seal so that the sample could be transferred into a vacuum system. The trap was cooled with solid CO_2, and the other end went by tubing directly to a hood. The glass from the copper-to-glass seal to the CO_2 (s) level was maintained at about 75° by a heating tape to prevent condensation upstream from the trap. Good conversion of the Xe to XeF_4 was attained with a flow rate such that the residence time in the reactor was one minute. The apparatus is similar to that of Holloway and Peacock [5], except that our apparatus had only one trap. This procedure yielded the material described by Gunn and Williamson [6] for which the chemical analysis was close to theoretical for XeF_4. Our X-ray studies of material prepared in this way detected crystals only of the structure described here, except when samples had been exposed to water.

In some of our earlier work we attempted quick transfers of the material in damp air into capillaries, but the resulting samples survived only long enough for a few preliminary X-ray patterns. It was only when the capillaries were loaded by sublimation under vacuum that we obtained stable specimens. The capillaries were thin-walled vitreous silica 0.5 mm. in diameter. During the investigation of the final crystal, it is estimated to have undergone about ten hours of irradiation with no evidence of decomposition, and, in fact, the crystal continued to grow at the expense of other crystals in the capillary. A few weeks after the experiment, the crystal disappeared by sublimation to regrow in another location in the capillary. Four months later it was still there. Photographs of the crystal taken the day following the intensity measurements are shown in Plate I. The crystal diameter ranged from 0.13 to 0.24 mm. in various directions. Eleven faces of the pseudo-cubic dodecahedron were developed; the twelfth surface was attached to the curved surface of the capillary.

Molybdenum Kα X-rays were produced with a General Electric XRD-5 unit operated at 25 ma. and 40 kvp. A 0.001-in. zirconium foil was used to filter the diffracted radiation just before it entered the scintillation counter. The range of intensities measured was from 1 to 14,000 counts per second. The counter was checked and found to be linear over this range.

The cell dimensions were measured with a take-off angle of 2° using the resolved Kα_1 peaks of Mo (λ = 0.70926 A.). The crystal was set on the goniostat with the a^* axis perpendicular to the phi circle; this axis coincides roughly with the axis of the capillary.

The intensities were measured using the stationary technique and counting each reflection for twenty seconds with a take-off angle of 4°. A fixed-time count is appropriate for approximately equal weighting of the data in the least-squares analysis. The background, plotted as a function of the diffraction angle 2θ, was

PLATE I.—Two views of the crystal of XeF_4 used in this structure determination. The two views are approximately 75° rotation apart from each other. The *a* axis is approximately parallel to the long edge of the crystal.

ordinarily applied to the data. If the reflection was a multiple of a strong reflection, the background was checked near the reflection. All of the 293 independent reflections up to a 2θ angle of 50° ($\sin \theta/\lambda \sim 0.59$) were measured; 35 of these were below the detection limit and were recorded as zero. The crystal grew about 30 per cent during the measurements (two days), and the data were normalized by repeated measurement of a few standard reflections. The data were corrected for the Lorentz-polarization factor using the formula:

$$I_{\text{cor}} = \frac{I \sin 2\theta}{1 + \cos^2 2\theta}.$$

The least-squares program of Gantzel, Sparks, and Trueblood [7] was used on an IBM 7090; this program minimizes the function $\Sigma(|F_{\text{obs}}| - |F_{\text{calc}}|)^2/\Sigma|F_{\text{obs}}|^2$ where F_{obs} and F_{calc} are the observed and calculated structure factors. The weighting factors were all unity. The program utilizes a full-matrix calculation for the parameter shifts. Our results are stated in terms of temperature factors of the form $\exp(-\beta_{11}h^2 - 2\beta_{12}hk - \ldots)$, although the program actually uses $\exp(-B_{11}h^2 - B_{12}hk - \ldots)$.

Scattering factors for the neutral xenon and fluorine atoms were obtained from Tables 3.3.1B and 3.3.1A, respectively, as given in the International Tables [8]. Due to an oversight the xenon scattering factors were not corrected for the dispersion correction $\Delta f'$ which is approximately -0.5 electrons.

STRUCTURE DETERMINATION

Reflections are strong when $h + k + l$ is even and weak when it is odd, showing that the xenon atoms are at 0, 0, 0 and $\frac{1}{2}, \frac{1}{2}, \frac{1}{2}$. Trial coordinates for fluorine atoms were estimated by some simple calculations, which in principle were equivalent to making projections of the fluorine electron density down the a and c axes with use of only a few terms in which the effect of the fluorine atoms was large. The electron densities were not actually calculated, but were roughly approximated graphically. For example, reflections 060 and 110 were judged to be stronger than average, while 031 and 200 were weaker than average. In these cases the phases are fixed by xenon. Reflections 012, 014, and 520 were judged to be strong among reflections depending only on fluorine. In these cases phases were chosen in all permutations. These calculations resulted in six coordinates for the two fluorine atoms which in five cases were within 0.05 of the final values. For $F(2)$ the trial value of y was 0.18, in error by 0.15. Refinement by least squares quickly corrected this error.

Eight cycles of least-squares refinement using isotropic temperature factors brought the unreliability factor $R = \Sigma||F_{\text{obs}}| - |F_{\text{calc}}||/\Sigma|F_{\text{obs}}|$ to 0.11. Four cycles using anisotropic temperature factors then diminished R to 0.089. Two obvious blunders in data taking were corrected by remeasurement of their intensities, and three more cycles of least squares brought R to 0.076.

Some of the low-angle data appeared to suffer from extinction and/or absorption, so the seven reflections with $\sin \theta/\lambda$ less than 0.17 were deleted from the refinement. A final set of refinements of five cycles reduced R to our final value of 0.059 for 286 data. The results in Table 1 and Table 2 are from this last calculation. Table 1 lists the final parameters. Table 2 lists the observed and calculated structure factors.

Some additional calculations were performed with the 96 non-zero, odd $h + k + l$ data. These reflections are the result of fluorine atoms exclusively. A refinement with isotropic temperature factors resulted in coordinates for fluorine

TABLE 1

CRYSTAL-STRUCTURE DATA FOR XeF_4

$a=5.050\pm0.003$ A.	$Z=2$
$b=5.922\pm0.003$A.	Space Group $P2_1/n$ (C_{2h}^5)
$c=5.771\pm0.003$ A.	Molecular Weight=207.30
$\beta=99.6\ \pm0.1°$	X-ray Density=4.04 gm/ml
$V=170.2$ A.^{3}	

ATOMIC POSITIONS

Xe: $0, 0, 0; \frac{1}{2}, \frac{1}{2}, \frac{1}{2}$
F: $\pm(x, y, z; \frac{1}{2}-x, \frac{1}{2}+y, \frac{1}{2}-z)$

F(1)	F(2)
$x=0.260\ \pm0.003$	$x=0.229\pm0.003$
$y=.146\ \pm .002$	$y=.033\pm .002$
$z=-0.153\pm0.002$	$z=0.297\pm0.002$

ANISOTROPIC TEMPERATURE PARAMETERS

	Xe	F(1)	F(2)
β_{11}....	0.0208±0.0007	0.044±0.006	0.044±0.006
β_{22}....	.0097± .0005	.025± .004	.021± .004
β_{33}....	.0120± .0005	.031± .004	.029± .005
β_{12}....	.0012± .0004	− .006± .004	.001± .004
β_{13}....	.0071± .0004	.023± .005	.002± .004
β_{23}....	0.0000±0.0003	0.004±0.004	0.000±0.004

atoms which were the same as those in Table 1 within 0.005 or less. The corresponding R was 0.18.

The data were not corrected for absorption. The dimensions of the crystal correspond to μR of about 0.9. In the approximation of spherical shape, absorption would be almost perfectly compensated by systematic errors in the thermal parameters. We estimate that to compensate for the absorption error, the temperature parameters of each atom in Table 1 should be increased by the following amounts:

β_{11}	β_{22}	β_{33}	β_{12}	β_{13}	β_{23}
0.0007	0.0005	0.0005	0.0000	0.0001	0.0000

DISCUSSION

The space-group symmetry requires the molecule to be planar, and within the accuracy of the determination it is square planar. Figure 1 shows the molecular packing; and Figure 2, the molecular dimensions before correction for thermal motion. If the fluorine atoms are assumed to ride on the xenon atoms, the Xe-F bond distances should be increased by 0.02 A. to the value 1.93 A.

In Table 3 are listed interatomic distances without correction for thermal motion. Each xenon has four fluorine neighbors in other molecules at an average distance of 3.25 A. Each fluorine atom has eight fluorine neighbors in other

TABLE 2

OBSERVED AND CALCULATED STRUCTURE FACTORS (EACH MULTIPLIED BY 10)

H,K= 0, 0

L	FOBS	FCAL
	550	567
	556	545
	417	405

H,K= 0, 1

L	FOBS	FCAL
	641	973*
	245	202
	692	724
	85	-70
	393	401
	0	-7

H,K= 0, 2

L	FOBS	FCAL
	730	1052*
	124	95
	571	616
	74	66
	577	585
	50	-53
	333	341

H,K= 0, 3

L	FOBS	FCAL
	541	586
	99	95
	690	724
	61	-61
	328	354
	0	14

H,K= 0, 4

L	FOBS	FCAL
	610	598
	140	-130
	498	528
	0	21
	450	463
	14	13

H,K= 0, 5

L	FOBS	FCAL
	473	478
	15	-24
	431	440
	0	-10

H,K= 0, 6

L	FOBS	FCAL
	460	456
	66	-73
	339	356
	0	19

H,K= 1, 0

L	FOBS	FCAL
-5	557	568
-3	611	605
-1	798	965*
1	758	837*
3	782	792
5	381	353

H,K= 1, 1

L	FOBS	FCAL
-6	344	363
-5	12	5
-4	771	740
-3	141	105
-2	615	551
-1	140	-98*
0	734	958*
1	145	-115*
2	807	943
3	88	69
4	415	404
5	0	10
6	356	339

H,K= 1, 2

L	FOBS	FCAL
-6	46	-41
-5	447	438
-4	64	56
-3	663	608
-2	175	130
-1	933	1047
0	366	-324
1	499	543
2	110	99
3	642	650
4	28	31
5	391	380
6	18	-20

H,K= 1, 3

L	FOBS	FCAL
-6	287	300
-5	0	13
-4	557	506
-3	17	-11
-2	746	703
-1	19	10
0	690	733
1	21	-25
2	541	548
3	0	-12
4	409	413
5	22	18

H,K= 1, 4

L	FOBS	FCAL
-5	336	333
-4	44	-40
-3	524	481
-2	33	38
-1	758	740
0	20	-38
1	430	427
2	38	46
3	467	468
4	33	-38
5	320	323

H,K= 1, 5

L	FOBS	FCAL
-4	398	377
-3	103	-93
-2	463	444
-1	57	58
0	504	497
1	46	43
2	439	435
3	72	-68
4	292	311

H,K= 1, 6

L	FOBS	FCAL
-3	335	330
-2	14	10
-1	398	383
0	23	-23
1	376	387
2	14	23
3	314	327

H,K= 2, 0

L	FOBS	FCAL
-6	305	335
-4	635	650
-2	912	977
0	417	364
2	810	813
4	472	428
6	245	253

H,K= 2, 1

L	FOBS	FCAL
-6	0	-13
-5	527	515
-4	103	-84
-3	688	646
-2	212	163
-1	874	831
0	13	4
1	609	649
2	133	-107
3	517	506
4	27	29
5	406	373

H,K= 2, 2

L	FOBS	FCAL
-6	344	367
-5	63	-64
-4	534	499
-3	30	26
-2	814	759
-1	129	89
0	599	599
1	48	-42
2	650	677
3	68	-58
4	362	355
5	20	25

H,K= 2, 3

L	FOBS	FCAL
-6	18	10
-5	460	449
-4	76	-71
-3	428	394
-2	78	75
-1	767	742
0	0	-13
1	626	644
2	59	-55
3	350	361
4	21	25
5	318	313

H,K= 2, 4

L	FOBS	FCAL
-5	29	20
-4	410	389
-3	0	-6
-2	554	509
-1	126	-102
0	542	540
1	87	83
2	473	479
3	24	-32
4	288	285

H,K= 2, 5

L	FOBS	FCAL
-4	0	-7
-3	390	373
-2	26	-26
-1	479	456
0	32	-30
1	422	421
2	0	13
3	300	312

H,K= 2, 6

L	FOBS	FCAL
-3	0	6
-2	373	368
-1	71	-69
0	309	321
1	33	42
2	310	323

H,K= 3, 0

L	FOBS	FCAL
-5	377	390
-3	626	621
-1	712	683
1	681	620
3	353	335
5	309	296

H,K= 3, 1

L	FOBS	FCAL
-6	355	384
-5	0	2
-4	404	402
-3	97	-83
-2	819	776
-1	31	28
0	609	582
1	62	54
2	423	428
3	21	-28
4	406	370
5	0	-9

H,K= 3, 2

L	FOBS	FCAL
-6	35	41
-5	430	444
-4	47	-55
-3	588	543
-2	118	-92
-1	525	487
0	155	135
1	604	629
2	40	-46
3	346	341
4	0	-13

H,K= 3, 3

L	FOBS	FCAL
-5	17	-8
-4	451	448
-3	0	-9
-2	528	490
-1	39	-38
0	485	469
1	0	16
2	431	446
3	0	-4
4	309	298

H,K= 3, 4

L	FOBS	FCAL
-4	0	15
-3	458	436
-2	24	-29
-1	382	367
0	0	1
1	467	472
2	30	-40
3	290	296

H,K= 3, 5

L	FOBS	FCAL
-3	51	57
-2	404	396
-1	53	-56
0	353	354
1	42	-32
2	284	309

H,K= 3, 6

L	FOBS	FCAL
-1	316	326
0	0	-4

H,K= 4, 0

L	FOBS	FCAL
-4	374	392
-2	411	417
0	672	610
2	329	314
4	263	261

H,K= 4, 1

L	FOBS	FCAL
-5	333	375
-4	50	48
-3	483	468
-2	81	-75
-1	489	451
0	0	-9
1	431	460
2	28	26
3	326	306
4	0	-11

H,K= 4, 2

L	FOBS	FCAL
-5	29	36
-4	400	422
-3	0	-9
-2	460	441
-1	67	-57
0	476	469
1	0	-1
2	291	305
3	0	18

H,K= 4, 3

L	FOBS	FCAL
-4	26	30
-3	486	488
-2	50	-49
-1	384	364
0	0	-8
1	315	344
2	0	3
3	289	287

H,K= 4, 4

L	FOBS	FCAL
-4	304	345
-3	23	-6
-2	379	385
-1	21	24
0	335	340
1	52	-50
2	237	259

H,K= 4, 5

L	FOBS	FCAL
-2	15	-7
-1	298	308
0	0	4

H,K= 5, 0

L	FOBS	FCAL
-3	345	375
-1	344	347
1	281	294

H,K= 5, 1

L	FOBS	FCAL
-4	352	395
-3	29	31
-2	314	318
-1	13	-10
0	328	353
1	22	-23
2	262	272

H,K= 5, 2

L	FOBS	FCAL
-4	0	15
-3	319	353
-2	31	25
-1	374	370
0	51	-48
1	222	258
2	0	4

H,K= 5, 3

L	FOBS	FCAL
-3	0	-9
-2	314	333
-1	27	5
0	291	315
1	0	-22

* Indicates a reflection given zero weight.

TABLE 3

DISTANCES IN XeF_4

Xe		F_1		F_2	
2 F_1	1.91 ± .02 A.*	Xe	1.91 ± .02 A.*	Xe	1.91 ± .02 A.*
2 F_2	1.91 ± .02 A.*	F_2	2.71 ± .03 A.*	F_1	2.71 ± .03 A.*
2 F_2	3.22 ± .02 A.	F_2	2.69 ± .03 A.*	F_1	2.69 ± .03 A.*
2 F_1	3.27 ± .02 A.	F_2	3.03 ± .03 A.	2 F_2	3.02 ± .01 A.
		F_2	3.08 ± .03 A.	F_1	3.03 ± .03 A.
		F_2	3.09 ± .03 A.	F_1	3.08 ± .03 A.
		2 F_1	3.16 ± .02 A.	F_1	3.09 ± .03 A.
		F_2	3.22 ± .03 A.	F_1	3.22 ± .03 A.
		F_1	3.24 ± .04 A.	F_1	3.26 ± .03 A.
		F_2	3.26 ± .03 A.	F_2	3.32 ± .04 A.
		Xe	3.27 ± .02 A.	Xe	3.22 ± .02 A.

* Values are intramolecular distances.

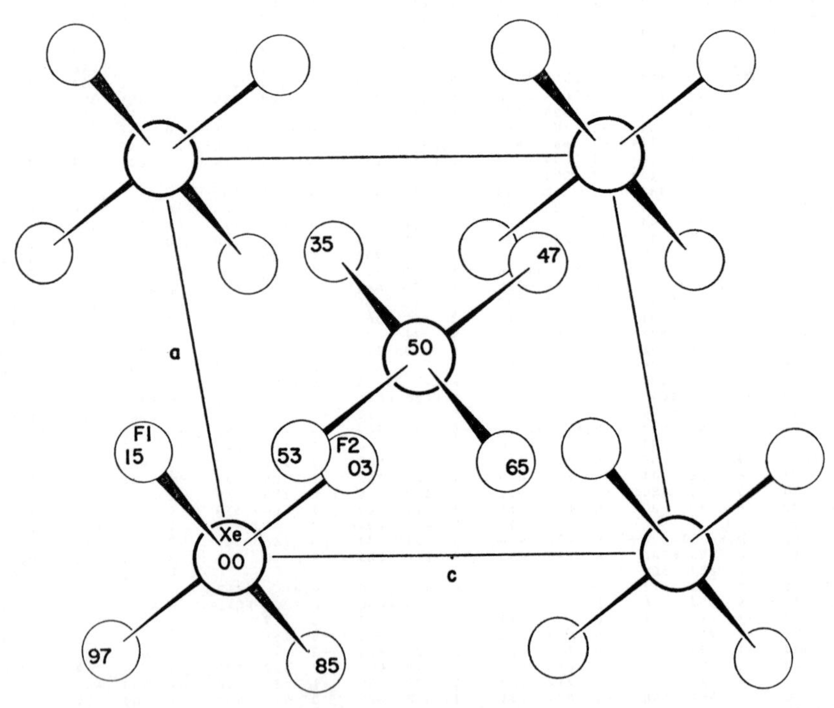

FIG. 1.—Molecular packing in XeF_4 as seen in projection down the *b* axis. The numbers on some of the atoms are *b* coordinates (× 100).

molecules at an average distance of 3.13 A. or 3.15 A., as well as one xenon neighbor in another molecule. The average intermolecular F-F distance infers a van der Waals radius of 1.57 A., which is considerably larger than the accepted value of 1.35 A. [9], perhaps because of the considerable thermal motion of the molecules. Using the smaller value for fluorine, one gets an upper limit of 1.9 A. for the van der Waals radius of xenon in this tetravalent state.

We have three independent sets of results for the structure of this crystal: the neutron-diffraction study of Burns, Agron, and Levy [see p. 211, this volume], the photographic X-ray study of Hamilton and Ibers [see p. 195, this volume],

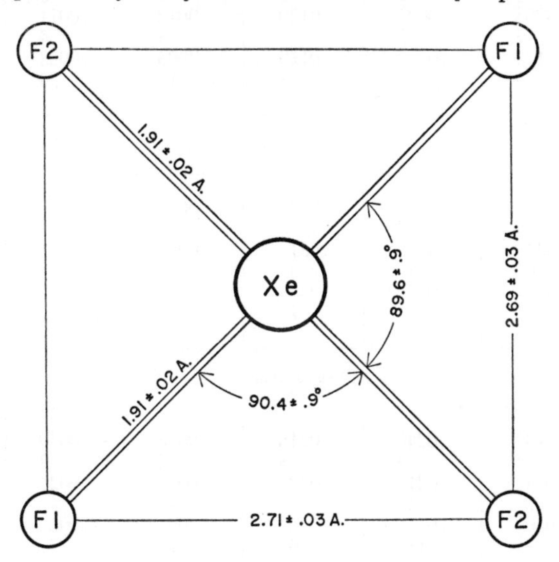

FIG. 2.—Molecular dimension in XeF_4. Distances have not been corrected for thermal vibrations in this figure.

and our own counter X-ray study. There is no significant disagreement with respect to the geometry of the structure: The three sets of coordinates agree in each case within two standard deviations or less. The thermal parameters of the fluorine atoms are in similar agreement. The agreement between the sets of thermal parameters for xenon is as good as for fluorine on an absolute scale, but is poorer than the ostensible precision of the measurements (Table 4). Systematic errors which are a function of θ (for example, absorption) will have equal effect on thermal parameters of heavy and light atoms. We attribute the disagreement to systematic errors which have an effect on the thermal parameters at a level of the order of 0.3 in terms of the equivalent isotropic B value, but we have not identified the precise nature of these errors. We are not surprised that such errors are present; rather, we did not expect them to be so small.

TABLE 4

COMPARISON OF THERMAL PARAMETERS IN DETERMINATIONS OF XeF_4 STRUCTURE

	β_{11}	β_{22}	β_{33}	β_{12}	β_{13}	β_{23}
			XE PARAMETERS			
Neutrons....	.0266	.0129	.0161	.0014	.0050	−.0002
	4	3	3	3	3	2
Counter.....	.0208	.0097	.0120	.0012	.0071	.0000
	7	5	5	4	4	3
Precession...	.0123	.0089	.0110	−.0003	.0053	.0006
	5	3	5	17	3	11
			F_1 PARAMETERS			
Neutrons....	.0437	.0284	.0329	−.0059	.0167	.0026
	4	3	3	3	3	3
Counter.....	.044	.025	.031	−.006	.023	.004
	6	4	4	4	5	4
Precession...	.034	.024	.033	−.015	.017	.004
	8	5	7	5	5	4
			F_2 PARAMETERS			
Neutrons....	.0437	.0248	.0216	−.0008	−.0030	−.0013
	4	3	3	3	2	2
Counter.....	.044	.021	.029	.001	.002	.000
	6	4	5	4	4	4
Precession...	.040	.034	.026	.009	−.006	−.005
	8	11	6	7	5	5

Standard deviations are indicated below the value.

REFERENCES

1. D. H. TEMPLETON, A. ZALKIN, J. D. FORRESTER, and S. M. WILLIAMSON, *J. Am. Chem. Soc.* **85**, 242 (1963).
2. S. SIEGEL and E. GEBERT, *ibid.* 240 (1963).
3. J. A. IBERS and W. C. HAMILTON, *Science* **139**, 106 (1963).
4. J. H. BURNS, P. A. AGRON, and H. A. LEVY, *ibid.* 1208 (1963).
5. J. H. HOLLOWAY and R. D. PEACOCK, *Proc. Chem. Soc.* 389 (1962).
6. S. R. GUNN and S. M. WILLIAMSON, *Science* **140**, 177 (1963).
7. P. GANTZEL, R. SPARKS, and K. TRUEBLOOD, private communication.
8. *International Tables for X-ray Crystallography*, vol. 3. Birmingham, England: Kynoch Press, 1962.
9. L. PAULING, *The Nature of the Chemical Bond*. 3d ed.; Ithaca, N.Y.: Cornell University Press, 1960.

THE CRYSTAL AND MOLECULAR STRUCTURE OF XENON TETRAFLUORIDE BY NEUTRON DIFFRACTION

John H. Burns, Paul A. Agron, and Henri A. Levy
Oak Ridge National Laboratory

Abstract

The crystal structure, molecular geometry, and thermal motion of xenon tetrafluoride were determined by three-dimensional neutron-diffraction analysis. The symmetry of the XeF_4 molecule was found to be D_{4h} to high precision. The average Xe-F bond length, corrected for thermal motion, is 1.95 (σ = 0.01)A. and the F(1)-Xe-F(2) angle is 90.0 (σ = 0.1)°. Thermal motion in the crystal is described as a translation of the rigid molecule plus a libration of approximately 5° about the three molecular axes. Each fluorine atom makes eight intermolecular contacts with other fluorine atoms and one close approach to another xenon atom.

INTRODUCTION

When this experiment was begun, the unit-cell dimensions and space group of crystalline XeF_4 were known [1], and shortly afterward there was reported an approximate crystal structure derived from visually estimated X-ray intensities [2]. This report was consistent with D_{4h} molecular symmetry but attached a high standard deviation to the parameters, indicating the need for a more precise determination. Such a study by use of counter-measured X-ray diffraction was soon available [3], and this work showed the XeF_4 molecule to be square planar. The neutron-diffraction analysis to be discussed here was carried out concurrently with these other studies; it provides a description of the structure that is not appreciably different from the final results of these workers [see Hamilton and Ibers, p. 195, this volume, for comparison] but is more precisely known.

The neutron-diffraction technique is inherently more precise for a structure of this type for two principal reasons. First, the X-ray intensities are considerably less sensitive to the parameters of the fluorine atoms than to those of xenon, which dominate the scattering because of their much larger atom form factor. The neutron-scattering factors, however, are nearly equal for these elements. Second, the X-ray absorption by xenon is high, making the exact shape and orientation of the crystal important; while the neutron absorption is low and easily measured, and a correction can be made.

EXPERIMENTAL PROCEDURES

Xenon tetrafluoride was synthesized from the elements [4] by D. F. Smith in the circulating-gas system previously described [5], and the volatile solid was sublimed into specially dried quartz tubes (~1 mm. diameter) and hermetically sealed. A crystal of suitable size for neutron-diffraction study (~25 mg.) was grown in one tube by effecting a small temperature gradient along it.

It was frequently observed during the X-ray examination of several small specimens that the crystals twin by reflection in $(10\bar{1})$. In the twinning process the [101] axis is unchanged, but the $[10\bar{1}]$ axis is shifted by 15.5°.

Preliminary examination of the large specimen by X-ray diffraction ensured that it was a single crystal and determined approximately its orientation. It was then mounted on the Oak Ridge automatic neutron diffractometer [6], its exact orientation ascertained, and all symmetry-independent reflections out to $\sin \theta/\lambda = 0.76$ recorded. During the period of data collection (about one month) the crystal grew at a nearly linear rate by transfer of material from elsewhere in the tube. This growth was monitored by frequent measurement of a reference reflection, and a normalization factor was applied to the intensities. Over one hundred scans were made across positions corresponding to space-group absences without finding significant intensity, thereby confirming the X-ray space group, $P2_1/n$. The unit-cell dimensions of Templeton *et al.* [3]—$a = 5.050$, $b = 5.922$, $c = 5.771$ A., $\beta = 99.6°$—were used throughout the determination.

Neutron absorption by the crystal was measured for several orientations and amounted to about 5 per cent. A correction was calculated for each reflection by the method of Busing and Levy [7]. The intensities were brought to an approximate absolute scale by comparison with a NaCl standard crystal and were converted to structure factors in the usual way.

STRUCTURE REFINEMENT

By the time that this neutron-diffraction data had been collected, the two X-ray studies were available, so that it was possible to proceed directly to structure refinement. The method of least squares was employed for this process, and all calculations were made on an IBM 7090 with the program of Busing, Martin, and Levy [8]. The structural parameters not fixed by symmetry, which were varied in the least-squares refinements, included positional coordinates for the fluorine atoms, individual anisotropic thermal parameters for all atoms, the neutron-scattering factor of xenon, and one over-all scale factor. The parameters of Templeton *et al.* provided starting values for coordinates and thermal parameters.

The observations used were the values of F^2_{obs}, where F_{obs} is the observed structure factor, and were weighted as the reciprocal of their variances. The latter were obtained from the expression

$$\sigma^2(F^2) = \left(\frac{sL}{A}\right)^2 [N + 2B + (0.03N)^2],$$

where N = net count in a peak, B = background count, s = intensity scale factor, L = Lorentz factor, and A = absorption factor. The justification for use of this expression with neutron-diffraction measurements has been given previously [9].

After a few cycles of least squares, it was apparent that some secondary extinction effects were present in the 24 strongest reflections. These were omitted from subsequent cycles, after which no further effect was detectable. This left a total of 599 observations to which 26 structural parameters were adjusted.

The final parameters and their standard deviations are given in Table 1. The scale factor from the least-squares refinement was applied to the observations to

TABLE 1

FINAL PARAMETERS AND THEIR STANDARD DEVIATIONS

	Xe		F(1)		F(2)	
	Parameter	σ	Parameter	σ	Parameter	σ
f^*.....	0.476	0.003	(0.55)	....	(0.55)	....
x.....	(.0)†	....	.2643	0.0002	.2356	0.0002
y.....	(.0)	....	.1481	.0002	.0297	.0002
z.....	(.0)	....	− .1536	.0002	.3002	.0002
β_{11}‡...	.0266	.0004	.0437	.0004	.0437	.0004
β_{22}....	.0129	.0003	.0284	.0003	.0248	.0003
β_{33}....	.0161	.0003	.0329	.0003	.0216	.0003
β_{12}....	.0014	.0003	− .0059	.0003	− .0008	.0003
β_{13}....	.0050	.0003	.0167	.0003	− .0030	.0002
β_{23}....	−0.0002	0.0002	0.0026	0.0003	−0.0013	0.0002

* f = neutron scattering factor in Fermi units. The value of 0.55 for fluorine is from Bacon [17].
† The numbers in parentheses were not varied in the least-squares refinement.
‡ The β's are the coefficients in the asymmetric temperature factor expression:

$$\exp - [\beta_{11}h^2 + \beta_{22}k^2 + \beta_{33}l^2 + 2\beta_{12}hk + 2\beta_{13}hl + 2\beta_{23}kl] .$$

bring them on an absolute scale. Values of F^2_{calc}, F^2_{obs} on an absolute scale, the standard deviation of F^2_{obs} (each multiplied by 10^3), and the calculated signs of the structure factors are presented in Table 2. Those reflections affected by extinction and omitted from the final refinement are denoted by an "E."

The quality of fit of the model to the observations may be judged by either of the following quantities. The discrepancy factor,

$$R = \Sigma |F^2_{\text{obs}} - F^2_{\text{calc}}| / \Sigma F^2_{\text{obs}} ,$$

reached 0.067, based on all reflections. The standard deviation of an observation of unit weight,

$$\Big[\sum_i w_i (F^2_{\text{obs}} - F^2_{\text{calc}})^2 / (m - n) \Big]^{1/2} ,$$

where w_i is the weight, m is the number of observations, and n is the number of variables, was 1.03 at the end of the refinement. The expectation value of this

TABLE 2

OBSERVED AND CALCULATED VALUES OF THE SQUARED STRUCTURE FACTORS ON AN ABSOLUTE SCALE

K	CALC	OBS	SIG
*** 0 K 0 ***			
E 2	4835	4615	141 +
4	480	488	25 +
6	2554	2613	89 +
8	356	319	29 +
*** 0 K 1 ***			
1	963	968	31 +
2	675	658	25 -
3	264	250	22 -
4	2349	2355	82 +
5	195	245	34 +
6	1375	1370	53 +
7	531	502	31 +
8	25	5	22 +
*** 0 K 2 ***			
E 0	1999	2021	63 -
E 1	2623	2525	79 -
2	145	146	16 -
3	915	866	38 -
4	102	121	46 +
5	148	89	56 +
6	16	6	19 +
7	206	220	25 -
8	96	98	23 +
*** 0 K 3 ***			
1	1685	1641	55 +
2	490	482	24 -
E 3	8358	8001	247 +
4	101	110	28 -
5	1321	1344	62 +
6	126	97	59 -
7	33	80	23 +
8	87	86	22 -
*** 0 K 4 ***			
0	54	81	16 +
1	841	802	34 +
2	1877	1962	65 +
3	793	735	40 +
4	1700	1641	63 +
5	45	0	21 +
6	48	90	24 +
7	188	166	25 +
*** 0 K 5 ***			
1	269	256	24 -
2	710	687	34 +
3	130	135	24 -
4	101	114	25 -
5	0	13	22 -
6	45	42	22 -
*** 0 K 5 ***			
7	77	79	23 +
*** 0 K 6 ***			
0	2090	2167	75 +
1	6	7	18 +
2	426	448	29 +
3	142	154	25 -
4	84	107	22 +
5	348	294	39 -
6	334	364	29 +
*** 0 K 7 ***			
1	1530	1421	56 +
2	2	47	21 -
3	252	292	29 +
4	49	51	21 +
5	251	241	27 +
*** 0 K 8 ***			
0	5	-16	19 -
1	20	5	21 -
2	8	28	18 -
3	6	31	18 +
*** 1 K -8 ***			
1	162	114	23 +
2	16	27	21 +
3	529	400	31 +
*** 1 K -7 ***			
0	407	436	29 +
1	214	228	25 -
2	748	816	40 +
3	113	115	23 -
4	636	708	42 +
5	0	6	19 -
*** 1 K -6 ***			
1	14	44	18 -
2	543	586	30 -
3	111	138	22 -
4	5	15	22 +
5	42	28	29 +
6	3	-4	21 +
*** 1 K -5 ***			
0	2582	2630	87 +
1	4	11	19 +
2	51	30	19 +
3	124	135	23 +
4	13	34	19 -
*** 1 K -5 ***			
5	369	394	46 +
6	456	507	34 +
7	149	127	30 +
*** 1 K -4 ***			
E 1	7407	7130	221 +
2	481	500	29 +
3	823	816	41 +
4	353	346	31 -
5	798	751	52 +
6	102	99	22 -
7	612	634	38 +
*** 1 K -3 ***			
0	96	114	13 -
1	1133	1121	40 +
2	44	18	16 +
3	22	27	17 -
4	116	124	30 +
5	2203	1979	89 -
6	60	39	22 +
7	119	84	23 -
8	162	99	41 +
*** 1 K -2 ***			
E 1	1826	1802	58 -
2	1508	1445	51 +
3	1540	1377	53 +
4	192	121	53 +
5	56	-5	21 +
6	65	86	19 +
7	4	28	22 -
8	153	132	27 +
*** 1 K -1 ***			
0	943	962	31 +
1	471	474	19 -
E 2	8410	6942	213 +
3	15	11	12 +
E 4	7178	6463	203 +
5	728	804	41 +
6	188	149	30 +
7	12	16	22 +
8	86	92	24 +
*** 1 K 0 ***			
1	938	955	31 +
E 2	6204	5751	176 -
3	1144	1120	40 +
4	53	47	18 -
5	629	655	33 +
6	89	82	21 -
7	200	243	25 +
*** 1 K 0 ***			
8	566	636	36 -
*** 1 K 1 ***			
0	156	156	10 +
1	820	844	29 -
2	1549	1604	52 -
3	34	68	15 -
4	1104	1159	45 -
5	637	687	35 +
6	410	460	30 +
7	4	8	22 +
8	209	178	29 +
*** 1 K 2 ***			
E 1	7700	7170	218 +
2	908	883	34 -
3	60	62	19 +
4	372	388	33 -
5	835	793	41 +
6	267	266	31 -
7	919	907	42 +
8	166	158	25 -
*** 1 K 3 ***			
E 0	5787	5737	177 +
1	653	681	28 -
2	2497	2579	85 +
3	36	-17	39 +
4	888	943	45 +
5	1490	1461	61 +
6	652	723	50 +
7	96	103	22 +
8	134	126	27 +
*** 1 K 4 ***			
1	1096	1091	42 -
2	275	312	27 -
3	12	17	19 +
4	391	409	30 +
5	5	17	21 +
6	190	173	24 +
7	7	-16	22 +
*** 1 K 5 ***			
0	89	106	21 -
1	48	29	18 -
2	466	393	28 +
3	253	237	24 -
4	768	736	40 +
5	302	321	38 -
6	31	2	22 +
7	144	133	28 -
*** 1 K 6 ***			
1	1583	1569	58 +
2	183	160	24 +
3	1726	1621	64 +
4	4	34	22 -
5	421	375	41 +
6	2	-23	21 -
*** 1 K 7 ***			
0	14	27	21 +
1	135	142	23 +
2	0	11	21 +
3	136	101	24 +
4	1	8	19 +
*** 1 K 8 ***			
1	5	48	18 +
2	8	11	18 -
*** 2 K -8 ***			
0	576	699	35 +
1	66	58	22 +
2	496	497	31 +
3	0	-10	19 +
*** 2 K -7 ***			
1	39	10	21 -
2	33	36	22 +
3	246	249	29 +
4	70	68	22 -
5	29	52	21 +
*** 2 K -6 ***			
0	410	420	27 -
1	11	-16	19 -
2	136	120	23 +
3	101	98	27 +
4	481	461	35 +
5	186	166	29 +
6	2	-5	21 +
*** 2 K -5 ***			
1	1821	1861	65 +
2	785	758	35 -
3	2058	2093	75 +
4	222	215	30 +
5	844	831	72 +
6	77	56	22 +
7	94	96	22 +
*** 2 K -4 ***			
0	2091	2179	74 +
*** 2 K -4 ***			
1	1153	1128	44 -
2	58	86	27 +
3	850	826	41 -
4	12	7	19 +
5	0	31	19 +
6	586	665	39 +
7	138	160	28 -
*** 2 K -3 ***			
1	233	239	17 +
2	73	89	19 +
3	2026	1998	72 -
4	10	23	19 +
5	6	8	19 +
6	90	41	38 +
7	540	502	42 +
8	26	47	22 +
*** 2 K -2 ***			
E 0	8548	7773	237 +
E 1	2152	2139	69 +
E 2	2565	2482	81 +
3	716	695	35 +
4	429	359	47 +
5	89	81	19 -
6	840	770	68 +
7	221	166	39 +
8	262	245	31 +
*** 2 K -1 ***			
1	1550	1535	50 +
2	950	956	36 +
E 3	3833	3669	118 +
4	1675	1805	68 -
5	633	621	38 +
6	1093	1093	50 -
7	121	109	46 +
8	7	18	22 -
*** 2 K 0 ***			
E 0	7715	6963	212 -
1	26	36	10 +
2	58	58	13 -
3	0	-7	13 +
4	574	582	33 +
5	107	127	22 -
6	91	87	25 -
7	29	63	22 -
8	56	67	24 +
*** 2 K 1 ***			
1	121	126	12 +
2	165	154	15 -
*** 2 K 1 ***			
E 3	3097	3048	98 +
4	1627	1686	61 +
5	622	638	31 +
6	801	801	38 +
7	33	50	25 +
8	28	50	22 +
*** 2 K 2 ***			
E 0	6691	6294	194 +
1	1470	1481	51 -
E 2	3811	3813	121 +
3	466	513	29 -
4	1641	1692	63 +
5	134	142	23 +
6	717	741	39 +
7	88	121	23 -
8	122	122	25 +
*** 2 K 3 ***			
1	53	47	21 +
2	646	700	34 +
3	777	795	39 -
4	167	170	25 +
5	12	13	19 +
6	104	148	23 +
7	260	261	46 +
*** 2 K 4 ***			
0	465	497	28 +
1	295	296	24 -
2	18	2	19 -
3	361	391	28 -
4	31	30	19 -
5	85	89	30 -
6	251	260	25 +
7	131	98	22 -
*** 2 K 5 ***			
1	2607	2565	86 +
2	345	360	27 -
3	1441	1415	56 +
4	6	-5	22 +
5	559	579	48 +
6	2	-36	22 +
*** 2 K 6 ***			
0	0	7	19 +
1	0	17	19 -
2	128	132	25 +
3	113	107	22 +
4	171	192	27 +
5	268	245	30 +
*** 2 K 7 ***			
1	155	165	22 -
2	0	6	19 -
3	2	-36	21 +
4	17	16	19 -
*** 3 K -8 ***			
1	212	241	24 +
2	28	56	21 -
*** 3 K -7 ***			
0	215	218	24 +
1	168	159	25 +
2	2	-38	21 +
3	30	-41	21 +
4	11	18	19 -
*** 3 K -6 ***			
1	843	858	39 +
2	634	724	35 +
3	937	966	45 +
4	18	-8	22 -
5	155	194	30 +
6	15	4	19 -
*** 3 K -5 ***			
0	39	41	18 -
1	17	4	18 +
2	1081	1083	45 +
3	10	4	21 -
4	1221	1201	52 +
5	303	388	59 -
6	8	38	21 +
*** 3 K -4 ***			
1	918	905	39 -
2	336	297	27 -
3	480	497	33 +
4	155	144	28 +
5	87	124	23 +
6	5	-13	22 +
7	1	2	22 -
*** 3 K -3 ***			
0	1157	1160	45 +
1	950	924	39 -
2	791	808	39 +
3	2	-18	21 +
4	802	774	44 +
5	1438	1446	62 +
6	372	340	36 +
7	37	25	22 +

The Miller Index K is the running index. CALC, OBS, and SIG refer to square of the calculated, square of the observed, and the standard deviation of the square of the observed structure factors, respectively (each multiplied by 10^3). The calculated sign of F is given. Those reflections preceded by E were affected by extinction. Negative values of F^2, which resulted from background counts larger than gross counts, are all insignificantly different from zero.

TABLE 2—*Continued*

K	CALC	OBS	SIG
*** 3 K -2 ***			
E 1	5954	5588	175 +
2	1245	1212	47 -
3	150	156	30 +
4	36	8	21 -
5	882	887	45 +
6	0	33	22 -
7	624	567	42 +
8	95	113	25 -
*** 3 K -1 ***			
0	1148	1133	41 +
1	84	73	16 +
2	394	354	24 -
3	123	48	34 -
4	650	767	38 -
5	412	412	34 -
6	434	437	38 +
7	14	63	22 -
8	195	220	27 +
*** 3 K 0 ***			
1	395	400	22 +
E 2	2812	2778	91 +
3	25	78	19 +
4	9	1	21 +
5	87	127	24 +
6	34	21	22 +
7	252	292	35 +
8	249	240	25 +
*** 3 K 1 ***			
0	1857	1867	63 +
1	573	547	27 +
E 2	4797	4687	148 +
3	90	114	22 +
4	2945	2975	98 +
5	274	283	25 -
6	194	180	28 +
7	0	-1	22 -
8	102	85	22 +
*** 3 K 2 ***			
1	195	186	21 -
2	272	292	24 -
3	1019	1087	44 +
4	293	332	24 -
5	110	115	22 +
6	285	258	30 -
7	1	11	21 +
*** 3 K 3 ***			
0	671	730	35 -
1	183	184	23 -
2	40	45	18 -
3	18	11	18 +
4	56	78	21 +
5	443	444	33 +
6	15	11	22 +
7	24	34	21 +
*** 3 K 4 ***			
1	2423	2375	81 +
2	54	92	21 +
3	846	880	41 +
4	227	250	28 -
5	501	554	40 +
6	130	147	27 -
*** 3 K 5 ***			
0	1146	1195	48 +
1	44	65	21 +
2	260	284	27 +
3	170	159	28 +
4	57	40	22 +
5	137	133	24 +
*** 3 K 6 ***			
1	79	99	21 -
2	19	1	21 -
3	107	108	25 -
4	7	23	19 +
*** 3 K 7 ***			
0	208	196	24 +
1	46	50	19 -
2	206	182	23 +
*** 4 K -8 ***			
0	0	-4	18 +
1	84	69	21 -
*** 4 K -7 ***			
1	529	523	30 +
2	87	79	22 -
3	20	-2	19 +
4	34	33	21 +
*** 4 K -6 ***			
0	2121	2208	78 +
1	2	21	21 +
2	229	243	28 +
3	70	72	22 -
4	7	31	21 +
5	54	48	21 -
*** 4 K -5 ***			
1	85	67	21 +
2	396	406	28 +
3	29	7	22 -
4	228	233	31 -
5	1	-13	22 +
6	66	59	22 -
*** 4 K -4 ***			
0	0	21	18 +
1	735	706	34 +
2	563	523	30 +
3	385	423	30 +
4	399	439	34 +
5	48	33	23 -
6	9	-16	22 +
*** 4 K -3 ***			
1	464	448	28 +
2	5	19	18 +
E 3	3958	3824	125 +
4	0	39	21 -
5	515	463	51 +
6	73	67	26 -
7	5	18	21 +
*** 4 K -2 ***			
0	509	503	29 -
1	838	810	38 -
2	78	47	21 +
3	263	235	27 -
4	611	607	35 +
5	13	5	22 +
6	20	40	23 +
7	158	183	28 -
*** 4 K -1 ***			
1	16	74	18 +
2	586	577	33 -
3	19	-7	18 -
4	528	505	33 +
5	162	188	29 +
6	338	313	48 +
7	121	154	24 +
*** 4 K 0 ***			
E 0	7116	6834	213 +
1	41	53	21 -
2	1351	1419	53 +
3	2	-13	18 -
4	239	212	25 +
5	116	118	27 +
6	809	742	46 +
7	22	7	22 +
*** 4 K 1 ***			
1	1185	1153	46 +
2	4	-17	19 +
3	0	-25	18 +
4	598	636	34 -
5	120	137	25 +
6	251	256	30 -
7	251	250	24 +
*** 4 K 2 ***			
0	226	196	22 -
1	367	329	24 +
2	88	91	19 -
3	76	96	21 +
4	9	19	19 -
5	81	48	22 -
6	16	36	22 +
7	9	0	19 +
*** 4 K 3 ***			
1	537	557	30 +
2	296	308	25 +
3	1291	1270	51 +
4	84	113	23 +
5	283	274	29 +
6	37	25	13 +
*** 4 K 4 ***			
0	344	361	27 +
1	50	47	21 -
2	837	882	41 +
3	85	103	24 -
4	594	576	38 +
5	50	45	22 -
*** 4 K 5 ***			
1	122	118	24 -
2	98	52	23 +
3	14	7	21 -
4	4	-19	19 +
*** 4 K 6 ***			
0	122	95	21 +
1	0	23	18 -
2	49	53	19 +
3	43	53	19 -
*** 5 K -7 ***			
0	16	-7	19 +
1	72	98	19 -
2	258	249	25 +
*** 5 K -6 ***			
1	21	31	21 +
2	370	423	30 -
3	17	-5	21 +
4	13	10	19 +
*** 5 K -5 ***			
0	788	802	38 +
1	51	65	22 -
2	2	28	21 +
3	10	12	21 -
4	2	7	21 -
5	106	114	22 +
*** 5 K -4 ***			
1	1883	1916	69 +
2	69	34	22 +
3	26	24	21 +
4	63	36	22 -
5	154	120	29 +
6	0	30	19 +
*** 5 K -3 ***			
0	337	335	25 +
1	352	365	25 +
2	206	222	23 +
3	2	-5	21 -
4	26	2	22 +
5	534	511	42 -
6	65	73	22 +
*** 5 K -2 ***			
1	134	160	21 -
2	446	457	28 +
3	326	321	28 +
4	1	-39	22 -
5	10	12	22 +
6	25	1	21 -
*** 5 K -1 ***			
0	32	44	18 +
1	2	22	18 -
2	1228	1214	48 +
3	75	76	23 +
4	1290	1367	56 +
5	70	90	23 +
6	33	73	21 +
*** 5 K 0 ***			
1	331	309	24 +
2	624	645	34 -
3	732	696	35 +
4	2	11	22 -
5	309	328	35 +
6	18	11	21 -
*** 5 K 1 ***			
0	0	6	18 +
1	187	182	23 -
2	97	96	22 -
3	63	91	21 -
4	13	33	21 -
5	34	70	22 +
6	79	115	21 +
*** 5 K 2 ***			
1	640	679	34 +
2	32	30	21 +
3	16	-7	22 +
4	77	70	23 +
5	125	119	29 +
6	84	98	22 +
*** 5 K 3 ***			
0	1356	1361	56 +
1	23	4	22 +
2	643	665	36 +
3	6	47	21 -
4	201	207	25 +
5	72	89	21 -
*** 5 K 4 ***			
1	17	7	21 -
2	1	-5	21 +
3	1	15	19 +
4	81	76	21 -
*** 5 K 5 ***			
0	9	8	19 -
1	12	21	19 +
2	10	0	19 +
*** 6 K -6 ***			
0	88	89	19 -
1	0	-7	19 +
2	87	89	22 +
*** 6 K -5 ***			
1	80	25	23 +
2	112	72	22 -
3	409	422	29 +
*** 6 K -4 ***			
0	371	388	29 +
1	243	256	25 -
2	67	19	22 +
3	107	115	27 -
4	87	55	21 +
*** 6 K -3 ***			
1	148	143	25 +
2	35	-2	22 -
3	120	82	22 -
4	3	-31	21 -
5	35	47	19 +
*** 6 K -2 ***			
0	1008	1020	45 +
1	138	135	28 +
2	182	213	27 +
3	24	27	22 +
4	8	19	21 +
5	5	50	19 -
*** 6 K -1 ***			
1	492	483	31 +
2	133	89	23 +
3	383	372	31 +
4	112	65	27 -
5	77	89	23 +
*** 6 K 0 ***			
0	172	195	27 -
1	13	-2	21 +
2	21	5	22 +
3	0	-15	22 -
4	87	59	22 +
5	62	24	21 -
*** 6 K 1 ***			
1	1	-1	21 +
2	0	22	21 +
3	225	241	30 +
4	81	68	24 +
5	67	114	21 +
*** 6 K 2 ***			
0	496	482	34 +
1	45	35	21 -
2	426	416	34 +
3	7	4	21 -
4	268	246	27 +
*** 6 K 3 ***			
1	30	21	19 +
2	66	61	21 -
3	0	-17	19 +
*** 6 K 4 ***			
0	4	-17	18 +
1	3	-6	18 +
*** 7 K -4 ***			
1	11	8	18 -
2	7	15	18 -
*** 7 K -3 ***			
0	17	17	19 +
1	68	50	21 -
2	57	61	19 +
3	0	-12	19 -
*** 7 K -2 ***			
1	324	303	29 +
2	91	99	21 -
3	47	25	19 +
*** 7 K -1 ***			
0	200	230	25 +
1	1	-6	19 -
2	3	-53	19 +
3	27	38	19 -
*** 7 K 0 ***			
1	45	63	21 +
2	53	61	21 +
3	1	17	18 +
*** 7 K 1 ***			
0	179	182	27 +
1	25	1	19 +
2	230	203	28 +
*** 7 K 2 ***			
1	17	-24	19 +

quantity is unity, provided the errors are normally distributed and the weights are properly assigned. Both of these criteria indicate a satisfactory agreement between the model and the data.

The parameters derived from X-ray diffraction analyses [pp. 196 and 211, this volume] are generally in agreement within their limits of error with the neutron diffraction results. However, the temperature factors for xenon are significantly lower in the X-ray case, indicating a systematic error in one or both of the methods, the source of which is presently unknown.

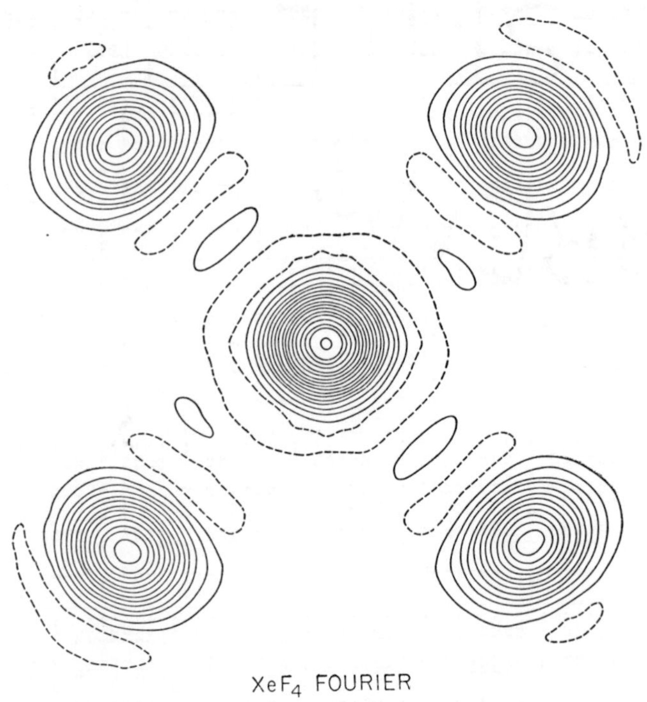

FIG. 1.—Fourier map of XeF_4 in the molecular plane. The contour interval is 0.2 Fermi units/A.^{3}. The zero contour is omitted, and negative contours are dotted. The lowest positive contour is 0.1.

When the final parameters were obtained, the calculated phases of the structure factors were used with the complete set of F_{obs} to produce a Fourier map (Fig. 1). A program for the IBM 7090 written by G. M. Brown was applied to calculate a section of the three-dimensional map lying in the plane of the XeF_4 molecule. It is seen that the atoms are clearly defined, although there are small negative diffraction rings due to the finite termination of the Fourier series.

STRUCTURE OF THE XeF_4 MOLECULE

All of the pertinent distances, angles, and thermal displacements in the crystal with their standard deviations were calculated from the final least-squares parameters, their associated variance-covariance matrix, and the unit-cell dimensions and their errors [3]. For this computation, a program by Busing and Levy [10] was used. The results are presented in Tables 3 and 4 and in Figures 2 and 3.

The distances between observed atomic positions, Xe-F(1) = 1.932 (σ = 0.002) A. and Xe-F(2) = 1.939 (σ = 0.002) A., are shorter than the true bond lengths because of thermal motion [11]; a correction was made [12] on the assumption that the fluorine atom "rides" on the heavier xenon, yielding bond lengths of 1.95 (σ = 0.01) A. for each. (A larger uncertainty is assigned to the bond lengths because of their dependence on the validity of the riding model.) The XeF_4 molecule is planar by crystal symmetry, the F(1)-Xe-F(2) angle was found to be 90.0 (σ = 0.1)°, and the bond lengths are equal; so the molecular symmetry is clearly D_{4h}.

TABLE 3

ROOT-MEAN-SQUARE COMPONENTS OF THERMAL DISPLACEMENT ALONG MOLECULAR AXES (A.)

Axis	Xe	F(1)	F(2)
Xe-F(1)	0.170	0.178	0.242
Xe-F(2)	.174	.256	.178
Plane normal	0.151	0.240	0.219

σ = 0.002 for all values in table.

TABLE 4

INTERATOMIC DISTANCES

DISTANCE, A.	$10^3\sigma$	NUMBER OF ATOMS ABOUT		
		Xe	F(1)	F(2)
1.932*†	2	2	1	
1.939*†	2	2		1
2.736	3		1	1
2.738*	3		1	1
2.986	3		1	1
3.025	2			2
3.044	2		1	1
3.096	2		1	1
3.158	2		2	
3.209	2		1	1
3.218	2	2		1
3.236	4		1	
3.244	4			1
3.248	2	2	1	
3.257	2		1	1

* The distances are intramolecular.

† Values are not corrected for thermal motion effects.

The components of thermal displacement along each of the principal axes of thermal motion were calculated for each atom, as well as the orientation of these axes relative to a Cartesian system passing through the three independent atoms of the molecule. Figure 2 provides a pictorial representation of this thermal motion. The ellipses are drawn through the end points of lines which are proportional to the principal-axis displacements of the atoms projected onto the plane

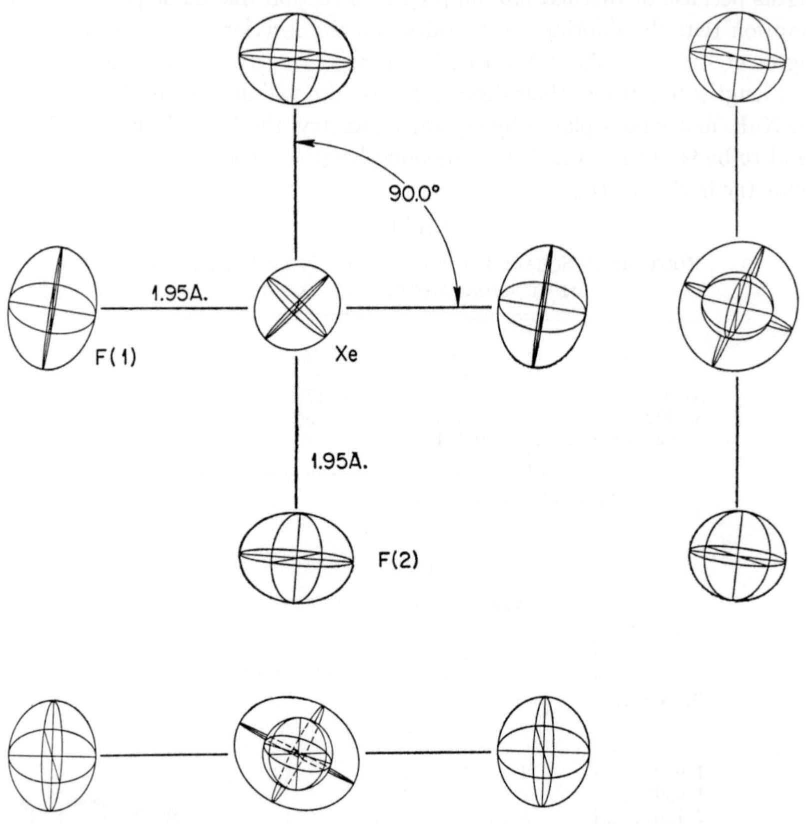

FIG. 2.—Molecular geometry and thermal motion in XeF_4

of the drawing. It is apparent that the principal axes of motion of the fluorine atoms lie almost along the two bond directions and along a normal to the plane of the molecule and that the motions perpendicular to the bond in each case are larger than along it. This anisotropy of thermal motion is also seen in the Fourier map (Fig. 1). Table 3 contains the root-mean-square components of thermal displacement of each of the atoms along the Cartesian axes.

The large motion of the fluorine atoms normal to the bonds suggests that the molecule is librating as a rigid body. If the translational motion of the molecule is taken to be that of the xenon atom and is subtracted from the fluorine atom displacements, the remainder can be ascribed to libratory motions with root-mean-square amplitudes of approximately 5° about each of the molecular axes.

MOLECULAR PACKING

The crystal structure of XeF_4 is pictured in Figure 3. The xenon atoms form a body-centered array, while the XeF_4 molecule at the origin makes a dihedral angle of 55.2° with the one at the body center. All of the unique interatomic distances are shown in Figure 3 and listed in Table 4 with their standard deviations. It is seen that each fluorine makes eight contacts with fluorines of other

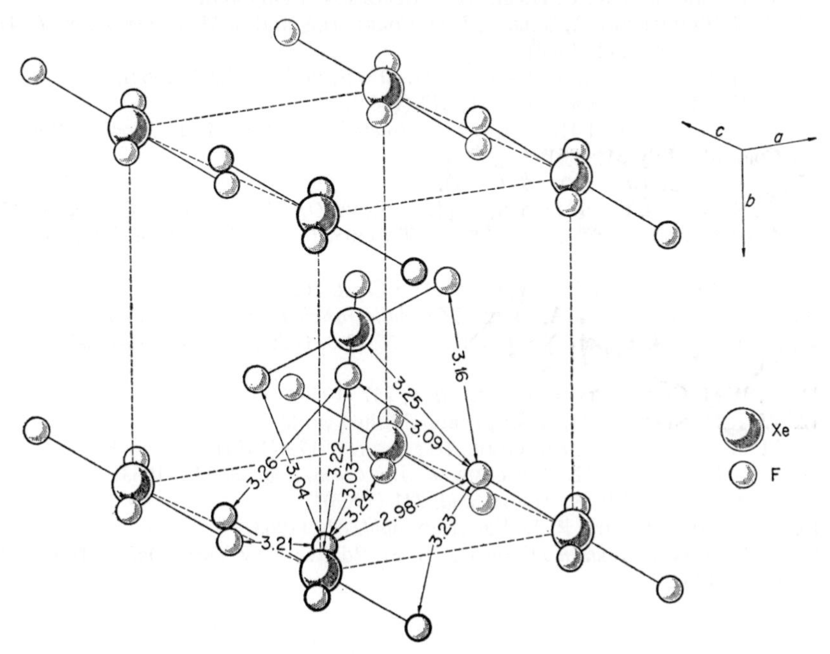

Fig. 3.—Crystal structure and interatomic distances

molecules at distances of 2.99 to 3.26 A., as well as two intramolecular contacts at 2.74 A. Atoms F(1) and F(2) make close xenon contacts at 3.25 and 3.22 A., respectively. The non-bonded fluorine-fluorine distances are normal for molecular crystals [13], and the total of twelve neighboring atoms about each fluorine represents efficient packing. If a van der Waals radius of 1.5 A. is taken for fluorine, the corresponding radius for xenon would be about 1.7 A., considerably shorter than half the interatomic distance in solid xenon, which is about 2.2 A.

CONCLUSION

Two features of the XeF_4 molecular geometry which are important to any explanation of the chemical binding were determined by this neutron-diffraction study. First, the square-planar symmetry, predicted for the free molecule by the molecular-orbital calculations of Lohr and Lipscomb [14], was found to exist to a surprisingly high precision in solid XeF_4. Also of interest is the smaller Xe-F bond length of 1.95 A. in XeF_4 compared to 2.00 A. in XeF_2 [15]. A similar variation in bond distance in the sequence ClO_2^-, ClO_3^-, ClO_4^- has been cited recently by Pimentel and Spratley [16] and attributed to the effect of increasing electric charge on the central atom. The same explanation may well apply here.

REFERENCES

1. S. Siegel and E. Gebert, *J. Am. Chem. Soc.* **85**, 240 (1963).
2. J. A. Ibers and W. C. Hamilton, *Science* **139**, 106 (1963).
3. D. H. Templeton, A. Zalkin, J. D. Forrester, and S. M. Williamson, *J. Am. Chem. Soc.* **85**, 242 (1963).
4. H. H. Claassen, H. Selig, and J. G. Malm, *ibid.* **84**, 3593 (1962).
5. D. F. Smith, *J. Chem. Phys.* **38**, 270 (1963).
6. W. R. Busing and H. A. Levy, *Abstracts, Am. Cryst. Assn. Meeting*, Boulder, Colorado, July 31, 1961.
7. ———, *Acta Cryst.* **10**, 180 (1957).
8. W. R. Busing, K. O. Martin, and H. A. Levy, *ORFLS, A Fortran Crystallographic Least-Squares Program*, Report No. TM-305. Oak Ridge National Laboratory, 1962.
9. S. W. Peterson and H. A. Levy, *Acta Cryst.* **10**, 70 (1957).
10. W. R. Busing and H. A. Levy, *A Crystallographic Function and Error Program for the IBM 704*, Report No. ORNL CF-59-12-3. Oak Ridge National Laboratory, 1959.
11. D. W. J. Cruickshank, *Acta Cryst.* **9**, 757 (1956).
12. W. R. Busing and H. A. Levy, unpublished work.
13. M. Atoji and W. N. Lipscomb, *Acta Cryst.* **7**, 173 (1954).
14. L. L. Lohr and W. N. Lipscomb, *J. Am. Chem. Soc.* **85**, 240 (1963).
15. H. A. Levy and P. A. Agron, *ibid.* 241 (1963).
16. G. C. Pimentel and R. D. Spratley, *ibid.* 826 (1963).
17. G. E. Bacon, *Neutron Diffraction*, p. 31. 2d ed.; New York: Oxford University Press, 1962.

THE CRYSTAL AND MOLECULAR STRUCTURE OF XENON DIFLUORIDE BY NEUTRON DIFFRACTION

HENRI A. LEVY AND PAUL A. AGRON
Oak Ridge National Laboratory

ABSTRACT

The crystal and molecular structure of xenon difluoride and its thermal motion were determined by a three-dimensional neutron-diffraction study. The crystal is a body-centered tetragon, space group $I4/mmm$, with two linear molecules per unit cell aligned along the tetrad axes. The thermal parameters suggest that the molecules undergo precession with a half angle of 7°. The Xe-F bond distance, after correction for this thermal motion, is $2.00 \pm .01$ A.

As accurate molecular parameters for the new compounds of xenon [1] are of interest in clarifying the nature of their chemical binding, a neutron-diffraction study of XeF_2 at room temperature was undertaken. A preliminary description of the work was reported earlier [2], as well as a discussion in relation to other measurements [3].

STRUCTURE DETERMINATION AND REFINEMENT

A sample of xenon difluoride furnished by D. F. Smith, who recently reported [4] preparation and characterization of the compound, yielded several crystals when distilled into thin-walled vitreous silica capillaries. One having the shape of an irregular hexagonal platelet ($\sim 1.5 \times 1.0 \times 0.5$ mm.), well separated from other crystals, was selected for neutron-diffraction measurements. These were made on the Oak Ridge Automatic Neutron Diffractometer [5] with a neutron wave length of 1.078 A.

The tetragonal body-centered lattice indicated by preliminary X-ray precession photographs was confirmed by neutron measurements on 67 non-extinguished equivalent pairs of reflections (hkl, khl) and at positions of 53 reflections required to be absent by body centering. Intensities for constant l index decreased monotonically with increasing scattering angle, and those for $l = 0, 3, 4,$ 7, and 10 were outstandingly strong; thus placement of xenon at the origin and fluorine on tetrad axes at $\pm$ (00z), $z \approx \frac{2}{7}$, of space group $I4/mmm$ was indicated. Absorption effects were indicated to be less than 1 per cent by measurement of the transmission of neutrons through the 1-mm. dimension of the crystal. The specimen was found to undergo a steady growth, ascribed to sublimation from other crystals in the tube, during the ten days of data collection. This necessitated a correction to the measured intensities to establish a consistent scale, based on periodic repetitions of the (020) reflection. The factors ranged from 1.38 to 0.807 and were judged accurate to better than 4 per cent. Corresponding crystal weights, deduced from the final scale factor and the beam intensity, were 1.53 to 2.62 mg. The intensities were brought to a preliminary absolute scale by comparison with a standard crystal of NaCl and were converted to structure factors in the usual way.

Measurements of intensity numbered 334, exclusive of space-group absences, of which 91 were symmetrically non-equivalent. The structure was refined by iterative least squares, using a full matrix program for the IBM 7090 [6]. The observations were taken as F^2 and were weighted as σ^{-2} where σ^2, the variance of the squared structure factor, was assigned the value

$$\sigma^2(F^2) = K^2[T + k^2B + .0016(T - kB)^2] .$$

Here T is the total integrated neutron count, B, the background count; k, the ratio of background-to-peak counting times; and K, the factor to convert counts to F^2. The linear terms in T and B represent the statistical error of a Poisson distribution, while the quadratic term represents an assumed independent relative error of 4 per cent; this value was suggested by the concordance of repeated measurements of (020). At the final stage of refinement, values of equivalent observations were averaged and their variances were adjusted under the assumption that the individual values were independent. Initially the neutron-scattering amplitude of xenon was determined from the data [2], but as a more precise value became available from the study [7; see also Burns, Agron, and Levy, this volume, p. 213] of XeF_4, it was used in the final cycle.

The standard deviation of an observation of unit weight reached the value of 1.62, and the reliability index,

$$\Sigma(F^2_{obs} - F^2_{calc})/\Sigma F^2_{obs} ,$$

reached 0.090; both of these as well as a detailed comparison of F^2_{obs} versus F^2_{calc} (Table 1) indicate a satisfactory refinement. The resulting parameter values, with their least-squares standard errors, are listed in Table 2.

RESULTS

Figure 1 shows the structure of the xenon difluoride crystal. The linear triatomic molecules are aligned on the tetrad axes in a body-centered array. The intramolecular xenon and fluorine positions are separated by 1.984 A. (least-squares standard error, 0.004 A.).

The temperature factor coefficients yield the root-mean-square thermal displacements listed in Table 3. It is evident that the fluorine atoms undergo appreciably larger displacements than do the xenon atoms, particularly in the direction perpendicular to the molecular axis. These displacements suggest that the principal mode of motion is a precession of the molecule as a rigid body with a

TABLE 1

CALCULATED AND OBSERVED VALUES OF THE SQUARED STRUCTURE FACTORS ON AN ABSOLUTE SCALE

h	*k*	*l*	$F^2 \times 10^3$		S.D.	Sign*
			Calc.	Obs.		
0	0	2	1028	1197	59	−
		4	4397	4373	149	+
		6	55	230	68	+
		8	160	96	56	+
		10	801	692	64	+
1	0	1	232	190	34	+
		3	3950	3968	96	+
		5	491	426	47	−
		7	3812	3812	105	+
		9	242	221	45	−
1	1	0	7984	7258	220	+
		2	712	714	54	−
		4	3577	3533	130	+
		6	60	130	68	+
		8	146	58	59	+
		10	657	496	56	+
2	0	0	6431	6221	187	+
		2	486	521	40	−
		4	2915	3003	85	+
		6	62	31	50	+
		8	132	198	46	+
		10	539	668	52	+
2	1	1	213	170	38	+
		3	2638	2739	78	+
		5	222	257	54	−
		7	2476	2443	69	+
		9	113	102	42	−
2	2	0	4192	4173	148	+
		2	215	181	61	−
		4	1945	2003	105	+
		6	64	100	62	+
		8	109	161	57	+
3	0	1	190	195	43	+
		3	1774	1698	70	+
		5	92	250	52	−
		7	1619	1681	62	+
		9	49	36	42	−
3	1	0	3393	3467	95	+
		2	138	148	49	−
		4	1593	1712	69	+
		6	63	119	38	+
		8	98	37	38	+
3	2	1	164	399	56	+
		3	1201	1245	63	+
		5	32	53	45	−
		7	1065	1115	45	+
3	3	0	1479	1567	104	+
		2	14	0	74	−
		4	729	642	62	+
		6	54	90	57	+
		8	63	185	60	+
4	0	0	1816	1908	76	+
		2	28	0	52	−
		4	884	827	55	+
		6	57	139	51	+
		8	71	0	46	+
4	1	1	139	117	53	+
		3	819	858	55	+
		5	8	16	40	−
		7	705	757	45	+
4	2	0	1207	1174	58	+
		2	6	75	46	−
		4	602	570	45	+
		6	50	0	40	+
4	3	1	95	65	41	+
		3	389	424	46	+
		5	0	6	40	+
4	4	0	369	344	51	+
		2	3	56	57	+
		4	198	90	56	+
5	0	1	95	91	41	+
		3	389	413	46	+
		5	0	69	41	+
5	1	0	662	652	49	+
		2	0	86	42	+
		4	343	361	39	+
		6	40	4	38	+
5	2	1	77	151	43	+
		3	270	212	38	+
		5	2	0	38	+
5	3	0	305	257	52	+
		2	4	0	41	+
		4	166	8	51	+
5	4	1	40	162	50	+
6	0	0	252	157	42	+
		2	5	117	37	+
		4	139	316	50	+
6	1	1	50	183	36	+
		3	133	0	66	+
6	2	0	174	60	38	+
		2	7	145	49	+

* Sign of the structure amplitude.

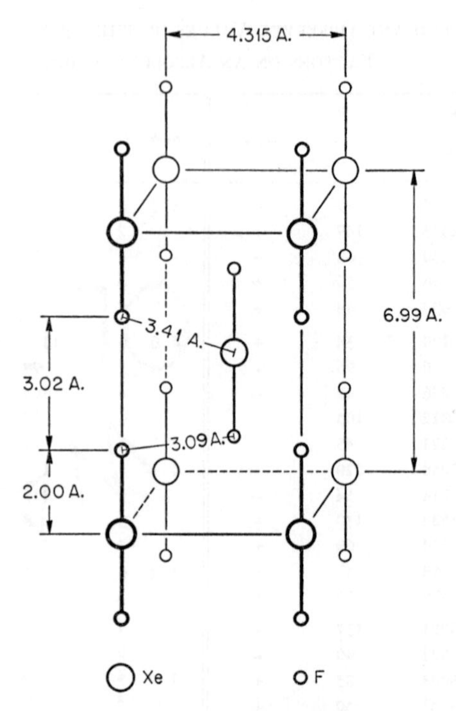

FIG. 1.—Crystal structure and interatomic distances of XeF_2

TABLE 2

PARAMETERS OF THE STRUCTURE OF XeF_2

	Xe	F
z	(0.0)*	0.2838±0.0004
β_{11}	0.0341±0.0020	.0635± .0022
β_{33}	0.0083±0.0006	0.0087±0.0004
f, fermi units	(0.476)*	(0.550)*

* Values in parentheses were not varied. The quoted errors are least-squares standard errors. Values of f, the neutron-scattering amplitudes, are taken from references 7 and 8, respectively.

TABLE 3

ROOT-MEAN-SQUARE COMPONENTS OF THERMAL DISPLACEMENTS, A.

	Perpendicular to c	Parallel to c
Xenon	0.179	0.143
Fluorine	0.245	0.147

half angle estimated as 7°, or alternatively as librations about axes perpendicular to the molecular axis of root-mean-square amplitudes about 5°. Similar amplitudes are indicated [7; see also Burns, Agron, and Levy, this volume, p. 219] in XeF_4. Such motion would cause the Xe-F distance to appear foreshortened [9]. A correction for this effect of thermal motion, computed [10] on the reasonable assumption that a fluorine atom "rides" on xenon, yielded the value 2.00 ± .01 A. for the mean separation of xenon and fluorine. This is significantly larger than the corresponding distance [7; see also this volume, p. 217] 1.95 ± .01A. in XeF_4.

Interatomic distances of interest are listed in Table 4 and in Figure 1. Each fluorine atom has one fluorine neighbor at 3.02 A. and four at 3.09 A. Xenon has eight non-bonded fluorine neighbors at 3.41 A., somewhat larger than the corresponding contacts [7; see also this volume, p. 217] in XeF_4.

TABLE 4

INTERATOMIC DISTANCES

DISTANCE (A.)	Number of Atoms About	
	Xe	F
1.98*...........	2 F	1 Xe
3.02............	....	1 F
3.09............	....	4 F
3.41............	8 F	1 Xe

* Intramolecular distance not corrected for thermal motion.

The cell dimensions used in the calculation of distances, $a = 4.315$ A., $c = 6.990$ A., are those of Siegel and Gebert [11] of Argonne National Laboratory, who kindly transmitted them to us.

REFERENCES

1. N. BARTLETT, *Proc. Chem. Soc.* 218 (1962); H. H. CLAASSEN, H. SELIG, and J. G. MALM, *J. Am. Chem. Soc.* **84,** 3593 (1962); C. L. CHERNICK *et al.*, *Science* **138,** 136 (1962).
2. H. A. LEVY and P. A. AGRON, *J. Am. Chem. Soc.* **85,** 214 (1963).
3. P. A. AGRON *et al.*, *Science* **139,** 842 (1963).
4. D. F. SMITH, *J. Chem. Phys.* **38,** 270 (1963).
5. W. R. BUSING and H. A. LEVY, Am. Cryst. Assn. Meeting, Boulder, Colorado, July 31, 1961.
6. W. R. BUSING, K. O. MARTIN, and H. A. LEVY, *ORFLS, A Fortran Crystallographic Least Squares Program,* Report No. TM-305. Oak Ridge National Laboratory.
7. J. H. BURNS, P. A. AGRON, and H. A. LEVY, *Science* **139,** 1208 (1963).
8. G. E. BACON, *Neutron Diffraction,* p. 31. New York: Clarendon Press, 1962.
9. D. W. J. CRUICKSHANK, *Acta Cryst.* **9,** 757 (1956).
10. W. R. BUSING and H. A. LEVY, submitted to *Acta Cryst.*
11. S. SIEGEL and E. GEBERT, *J. Am. Chem. Soc.* **85,** 240 (1963).

THE CRYSTAL STRUCTURE OF $XeF_2 \cdot XeF_4$ BY X-RAY DIFFRACTION

JOHN H. BURNS, R. D. ELLISON, AND HENRI A. LEVY
Oak Ridge National Laboratory

A previous report [1] on the crystalline phase whose structure is described below gave the unit cell, $a = 6.64$ A., $b = 7.33$ A., $c = 6.40$ A., $\beta = 92°40'$, the probable space group, $P2_1/c$, and the arrangement of xenon atoms as a face-centered array. The calculated density of the phase, based on the assumption that the unit cell contains four XeF_4 molecules was 10 per cent higher than that of the readily grown crystals XeF_4 found earlier [2]; it has consequently been called "the high-density form of XeF_4." By determination of the crystal structure, we have shown that this phase is, in fact, a new compound $XeF_2 \cdot XeF_4$ (calculated density = 4.02 gm/cc).

Crystals were grown initially by condensation of vapor containing predominantly XeF_4, but subsequent controlled mixing of XeF_2 and XeF_4 vapors yielded larger quantities of the compound. The crystal data given above were derived from precession-camera photographs; the intensity data for structure determination were obtained from a single crystal with Mo $K\alpha$ X-rays, a goniostat, and a scintillation-counter detector. A virtually complete set of independent reflections was recorded out to the limit of measurable intensity. These data yielded some 574 structure amplitudes, F_{obs} (hkl).

There are two ways of achieving a face-centered arrangement of xenon atoms in space group $P2_1/c$: either they occupy positions $2(a)$: $0, 0, 0; 0, \frac{1}{2}, \frac{1}{2}$ and positions $2(d)$: $\frac{1}{2}, 0, \frac{1}{2}; \frac{1}{2}, \frac{1}{2}, 0$; or else they are in general positions $4(e)$: $\pm(x, y, z; x, \frac{1}{2} - y, \frac{1}{2} + z)$ with $x = z = \frac{1}{4}$, $y = 0$. In either case the xenon atoms dominate the X-ray scattering, and the signs of structure factors $F(hkl)$ with h, k, l unmixed (all odd or all even) are determined by the xenon positions above. Electron-density maps computed by Fourier series summation using only unmixed-index reflections should show the structure, although obscured by additional (false) symmetry. Such maps were computed, but no reasonable assignment of fluorine-atom positions was attained. Five of the strongest mixed-index reflections were then included, their signs being determined by Sayre's squaring method [3]; the resulting map with xenon atoms in positions 2(*a*) and 2(*d*) proved to be interpretable. The placement of xenon atoms in positions 4(*e*), which would require all molecules to be equivalent, did not produce an interpretable map. Thus it was found that two XeF_2 molecules are centered in positions 2(*a*) and two XeF_4 molecules are centered in positions 2(*d*).

After this structure had been shown by structure-factor calculations to be approximately correct, it was refined by iterative least-squares method [4], in which positional parameters and individual anisotropic temperature factors were varied. After several least-squares cycles and the omission of the 35 strongest reflections, which were affected by extinction, the agreement factor, $R = \Sigma|F^2_{obs} - F^2_{calc}|/\Sigma F^2_{obs}$, reached 0.050.

Since the structure is composed of molecules of XeF_2 and XeF_4, it is of interest to compare their configurations with those obtained from studies of the separate phases. The geometries of the XeF_2 and XeF_4 molecules were determined by neutron diffraction [5] and by X-ray diffraction [2, 6]. In pure crystalline XeF_2 the molecule is linear with an Xe-F bond distance of 2.00 A., corrected for thermal motion [7]. In pure crystalline XeF_4 the Xe-F bond lengths, corrected for thermal motion, are significantly shorter, 1.95 A., and the F-Xe-F angle is 90.0°. In the phase $XeF_2 \cdot XeF_4$, the difluoride molecule was found to be linear with an Xe-F bond length of 2.01 A., and the tetrafluoride molecule to be planar with bond lengths of 1.94 A. and 1.97 A. Each distance has been corrected [7] for thermal motion on the reasonable assumption that fluorine "rides" on xenon. The F-Xe-F bond angle in the tetrafluoride molecule was found to be 89.0°. We estimate the standard errors in these determinations, considered as measures of accuracy, to be 0.02 A. in bond length and 0.8° in bond angle. Thus the hypothesis that XeF_4 is square planar in the present phase is consistent with this study, and the two component molecules retain essentially the same geometry as in their separate crystals.

The close intermolecular fluorine-to-xenon distances of interest are 3.41 A. in XeF_2, 3.22 and 3.25 A. in XeF_4. In $XeF_2 \cdot XeF_4$ there are intermolecular fluorine-to-xenon contacts of 3.28, 3.35, 3.35, 3.37, 3.42 A. Thus there are no unusually short distances between molecules to indicate the presence of strong bonds. Indeed, one must conclude that the forces holding together this compound are not qualitatively different from those in the separate crystals. Therefore, the phase $XeF_2 \cdot XeF_4$ is appropriately described as a molecular addition compound.

REFERENCES

1. J. H. Burns, *J. Phys. Chem.* **67**, 536 (1963).
2. S. Siegel and E. Gebert, *J. Am. Chem. Soc.* **85**, 240 (1963).
3. D. Sayre, *Acta Cryst.* **5**, 60 (1952).
4. W. R. Busing, K. O. Martin, and H. A. Levy, *ORFLS, A Fortran Crystallographic Least Squares Program*, Report No. TM-305. Oak Ridge National Laboratory, 1962.

5. H. A. Levy and P. A. Agron, *J. Am. Chem. Soc.* **85,** 241 (1963); J. H. Burns, P. A. Agron, and H. A. Levy, *Science* **139,** 1208 (1963).
6. J. A. Ibers and W. C. Hamilton, *ibid.* 106 (1963); D. H. Templeton *et al.*, *J. Am. Chem. Soc.* **85,** 242 (1963).
7. W. R. Busing and H. A. Levy, submitted to *Acta Cryst.* (1963); W. R. Busing and H. A. Levy, *A Crystallographic Function and Error Program for the IBM 704*, Report No. 59-12-3. Oak Ridge National Laboratory, 1959; see also D. W. J. Cruickshank, *Acta Cryst.* **9,** 757 (1956).

DETERMINATION OF THE CRYSTAL STRUCTURE OF XENON TRIOXIDE

DAVID H. TEMPLETON, ALLAN ZALKIN, J. D. FORRESTER, AND STANLEY M. WILLIAMSON
University of California, Berkeley

ABSTRACT

Xenon trioxide was characterized by determination of its crystal and molecular structure using X-ray diffraction through single crystals. Four molecules of XeO_3 occupy an orthorhombic cell in space group $P2_12_12_1$ with dimensions $a = 6.163 \pm 0.008$ A., $b = 8.115 \pm 0.010$ A., $c = 5.234 \pm 0.008$ A. The crystal structure is closely related to that of the isoelectronic HIO_3. The XeO_3 molecule has trigonal pyramidal shape with average Xe-O bond length 1.76 A. (corrected for thermal motion) and average O-Xe-O bond angle 103°.

INTRODUCTION

This paper is a more detailed version of an earlier report [1] of the identification of XeO_3 by determination of its crystal and molecular structure. The preparation of this material by hydrolysis of XeF_4 and its chemical analysis have been described by Williamson and Koch [2]. It was independently prepared from XeF_6 and identified by Smith [3].

EXPERIMENTAL

The material used resulted from the hydrolysis of XeF_4, after which the solution was evaporated to dryness. A small amount of the white powder was placed on a glass slide in air and was allowed to pick up water from the atmosphere until it completely dissolved. Crystals were then grown by focusing a microscope lamp on the saturated drops on the slide to provide gentle heating. With a microscope of about 40 power, crystals were observed to develop as fine needles which grew in a few minutes to elongated rods. The *b* axis coincides with the long dimension of the crystal and is an extinction direction of light with crossed polarizers.

Crystals, 0.1 to 0.5 mm., were separated from the solution and stored on a glass slide (warm to the touch) in air, where they were stable for short periods. After a selection of crystals had been made, they were individually transferred into thin-walled capillaries 0.5 mm. in diameter and stored with one end open in a desiccator charged with $CaSO_4$ (Drierite). When a crystal was needed for X-ray examination, a capillary was removed from the desiccator and the end sealed off. Each crystal stuck to the side of the capillary with its long direction, the *b* axis, parallel to the direction of the capillary.

Several crystals were required for the work because of decomposition during the irradiation. The first five or six crystals were used in learning how to handle the substance and to take Weissenberg patterns to establish the crystal symmetry. An additional four or five crystals were required to get accurate cell dimensions and to take intensity data. These measurements were made with a General Electric XRD-5 apparatus equipped with a goniostat and scintillation counter and with Mo Kα X-rays. The tube was operated at 40 kv. and from 1 to 20 ma., the milliamperage being boosted as the crystal decayed. The cell dimensions were measured with the resolved Kα_1 ($\lambda = 0.70926$ A.) X-ray using a take-off angle of 2°. The intensity data were obtained using ten-second counts with take-off angle of 4°. Ultimately, 482 independent reflections were observed, 16 of which were below the detection limit and recorded as zero. The intensities were measured only for 2θ below 60°.

The crystals contained a film of moisture on their surfaces, probably acquired during their exposure to the room before sealing. During irradiation the crystal would decompose, and the resulting gas would form bubbles in this film. Where the crystal contacted the capillary, these bubbles caused the crystal to move out of alignment, necessitating frequent realignment of the crystal. The crystal could be observed to break up during the measurements. The decomposition rate decreased markedly when the irradiation was stopped; therefore, to diminish this damage a beam stop which exposed the crystal to X-rays only during the actual ten seconds of counting was used. Even with these precautions, a crystal would only last for about four to six hours of measurements.

The (200) reflection was used as the standard reflection with which to adjust the data. The data were taken in an order dictated primarily by the convenience of setting the angles on the goniostat. The actual order has been retained in reporting the observed and calculated values in Table 3. The intensities fell off with time principally because the fragments were rotated to each other and therefore not all in diffraction position, rather than because of destruction of the lattice. This rotation was not isotropic, and therefore the method of normalization is not correct. It was corrected by introduction of additional scaling factors.

The Fourier and least-squares calculations were done on an IBM 7090 computer. We used a Fourier program of our own design and a least-squares program written by Gantzel, Sparks, and Trueblood [4]. The least-squares program minimizes the function $\Sigma w||F_{obs}|-|F_{calc}||^2/\Sigma w|F_{obs}|^2$ and utilizes a full matrix. For lack of any better estimate, we used unity for all of the weighting factors, w. Lorentz-polarization corrections were applied to the data. Absorption corrections were not made. Atomic scattering factors for neutral xenon and oxygen were taken from Tables 3.3.1B and 3.3.1A in the International Tables [5]. Anomalous dispersion corrections were omitted.

DETERMINATION OF THE STRUCTURE

A set of Weissenberg films was sufficient to determine a set of cell dimensions. A comparison of the a/b and c/b ratios with those in Crystal Data [6] showed a resemblance to HIO_3. A thorough check showed the extinctions to correspond to space group $P2_12_12_1$. As xenon is the principal diffractor in this crystal, it was possible to estimate its location by an investigation of first the $h00$, $0k0$, and $00l$ data, and then a set of $h0l$ and $hk0$ data. This position had parameters that were less than 0.04 different from that found for I in HIO_3.

An electron density projection using 55 non-zero $hk0$ terms with signs based on xenon only was calculated. The three largest peaks exclusive of the xenon

TABLE 1

CRYSTAL-STRUCTURE DATA FOR XeO_3

CELL DATA

Orthorhombic	
Space Group $P2_12_12_1$ (D_2^4)	
$a=6.163\pm.008$ A.	$Z=4$
$b=8.115\pm.010$ A.	$V=261.8$ A.^{3}
$c=5.234\pm.008$ A.	Molecular Weight$=179.30$
	X-ray Density$=4.55$ gm/ml

ATOMIC PARAMETERS—ALL ATOMS IN THE GENERAL POSITION

4a $x, y, z; \frac{1}{2}-x, \bar{y}, \frac{1}{2}+z; \frac{1}{2}+x, \frac{1}{2}-y, \bar{z}; \bar{x}, \frac{1}{2}+y, \frac{1}{2}-z$.

	x	y	z	B(A.$^{-2}$)
Xe........	$.9438\pm.0003$	$.1496\pm.0003$	$.2192\pm.0004$	$1.34\pm.04$
O_1........	$.537\pm.004$	$.267\pm.004$	$.066\pm.006$	$2.3\pm.5$
O_2........	$.171\pm.005$	$.096\pm.004$	$.406\pm.006$	$2.2\pm.5$
O_3........	$.142\pm.004$	$.454\pm.003$	$.142\pm.006$	$1.8\pm.4$

peak were observed to occur at positions not far different from those of oxygen in HIO_3. We estimated the third parameter of each from the iodic acid positions.

Least-squares refinement proceeded as follows. Four cycles of least-squares using 297 reflections and isotropic temperature factors resulted in $R = \Sigma||F_{obs}| - |F_{calc}||/\Sigma|F_{obs}| = 0.24$. More data were added to make a total of 386, and another four cycles reduced R to 0.19. With all 482 reflections and six more cycles, R was 0.17. The data were corrected for obvious blunders by remeasurement of a few reflections that showed poor agreement. The reflections were separated into twelve groups, and a scaling factor was determined for each from the sums of observed and calculated structure factors. Six cycles of refinement reduced R to 0.11.

A second rescaling was done and the number of scale factors increased to 15. The beginning and end of each group could be correlated with an event in the data taking, e.g., a new crystal was put on, the crystal was realigned, a fragment of the crystal dropped off.

The final result after six more iterations, with $R = 0.098$, is shown in Table 1 with all of the pertinent crystal-structure data. Attempts to refine the xenon atom with anisotropic temperature factors did not improve the agreement and are of doubtful validity considering the method of scaling. Any attempt at a thermal description other than isotropic is not warranted from these data. The standard deviations listed in Table 1 are those estimated by the method of least squares. The accuracy cannot be expected to be this good, at least for the thermal parameter of xenon.

The fifteen scale factors that were applied to the intensities are shown in Table 2. Observed and calculated structure factors are listed in Table 3, in the order in which they were measured. A list of interatomic distances is given in Table 4.

TABLE 2

SCALING FACTORS APPLIED TO THE RAW INTENSITIES*

No.	Scale Factor	No.	Scale Factor	No.	Scale Factor
1.....	0.62	6...	1.12	11...	0.71
2.....	0.49	7...	1.72	12...	0.47
3.....	1.17	8...	0.60	13...	1.00
4.....	1.95	9...	0.85	14...	0.85
5.....	1.41	10...	1.41	15...	0.97

* These correlate to Table 3.

STOICHIOMETRY OF XENON TRIOXIDE

Although the structures of HIO_3 and XeO_3 are closely related, there are significant differences that forced us to conclude that the structure we had just analyzed was xenon trioxide rather than "xenic acid." Iodic acid has one hydrogen bond as deduced by Wells [7] from the X-ray structure determination by Rogers and Helmholz [8]. Garrett [9] did a neutron-diffraction study of HIO_3 and confirmed the basic structure as well as the location of the hydrogen bond as suggested by Wells. In iodic acid this hydrogen-bond distance is 2.69 A., and the corresponding distance in xenon trioxide is 3.19 A., far too long to be considered as a hydrogen bond. This is consistent with the observation that though HIO_3 is a larger molecule than XeO_3, its crystal volume is smaller due to the contraction by hydrogen bonding. Furthermore, we find no evidence of a hydrogen bond elsewhere in the structure. The structures of the two substances are otherwise so similar that we could not believe that the xenon compound would not make a hydrogen bond if it contained hydrogen. Structural data for XeO_3 and HIO_3 are compared in Table 5.

This formulation as XeO_3 was shortly after confirmed by chemical analysis [1–3].

TABLE 3

Observed and Calculated Structure Factors for XeO_3

The order of the reflections is that of taking the data. The phase angle phi is given in terms of thousandths a circ .

SCALE FACTOR 1

K	L	FCB	FCA	PHI
0	0	130	139	000
0	0	8	11	000
0	0	34	34	500
0	0	46	53	500
1	1	38	36	250
2	2	35	37	000
3	3	18	18	750
4	4	47	54	500
5	5	45	54	750
6	6	10	12	500
2	1	158	170	500
4	2	101	104	000
6	3	31	33	500
8	4	15	16	000
3	1	39	35	750
6	2	77	83	500
9	3	23	26	750
4	1	84	84	000
8	2	21	22	500

SCALE FACTOR 2

K	L	FCB	FCA	PHI
5	1	24	26	750
0	2	61	52	000
6	1	71	76	000
7	1	4	1	250
8	1	70	73	500
9	1	14	14	250
0	1	4	1	000
1	1	9	9	750

SCALE FACTOR 3

K	L	FCB	FCA	PHI
1	2	41	39	250
2	4	31	32	500
3	6	53	57	750
3	2	27	35	750
6	4	39	48	000
5	2	6	6	250
7	2	43	35	250
9	2	28	22	750
0	3	6	4	000
8	3	58	59	000
7	3	14	12	750
5	3	60	62	250
4	3	50	56	500
2	3	97	101	000
4	6	15	19	000
1	3	53	52	750
2	6	8	8	000
1	4	49	48	750
3	4	71	72	250
5	4	5	2	250
7	4	38	42	750
9	4	13	17	250
8	5	25	29	500
7	5	14	15	250
6	5	23	22	000
4	5	22	21	000
3	5	32	32	250
2	5	47	47	500
1	5	62	60	250
1	6	37	34	250
5	6	6	3	750
3	7	11	12	750
2	7	8	11	000
1	7	43	43	750

SCALE FACTOR 4

K	L	FCB	FCA	PHI
1	1	42	50	164
1	2	43	42	170
3	3	50	41	871
4	4	33	27	286
5	5	34	26	741
2	1	28	35	835
4	2	82	79	954
6	3	34	29	894
8	4	37	31	751
3	1	16	25	506
6	2	64	60	560
9	3	38	32	785
4	1	92	112	742
8	2	18	18	415

1	5	1	38	49	783
2	10	2	37	36	999
1	6	1	63	82	277
1	7	1	12	15	098
1	8	1	17	22	253
1	9	1	15	21	306
1	10	1	47	54	745
1	11	1	16	17	653

SCALE FACTOR 5

H	K	L	FCB	FCA	PHI
2	9	2	32	33	550
2	7	2	27	28	327
2	5	2	60	61	980
2	3	2	45	45	649
4	6	4	35	28	226
2	1	2	83	84	448
4	2	4	61	54	756
2	1	1	82	80	090
4	2	2	28	35	251
6	3	3	39	42	028
2	2	1	115	108	526
4	4	2	23	27	837
6	6	3	22	25	135
2	3	1	88	80	493
4	6	2	15	15	639
2	4	1	67	60	056
4	8	2	23	24	261
2	5	1	6	10	849
2	6	1	51	46	976
2	7	1	61	54	996
2	8	1	65	58	499
2	9	1	33	28	450
2	10	1	9	7	261
2	11	1	29	22	542
4	9	2	38	41	498
4	7	2	16	19	447
4	5	2	73	77	991
4	3	2	21	25	568
4	1	2	75	85	504
6	1	3	33	37	423
6	2	3	31	33	561
6	4	3	27	27	841
6	5	3	17	15	771
1	0	1	140	183	250
2	0	2	90	99	500
3	0	3	70	70	750
4	0	4	20	16	000
5	0	5	12	7	750
2	0	1	41	44	750
4	0	2	12	16	500
6	0	3	29	32	250
3	0	1	52	47	250
6	0	2	31	41	000
4	0	1	0	3	750
8	0	2	33	45	000
5	0	1	20	19	750
6	0	1	7	6	750
7	0	1	52	49	750
8	0	1	9	5	750
1	0	2	63	57	000
2	0	4	62	60	000
3	0	6	56	48	000
3	0	2	41	54	000
6	0	4	29	27	500
5	0	2	23	28	000
7	0	2	17	20	000
1	0	3	117	104	750
2	0	6	22	20	500
2	0	3	54	54	250
4	0	6	6	2	500
4	0	3	54	52	250

SCALE FACTOR 6

H	K	L	FCB	FCA	PHI
5	0	3	12	14	250
7	0	3	31	42	250
1	0	4	14	16	500
3	0	4	48	52	500
5	0	4	48	49	500
7	0	4	17	18	500
1	0	5	41	41	250
3	0	5	17	19	250
4	0	5	60	62	750
6	0	5	39	39	750
1	0	6	19	18	000
5	0	6	48	47	000
1	0	7	12	14	750
3	1	1	108	103	236
6	2	2	21	29	310
3	2	1	35	31	841
6	4	2	22	29	557
3	3	1	40	37	352
6	6	2	17	22	946
3	4	1	57	49	708
3	5	1	109	91	742
3	6	1	55	44	301
3	7	1	21	19	210
3	8	1	23	17	124
3	9	1	56	43	240
3	10	1	32	24	742
6	7	2	12	14	616
6	5	2	40	48	011
6	3	2	17	22	369
6	1	2	41	54	529
4	1	1	80	73	029
8	2	2	14	14	428
4	2	1	18	19	454
4	3	1	115	105	493
4	4	1	23	18	205
4	5	1	4	7	343
4	6	1	25	20	786
4	7	1	77	67	002
4	8	1	16	13	537
4	9	1	29	28	484
4	10	1	8	9	255
8	3	2	13	17	286
8	1	2	14	16	579
5	1	1	93	85	262
5	2	1	15	17	066
5	3	1	35	32	238
5	4	1	17	12	328
5	5	1	78	71	757
5	6	1	10	13	681
5	7	1	19	18	267
5	8	1	20	16	967
5	9	1	45	39	244
6	1	1	45	39	963
6	2	1	38	38	010
6	3	1	68	64	522
6	4	1	29	26	404
6	5	1	9	5	247
6	6	1	25	19	589
6	7	1	50	44	995
6	8	1	30	25	984
7	6	1	35	32	722
7	5	1	35	33	756
7	4	1	40	38	255
7	3	1	15	15	109
7	2	1	20	20	172
7	1	1	37	34	267
8	1	1	16	13	894
8	2	1	49	49	989
8	3	1	17	14	525
1	5	2	56	53	991
1	6	2	56	54	241
1	7	2	31	28	313

SCALE FACTOR 7

H	K	L	FOB	FCA	PHI
8	4	1	28	25	471
1	1	2	38	41	349
2	2	4	39	42	698
3	3	6	21	18	698
1	2	2	125	122	760
2	4	4	43	46	420
1	3	2	48	48	709
2	6	4	33	35	086
1	4	2	69	63	288
2	8	4	28	29	803

SCALE FACTOR 8

H	K	L	FOB	FCA	PHI
4	10	0	0	7	500
0	1	0	0	0	082
0	2	0	58	52	500
0	4	0	118	113	500
0	6	0	76	74	000
0	8	0	30	28	000
0	10	0	63	63	500

SCALE FACTOR 9

H	K	L	FOB	FCA	PHI
2	0	0	152	139	000
4	0	0	9	11	000
6	0	0	36	34	500
8	0	0	48	53	500
4	1	0	92	99	000
8	2	0	15	17	000
5	1	0	58	57	750
6	1	0	54	54	000
7	1	0	19	22	750
8	1	0	9	9	000
7	2	0	49	52	750
3	1	0	96	84	750
6	2	0	18	17	000
8	3	0	0	4	000
5	2	0	14	12	750
7	3	0	37	36	250

SCALE FACTOR 10

H	K	L	FCB	FCA	PHI
2	1	0	80	64	000
4	2	0	25	22	500
6	3	0	24	28	000
8	4	0	28	39	000
7	4	0	23	25	250
5	3	0	93	93	250
3	2	0	63	62	250
6	4	0	30	30	000

SCALE FACTOR 11

H	K	L	FCB	FCA	PHI
7	5	0	0	1	750
4	3	0	53	54	000
5	4	0	14	13	250
6	5	0	53	58	500
7	6	0	21	21	250
1	1	0	45	40	750
2	2	0	65	63	500
3	3	0	100	104	250
4	4	0	0	10	500
5	5	0	0	8	250
6	6	0	23	30	500
6	7	0	17	19	000
5	6	0	0	10	250
4	5	0	95	95	500
3	4	0	35	36	750
6	8	0	9	9	500
5	7	0	58	64	750
2	3	0	11	11	000
4	6	0	12	11	000
5	8	0	9	13	750
3	5	0	0	4	250
4	7	0	21	23	000
5	9	0	23	26	250
1	2	0	185	171	250
2	4	0	72	69	500
3	6	0	12	15	750
4	8	0	0	7	000
4	9	0	37	43	000
3	7	0	75	71	750
2	5	0	79	94	500
4	10	0	0	7	500
3	8	0	40	39	250
1	3	0	41	38	250
2	6	0	69	63	000
3	9	0	35	34	250

SCALE FACTOR 12

H	K	L	FOB	FCA	PHI
3	10	0	0	4	250
2	7	0	13	18	000
1	4	0	90	91	750
2	8	0	21	23	000
2	9	0	36	35	000
1	5	0	9	10	750
2	10	0	45	46	500
2	11	0	21	25	500
1	6	0	70	66	750
1	7	0	37	34	750
1	8	0	70	65	250
1	9	0	21	20	250
1	10	0	0	3	250
1	11	0	18	15	250

SCALE FACTOR 13

H	K	L	FOB	FCA	PHI
0	0	2	146	130	500
0	0	4	69	63	000
0	0	6	40	36	500
3	1	2	63	65	305
6	2	4	36	32	771
3	1	2	60	65	305
6	2	4	37	32	771
3	2	2	53	57	712
6	4	4	29	25	132
3	4	2	36	35	367
3	5	2	12	14	008
3	6	2	32	34	147
3	7	2	57	59	268
3	8	2	30	29	789
3	9	2	24	27	705
3	10	2	22	18	493
6	5	4	36	32	494
6	3	4	21	20	822
6	1	4	29	32	061
5	1	2	49	51	221
5	2	2	26	28	391
5	3	2	73	75	755
5	4	2	21	22	558
5	5	2	7	6	442
5	6	2	20	22	950
5	7	2	47	51	256
5	8	2	8	12	111
7	5	2	11	12	506
7	4	2	26	26	683
7	3	2	33	34	794
7	2	2	39	43	271
7	1	2	26	28	149
1	1	3	54	57	614
2	2	6	33	37	211
1	2	3	57	54	318
2	4	6	21	21	840
1	3	3	68	62	926
2	6	6	14	18	641
1	4	3	71	64	213
1	5	3	26	24	245
1	6	3	64	56	787
1	7	3	47	40	528
1	8	3	29	24	680
1	9	3	27	23	906
1	10	3	51	40	250
2	5	6	13	16	995
2	3	6	35	39	741
2	1	6	25	28	308
2	1	3	52	54	630
4	2	6	41	46	237
2	2	3	79	79	981
4	4	6	25	26	759
2	3	3	71	67	954
2	4	3	53	49	599
2	5	3	44	41	218
2	6	3	44	39	404
2	7	3	45	41	539
2	8	3	41	35	030
2	9	3	26	24	853
2	10	3	20	18	766
4	3	6	11	11	572
4	1	6	11	18	441
3	1	3	57	63	712
3	2	3	49	49	466
4	1	4	41	45	978
5	1	5	27	29	268
5	2	5	46	45	012
5	3	5	0	11	111
5	4	5	25	26	477
4	7	4	15	12	940
4	5	4	48	48	507
4	3	4	17	14	087
3	4	3	44	41	129
3	5	3	68	64	244
3	7	3	23	26	623
3	8	3	33	27	547
4	1	3	42	45	524
4	2	3	22	22	840
4	3	3	62	66	985
4	4	3	37	37	704
4	5	3	9	5	311
4	6	3	29	31	294
4	7	3	48	51	511
4	8	3	0	11	212
4	9	3	20	26	978
5	1	4	29	32	724
5	2	4	18	14	856
5	3	4	41	41	261
5	4	4	42	38	013
5	5	4	0	9	969
5	6	4	26	30	478
6	2	5	11	16	102
6	1	5	20	22	873
7	1	4	26	26	594
7	2	4	29	25	780

SCALE FACTOR 14

H	K	L	FOB	FCA	PHI
1	1	5	41	40	059
1	2	5	23	23	928
1	3	5	61	60	489
1	4	5	40	38	701
1	5	5	13	12	701
1	6	5	32	29	284
1	7	5	42	42	015
1	8	5	19	18	069
2	1	5	40	40	188
2	2	5	37	37	410
2	3	5	29	31	428
2	4	5	34	34	166
2	5	5	42	43	753
2	6	5	34	33	837
2	7	5	21	22	086
5	1	3	44	47	772
5	2	3	42	43	508
5	3	3	20	18	682
5	4	3	24	25	911
5	5	3	61	61	251
5	6	3	18	18	064
5	7	3	16	15	804
5	8	3	24	25	484
7	1	3	28	26	832
7	2	3	21	21	567
7	3	3	25	27	585
7	4	3	28	28	807
1	1	4	69	69	968
1	2	4	67	68	227
1	3	4	32	33	161
1	4	4	39	37	826
1	5	4	49	49	487
1	6	4	35	32	655
1	7	4	22	19	839
1	8	4	41	41	268
1	9	4	31	29	023
2	9	4	23	24	101
2	7	4	33	33	801
2	5	4	31	32	518
2	3	4	51	53	181
2	1	4	48	53	902

SCALE FACTOR 15

H	K	L	FOB	FCA	PHI
3	1	4	41	41	865
3	2	4	27	34	183
3	3	4	54	52	250
3	4	4	51	47	937
3	5	4	25	24	497
3	6	4	41	39	580
3	7	4	40	37	773
3	8	4	21	17	349
3	7	5	19	21	052
3	6	5	27	26	409
3	5	5	27	28	765
3	4	5	28	30	588
3	3	5	25	25	437
3	2	5	51	52	993
3	1	5	35	37	169
4	1	5	27	24	027
4	2	5	7	16	294
4	3	5	35	34	502
4	4	5	44	40	235
4	5	5	14	11	752
4	6	5	32	30	773
1	6	6	18	18	113
1	5	6	50	46	007
1	4	6	19	20	382
1	3	6	24	22	547
1	2	6	27	26	691
1	1	6	45	46	484
3	1	6	22	23	403
3	2	6	21	19	583
3	4	6	33	35	468
3	5	6	25	25	994
1	3	7	41	43	008
1	2	7	14	17	485
1	1	7	24	27	507
2	1	7	31	32	723
2	2	7	11	11	860
3	3	2	112	96	702
2	5	0	124	94	500
2	0	5	43	35	750
2	0	7	39	31	250

DISCUSSION

The arrangement of molecules in the unit cell is shown in Figure 1. The dimensions found for the molecule are shown in Figure 2. No correction has been made for thermal motion. While we do not have an accurate description of this motion, we estimate that it affects bond distances by less than 0.01 A.

Standard deviations of coordinates correspond to 0.002 A. for xenon and 0.03 A. for oxygen. Because of the unorthodox method of scaling, the true accuracy may not be this good. Thus we have no basis for claiming any significant

TABLE 4

LIST OF DISTANCES LESS THAN 3.5 A. IN XeO_3

Xe– O_1.....	1.74±.03 A.*	O_1– Xe....	1.74±.03 A.*
O_2.....	1.76±.03 A.*	O_3.....	2.68±.04 A.*
O_3.....	1.77±.03 A.*	O_2.....	2.84±.04 A.*
O_1.....	2.80±.03 A.	Xe....	2.80±.03 A.
O_3.....	2.89±.03 A.	O_3.....	3.05±.04 A.
O_3.....	2.90±.03 A.	O_2.....	3.09±.04 A.
O_2.....	3.31±.03 A.	$2O_1$.....	3.17±.02 A.
		O_2.....	3.19±.04 A.
		O_2.....	3.22±.04 A.
		O_3.....	3.23±.04 A.
		O_2.....	3.32±.04 A.
		O_3.....	3.33±.04 A.
		O_3.....	3.43±.04 A.
O_2– Xe.....	1.76±.03 A.*	O_3– Xe....	1.77±.03 A.*
Xe.....	3.31±.03 A.	Xe....	2.89±.03 A.
O_3.....	2.73±.04 A.*	Xe....	2.90±.03 A.
O_1.....	2.84±.04 A.*	O_1.....	2.68±.04 A.*
O_3.....	2.91±.04 A.	O_2.....	2.73±.04 A.*
O_1.....	3.09±.04 A.	O_2.....	2.91±.04 A.
O_3.....	3.12±.04 A.	$2O_3$.....	3.03±.03 A.
O_1.....	3.19±.04 A.	O_1.....	3.05±.04 A.
$2O_2$.....	3.20±.04 A.	O_2.....	3.12±.04 A.
O_1.....	3.22±.04 A.	O_1.....	3.23±.04 A.
O_1.....	3.32±.04 A.	O_1.....	3.33±.04 A.
O_3.....	3.46±.04 A.	O_1.....	3.43±.04 A.
		O_2.....	3.46±.04 A.

* Intramolecular distances.

deviation of the dimensions of the molecule from threefold symmetry, and we report the average bond distance as 1.76 A. and the average bond angle as 103°. These dimensions may be compared with 1.82 A. and 97° reported for the isoelectronic iodate ion [10].

Each xenon has three close oxygen neighbors from other molecules, at an average distance of 2.86 A. The interaction of these atoms is analogous to the situation in HIO_3 which has been discussed in terms of "weak secondary bonds" [7, 8]. Whatever its nature, this interaction presumably is an important part of the explanation of the similarity of the two crystal structures.

TABLE 5

XeO_3 AND HIO_3 CRYSTAL-STRUCTURE COMPARISON*

CELL DIMENSIONS

	a	b	c	V	a/b	c/b	Mol. Wt.	X-ray Density
XeO_3....	6.163 A.	8.115 A.	5.234 A.	261.8 $A.^3$	.759	.645	179.3	4.55 gm/ml
HIO_3....	5.888 A.	7.733 A.	5.538 A.	252.2 $A.^3$	.761	.716	175.9	4.63 gm/ml

ATOMIC PARAMETERS

	x	y	z	x	y	z
	Heavy Atom:			O_1:		
XeO_3....	.9438	.1496	.2192	.537	.267	.066
HIO_3....	.914	.158	.204	.523	.243	.071
	O_2:			O_3:		
XeO_3....	.171	.096	.406	.142	.454	.389
HIO_3....	.198	.086	.334	.157	.448	.404
	H:					
HIO_3....	.322	.135	.234			

INTRAMOLECULAR DISTANCES AND ANGLES

	Heavy Atom					
	to O_1	to O_2	to O_3	O_1–O_2	O_1–O_3	O_2–O_3
XeO_3....	1.74 A.	1.76 A.	1.77 A.	2.84 A.	2.68 A.	2.73 A.
HIO_3....	1.82 A.	1.90 A.	1.78 A.	2.80 A.	2.77 A.	2.69 A.

ANGLES

	Xe O_1–I–O_2	Xe O_1–I–O_3	Xe O_2–I–O_3
XeO_3....	108°	100°	101°
HIO_3....	98°	100°	94°

SELECTED INTERMOLECULAR DISTANCES

	Distance That Is Hydrogen Bond in HIO_3	Heavy Atom		
		to O_1'	to O_3'	to O_3''
XeO_3....	3.19 A. O_2 .... O_1'	2.80 A.	2.90 A.	2.89 A.
HIO_3....	2.69 A. O_2–H ..... O_1'	2.50 A.	2.87 A.	2.77 A.

* The cell dimensions for HIO_3 are those given by Swanson *et al.* [11], and the HIO_3 structural data are those of the neutron-diffraction work of Garrett [9].

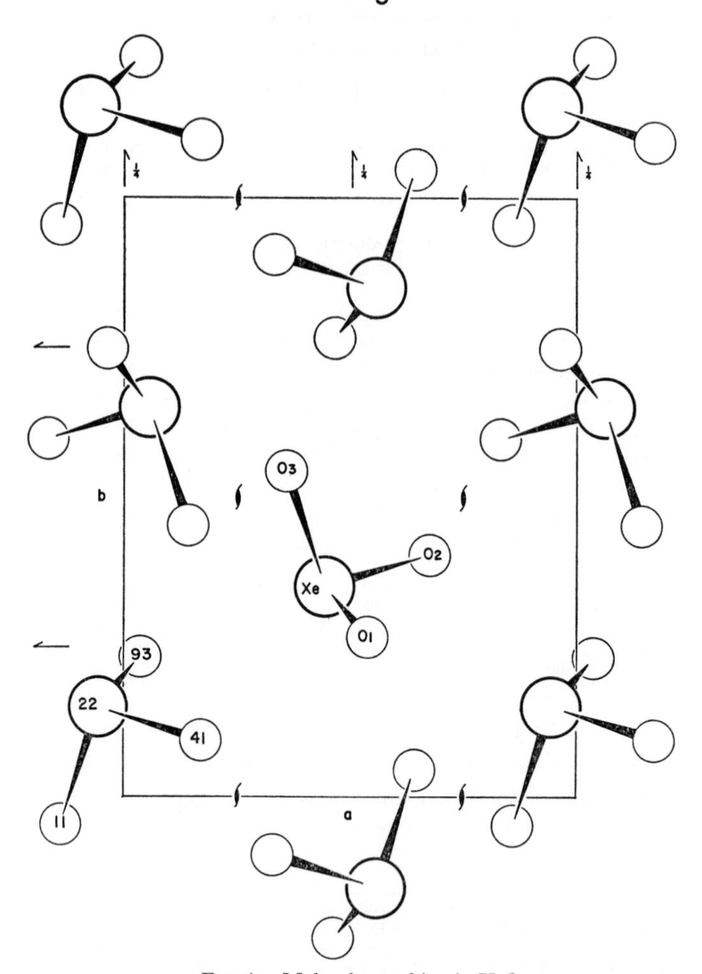

FIG. 1.—Molecular packing in XeO_3

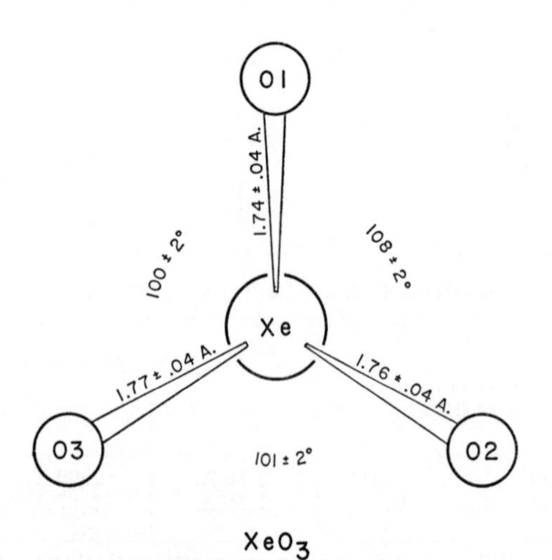

FIG. 2.—Molecular dimensions of XeO_3

Acknowledgments.—Crystals of xenon trioxide were first prepared at Berkeley by K. A. Maxwell. We thank Professor R. E. Connick for bringing them to our attention and Dr. C. W. Koch for assistance in the chemical preparation and analysis.

REFERENCES

1. D. H. Templeton, A. Zalkin, J. D. Forrester, and S. M. Williamson, *J. Am. Chem. Soc.* **85**, 817 (1963).
2. S. M. Williamson and C. W. Koch, *Science* **139**, 1046 (1963).
3. D. F. Smith, *J. Am. Chem. Soc.* **85**, 816 (1963).
4. P. Gantzel, R. Sparks, and K. Trueblood, private communication.
5. *International Tables for X-ray Crystallography*, vol. 3. Birmingham, England: Kynoch Press, 1962.
6. J. D. Donay and W. Nowacki, *Crystal Data, Memoir 30*. New York: The Geological Society of America, 1954.
7. A. F. Wells, *Acta Cryst.* **2**, 128 (1949).
8. M. T. Rogers and L. Helmholz, *J. Am. Chem. Soc.* **63**, 278 (1941).
9. B. S. Garrett, Report 1745, 97. Oak Ridge National Laboratory, 1954. Abstracted in *Structure Reports* **18**, 393 (1954).
10. J. A. Ibers, *Acta Cryst.* **9**, 225 (1956).
11. H. E. Swanson, N. T. Gilfrich, and G. M. Ugrinic, *Standard X-ray Diffraction Patterns*, N.B.S. Circular 539, **5**, 28 (1955).

AN ELECTRON-DIFFRACTION STUDY OF THE STRUCTURE OF GASEOUS XENON TETRAFLUORIDE

R. K. Bohn, K. Katada, J. V. Martinez, and S. H. Bauer
Department of Chemistry, Cornell University

INTRODUCTION

The X-ray and neutron-diffraction investigations of xenon tetrafluoride have given conclusive evidence that in the crystal the molecule is square planar [1–4], a result supported in theory on the nature of xenon-fluorine bonding [5–8].

The only available data regarding the structure of the gaseous species has been the recent interpretation of the gaseous infrared spectrum by Claassen, Malm, and Chernick [9]. The study by electron diffraction of gaseous xenon tetrafluoride was undertaken to obtain additional information regarding its structure. The physical properties of this compound [10] permitted application of standard electron-diffraction methods.

EXPERIMENTAL

A sample of xenon tetrafluoride was furnished by John Malm of Argonne National Laboratory. The substance was obtained in a nickel bulb which was attached directly to the electron-diffraction apparatus. Mass-spectral data and the infrared spectra of sublimed films agreed with results reported earlier [9, 10] and showed no detectable impurities.

The electron-diffraction apparatus used was described earlier [11]. Photographs were taken on Kodak process plates with an electron-beam energy of 45 kev. Gold foil was used for calibration; the sample-plate distance was $L \approx 17$ cm. The diffraction photographs were read on a microphotometer while being oscillated to reduce fluctuations due to emulsion granularity.

RESULTS

An experimental molecular scattering–intensity curve was obtained from the microphotometer data. The observed scattering intensities were corrected for non-nuclear scattering [12], and a radial-distribution function was obtained (Fig. 1).

Inspection of the radial-distribution curve shows a principal peak at 1.94 A. corresponding to the Xe-F distance. This peak yields a root-mean-square ampli-

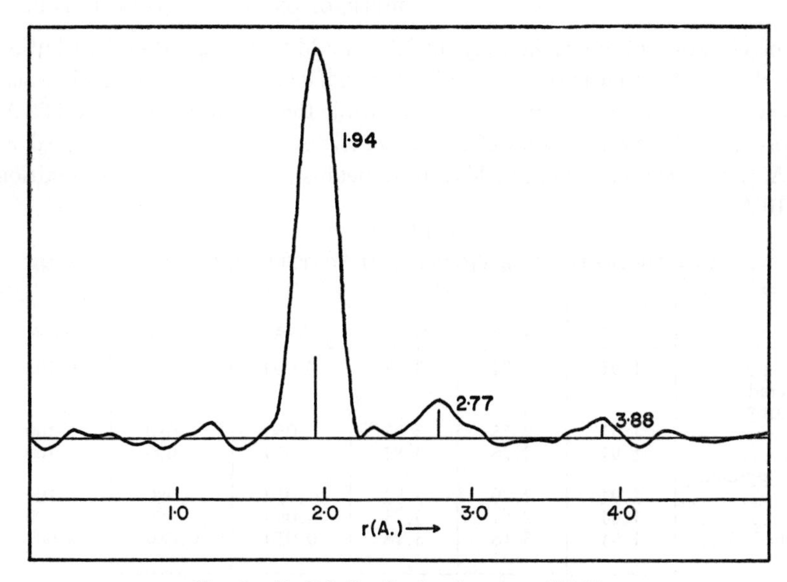

FIG. 1.—Radial-distribution function of XeF_4

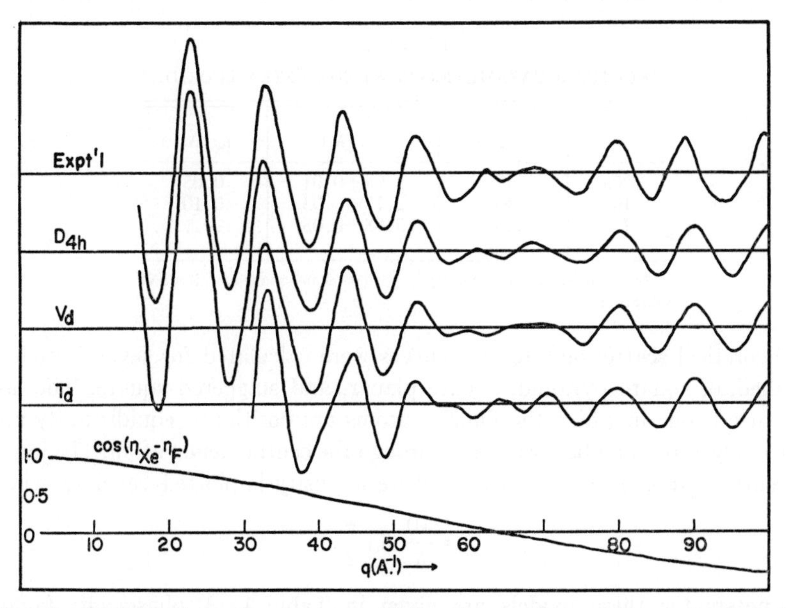

FIG. 2.—Scattering-intensity curves of XeF_4. The experimental scattering-intensity curve, the theoretical scattering-intensity curves for the square planar, D_{4h}, staggered square, V_d—with each of the Xe-F bonds making an angle of 10° with a plane through the xenon atom—and tetrahedral, T_d, models are shown. The phase-shift factor for Xe-F scattering is also shown.

tude of vibration of approximately 0.05 A. In addition, the radial-distribution curve shows two other maxima, at 2.77 A. and 3.88 A. The peak at 2.77 A. has an area corresponding to four F···F pairs, while the area of the peak at 3.88 A. is approximately one-third that of the one at 2.77 A. Approximating the peak at 2.77 A. with a gaussian curve yields a root-mean-square amplitude of vibration of 0.10 A.

TABLE 1

MODELS AND PARAMETERS FOR THEORETICAL SCATTERING-INTENSITY CURVES*

	$r_{Xe\text{-}F}$	$r_{F_1-F_2}$	$r_{F_1-F_3}$	$l_{Xe\text{-}F}$	$l_{F_1-F_2}$	$l_{F_1-F_3}$
Square planar...	1.94	2.74	3.88	0.050	0.080	0.100
Staggered square†						
a)...........	1.94	2.75	3.87	.050	.080	.100
b)...........	1.94	2.78	3.82	.050	.080	.100
Square pyramid‡						
c)...........	1.94	2.70	3.82	.050	.080	.100
d)...........	1.94	2.64	3.73	.050	.080	.100
Tetrahedral.....	1.94	3.18	3.18	0.050	0.080	0.080

* r_{ij} = interatomic distance in A.; l_{ij} = root mean square amplitude of vibration in A.

† The four fluorine atoms are located alternately and equidistantly above and below a plane through the xenon atom. The Xe-F bond makes an angle with the plane of 5° and 10° in (*a*) and (*b*), respectively.

‡ The Xe-F bond makes an angle with the plane of the four fluorine atoms of 10° and 15° in (*c*) and (*d*), respectively.

TABLE 2

STRUCTURAL PARAMETERS OF XENON TETRAFLUORIDE*

i	*j*	r_{ij} (A.)	l_{ij} (A.)
Xe	F	1.94±0.01	0.05
F_1	F_2	2.77± .03	0.10
F_1	F_3	3.88±0.05	....

* r_{ij} = interatomic distance; l_{ij} = root-mean-square amplitude of vibration.

Theoretical scattering-intensity curves were calculated for several models: tetrahedral, square pyramid, square planar, and staggered square. The last structure is one in which the fluorine atoms are positioned equidistantly and alternately above and below a plane through the central xenon atom. The intensity curves are shown in Figure 2, where intensity is plotted versus q, where

$$q = \frac{40}{\lambda} \sin \frac{\theta}{2}.$$

Parameters for these models are given in Table 1. A phase-shift factor, $\cos[\eta_{Xe}(q) - \eta_F(q)]$, was included in these calculations using the method of Bonham and Ukaji [13].* This is a correction for the failure of the Born approxima-

* Corrections to equation 6 in reference 13 were incorporated in our computations after discussion with Dr. Bonham.

tion due to the large difference in the atomic numbers of the xenon and fluorine atoms.

The experimental scattering-intensity curve is shown in Figure 2 along with the curves for the square-planar, D_{4h}, staggered-square, V_d, and tetrahedral, T_d, models for comparison. The theoretical intensity curves for the various models in the region of $q < 35$ do not differ significantly from one another.

The structural parameters obtained in this experiment are shown in Table 2.

DISCUSSION

The value 1.94 A. found for the Xe-F bond distance is in close agreement with the value 1.951 A. reported [4] in the neutron-diffraction study of solid xenon tetrafluoride.

Due to the fluctuations remaining in the radial-distribution function, the F···F distance of 2.77 A. found in this work can not be considered significantly different from the F···F distance of 2.74 A. as expected for a square-planar structure with an Xe-F distance of 1.94 A.

The scattering intensities are dominated by the contribution of the Xe-F pairs. This contribution becomes small only when the phase-shift factor is very small, i.e., $55 < q < 75$. However, the amplitudes of vibration for the F···F pairs are so large that the scattering intensities for these pairs are quite small in this region. Hence, the comparison of experimental with theoretical intensities is not very sensitive to the configuration of the molecule. The tetrahedral model, the staggered-square model with angles greater than 8° between the Xe-F bond and a plane through the xenon atom, and the square-pyramid model with angles greater than 15° between the Xe-F bonds and the plane of the F atoms can be eliminated in the comparison with the observed scattering curves.

The results of this electron-diffraction study show that gaseous xenon tetrafluoride is square planar to within experimental error. It must be emphasized, however, that conclusions as to the configuration of the molecule depend on the scattering-intensity contribution of the F···F pairs, which contribute approximately 13 per cent of the total molecular-scattering intensity.

Acknowledgments.—The authors wish to thank J. G. Malm of Argonne National Laboratory for furnishing the sample of XeF_4 and C. Lifshitz of Cornell University for the results of a mass-spectrum analysis. This work was supported, in part, by ARPA, Order No. 23–63 (Amd. No. 33), under ONR Contract Nonr–401 (41), to whom grateful acknowledgment is made.

REFERENCES

1. J. A. Ibers and W. C. Hamilton, *Science* **139**, 106 (1963).
2. S. Siegel and E. Gebert, *J. Am. Chem. Soc.* **85**, 240 (1963).
3. D. H. Templeton, A. Zalkin, J. D. Forrester, and S. M. Williamson, *ibid.* 242 (1963).
4. J. H. Burns, P. A. Agron, and H. A. Levy, *Science* **139**, 1208 (1963).

5. L. C. ALLEN, *ibid.* **138,** 892 (1962).
6. R. E. RUNDLE, *J. Am. Chem. Soc.* **85,** 112 (1963).
7. L. L. LOHR, JR., and W. N. LIPSCOMB, *ibid.* 240 (1963).
8. K. S. PITZER, *Science* **139,** 414 (1963).
9. H. H. CLAASSEN, J. G. MALM, and C. L. CHERNICK, *Abstract 149*, Div of Phys. Chem. American Chemical Society Meeting, Los Angeles, April, 1963. [Cf. this volume, p. 287.]
10. C. L. CHERNICK *et al.*, *Science* **138,** 136 (1962).
11. J. M. HASTINGS and S. H. BAUER, *J. Chem. Phys.* **18,** 13 (1950).
12. L. S. BARTELL, L. O. BROCKWAY, and R. H. SCHWENDEMAN, *ibid.* **23,** 1854 (1955).
13. R. A. BONHAM and T. UKAJI, *ibid.* **36,** 72 (1962).

PART 7 STUDIES OF ESR, NMR, MÖSSBAUER, IR, AND RAMAN SPECTRA AND RELATED EXPERIMENTS

In this section we have collected studies on xenon compounds using the electronic instrumentation which has become available to scientists in recent years. While absorption and Raman spectra investigations go back to an earlier period than measurements of nuclear magnetic resonance or the Mössbauer effect, the type of instrumentation now available even for these has extended both the useful range and the ease with which data can be accumulated.

It is perhaps worth pointing out that it is possible to accumulate substantial amounts of data with the use of these machines, but it is not always as easy to interpret this information in terms of chemical behavior. The papers included in this section do contain a significant amount of interpretative material.

The paper on hydrogen fluoride solutions of xenon compounds is included at this point, since a number of the other studies have involved such solutions.

RADIATION DAMAGE IN XENON TETRAFLUORIDE ELECTRON SPIN RESONANCE OF THE TRAPPED RADICAL XeF*

W. E. Falconer and J. R. Morton
National Research Council, Canada

Abstract

On γ-irradiation of XeF_4 at 77° K. the radical XeF is trapped in the crystal lattice. The electron spin resonance spectrum of XeF has been interpreted, and it appears that the unpaired electron occupies an antibonding σ orbital of predominantly fluorine $2p$- and xenon $5p$-character.

INTRODUCTION

Crystals that are subjected to the action of high-energy γ- or X-rays frequently suffer damage which affects the ions or molecules of which the crystal is composed. Quite often a fragment possessing an unpaired, or odd, electron may be trapped in the crystal, and this fragment will exhibit an electron spin resonance (ESR) spectrum [1]. In an ESR experiment, the irradiated sample is placed in a homogeneous magnetic field, which splits the energy level of the unpaired electron into its Zeeman components. Usually, the magnetic field strength is several thousand gauss, in which case transitions between the Zeeman levels may be induced by suitably polarized microwave radiation. The Boltzmann distribution of the spins between the two Zeeman levels ensures that there will be a net absorption of the microwave energy, and this absorption may be detected as an electron spin resonance. The resonance condition [1, 2] is

$$h\nu = g\beta H ,$$

where H is the magnetic field strength in gauss; ν, the frequency of the radiation; β, the Bohr magneton; and h, Planck's constant. The g-values of paramagnetic fragments produced by irradiation of crystals are almost always close to that corresponding to a free spin, 2.0023. According to the above equation, a magnetic field of 3200 gauss will induce an electron spin resonance at 9000 mc., a convenient frequency for spectrometer operation.

The most interesting and informative feature of ESR spectroscopy is the

* The research reported in this chapter is published in full in *J. Chem. Phys.* **39**, 427 (1963).

interpretation of the hyperfine pattern produced by the interaction of the unpaired electron with nearby magnetic nuclei [3]. The hyperfine interaction splits the Zeeman levels into a number of components, depending on the spin and the number of interacting nuclei. The spectrum need not, then, be a single absorption line, but may be many lined and may extend over several thousand gauss.

A study of paramagnetic fragments trapped in irradiated single crystals soon reveals that the ESR spectrum (hyperfine pattern *and* g-value) is anisotropic with respect to the magnetic-field direction. The trapped radicals are usually held by crystal forces in a fixed orientation in the crystal. Furthermore, all radicals are held in the *same* orientation, or in two or four sites (rarely more) related by the crystal symmetry. Twisting the irradiated crystal about an axis perpendicular to the homogeneous magnetic field of the spectrometer rotates the trapped radicals in the field, and this technique enables one to study the anisotropy of the spectrum. This anisotropy with respect to the field direction can be described by a g-tensor and one or more hyperfine interaction tensors. The deviation of the principal g-values from 2.0023 (free spin) may be interpreted in terms of spin-orbit interactions with excited states of the radical. The traceless or anisotropic part of the hyperfine tensor(s) arises from dipolar interaction between the nucleus and the electron and can give information on the p-character (and, in principle, d-character) of the orbital of the unpaired electron.

RADIATION DAMAGE IN XeF_4 CRYSTALS

Xenon tetrafluoride was prepared by the now well-established technique [4] of heating xenon and fluorine to 400° C. in a nickel or Monel vessel. Approximately 100 per cent excess fluorine was used, the total pressure in the 400-cc Monel vessel being 4 atm. at 400° C. After cooling the vessel, the product was distilled out and purified by resublimation. Mass-spectrometric measurements [5] established the presence of XeF_4. No ESR spectrum could be obtained from powdered samples of XeF_4 irradiated at 300° K.; however, on irradiation at 77° K. the powder turned blue and exhibited a powerful (though uninterpretable) spectrum. This spectrum and the blue color disappeared at 140° K. Sufficiently large single crystals of XeF_4 were eventually grown by sublimation inside quartz tubes at 280° K. X-ray–diffraction data [6] were identical with that reported by others for the XeF_4. The crystal used in the ESR experiments had the shape shown [7] in Figure 1, being an elongated hexagonal plate with prismatic ends. After irradiation with Co^{60} γ-rays at 77° K., the crystal was placed in one of three quartz tubes designed to enable the magnetic field of the spectrometer to explore successively the xy, xz, and yz planes. The magnetic-field strength was determined with the aid of an NMR probe in conjunction with a frequency counter, and the microwave frequency was measured with a transfer oscillator and the frequency counter.

The ESR spectra of the irradiated crystal of XeF_4 were examined at 77° K., and it was soon apparent that XeF had been trapped in the crystal. Figure 2 shows [7] the spectrum obtained when the magnetic field was parallel to the longitudinal axis of the crystal and, incidentally, parallel to the Xe-F bonds in the radical. It will be noted that all the lines in the spectrum are doublets. This small splitting, which reaches a maximum of 10 gauss for this orientation, is due to hyperfine interaction of the odd electron in XeF with a fluorine nucleus in a neighboring XeF_4 molecule—a so-called near-neighbor interaction. The strongest lines in the spectrum are due to XeF radicals containing xenon isotopes of zero spin. Of these, Xe^{132} is the most abundant, but masses 124 through 136 are included. The splitting between the lines due to Xe^{132} F (etc.) arises from hyper-

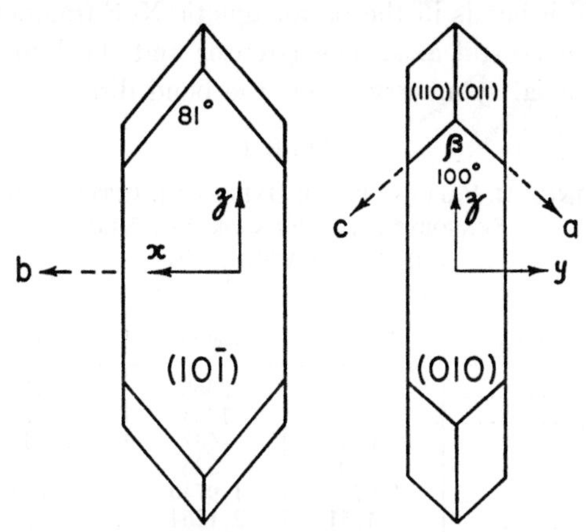

FIG. 1.—Diagram of the XeF_4 crystal, correlating the crystallographic *a*,*b*,*c* axis system with the *x*,*y*,*z* axis system referred to in the text.

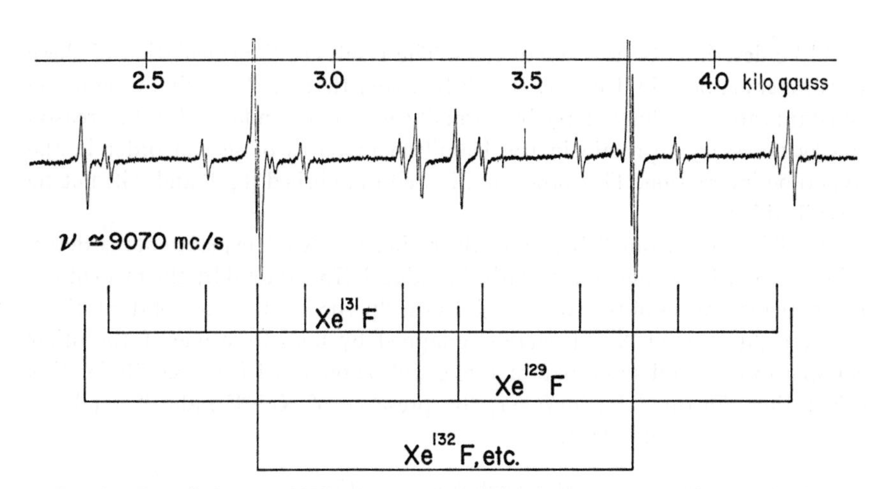

FIG. 2.—First derivative ESR spectrum of XeF in a γ-irradiated single crystal of XeF_4 (H parallel to z). The doublet structure of the lines is due to a near-neighbor interaction with F^{19} on an adjacent XeF_4 molecule.

fine interaction with the F^{19} nucleus ($I = \frac{1}{2}$). The Xe^{129} nucleus also has spin $I = \frac{1}{2}$, and so from $Xe^{129}F$ (26 per cent natural abundance) a four-line contribution is expected. Finally, from Xe^{131} F (21 per cent natural abundance) an eight-line pattern arises, because of the spin $I = \frac{3}{2}$ of Xe^{131}, together with the F^{19} spin of $\frac{1}{2}$.

The anisotropy of the spectra with respect to the magnetic-field direction showed that the largest hyperfine interactions occurred when H was parallel to the longitudinal axis of the crystal. Moreover, identical spectra were obtained when H explored the plane perpendicular to this axis. These considerations imply that the Xe-F bonds in the paramagnetic XeF fragments were parallel to the longitudinal crystal axis. The g-tensor and the hyperfine interaction tensors possessed axial symmetry about the bond direction, and each tensor

TABLE 1

PRINCIPAL VALUES OF THE HYPERFINE INTERACTION TENSORS* AND g-TENSORS† OF XEF IN IRRADIATED XEF_4

Species	Xe^{132} F	Xe^{129} F	Xe^{131} F
$Xe_{\parallel}$	....	2368	701
$Xe_{\perp}$	....	1224	....
$F_{\parallel}$	2649	2637	2653
$F_{\perp}$	540	526	....
$g_{\parallel}$	1.9740	1.9740	1.9740
$g_{\perp}$	2.1251	2.1264	....

* Units are mc.; errors: ±10 mc.
† Errors: ±0.0008.

could be described by two parameters. The results of the calculation of these parameters appear in Table 1. It will be noted that the hyperfine interaction constants are exceedingly large by normally accepted standards. For this reason it was necessary to retain in the Hamiltonian terms of second order in the hyperfine interaction. This procedure is well established [8, 9] and will not be described here.

It will be noted from Table 1 that the ratio Xe^{129}/Xe^{131}, experimentally determined as 3.378, is in agreement with the value 3.375 required by the ratio of the magnetogyric ratios of the two nuclei. In the "perpendicular" orientation, three of the eight lines of $Xe^{131}F$ were overlapped by lines from one of the other isotopic species, and so no parameters could be obtained from $Xe^{131}F$ for this orientation. Fortunately, however, the presence of $Xe^{129}F$ radicals rendered such information redundant.

DISCUSSION OF RESULTS

The hyperfine interaction constants of Table 1 can be correlated with certain parameters which may be derived from atomic wave functions, and in this way some idea of the nature of the orbital of the unpaired electron may be obtained. In order to do this, the principal hyperfine components must be broken down into an isotropic part **A**, and an anisotropic part **B.** These parameters are so defined that, in the case of axial symmetry, the hyperfine component along the unique direction is **A** + 2**B**, and perpendicular to it, **A** − **B.** Using the data in Table 1 it is seen that

$$\mathbf{A}_{F} = 1243 \text{ mc.}, \qquad \mathbf{B}_{F} = 703 \text{ mc. from } Xe^{132}F,$$

and

$$\mathbf{A}_{Xe} = 1605 \text{ mc.}, \qquad \mathbf{B}_{Xe} = 382 \text{ mc. from } Xe^{129}F.$$

The isotropic components **A** are dependent upon the atomic *s*-character of the orbital of the unpaired electron. The experimentally determined **A** is to be

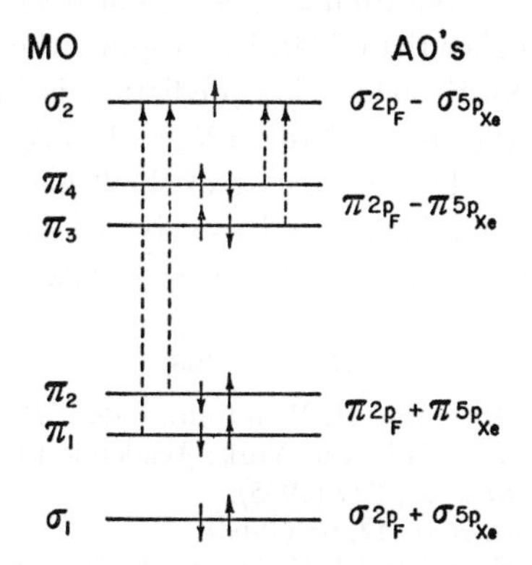

FIG. 3.—Schematic energy-level diagram for XeF. Dotted arrows represent transitions giving rise to the *g*-shifts.

compared with a theoretical **A,** calculated as if the odd electron were wholly in the valence *s* shell of the atom under consideration. The theoretical **A** is proportional to $\psi^2(0)$ for the atom and wave function in question, i.e., the probability that the electron is at the nucleus concerned. Having calculated $\psi^2(0)$ from available [10, 11] wave functions, the theoretical **A,** which is equal to $[8\pi g\beta\gamma/3h]\psi^2(0)$, was found to be 47,900 mc. in the case of the F^{19} 2*s* atomic orbital, and 33,030 mc. for the Xe^{129} 5*s* orbital. The observed values of $\mathbf{A}_F$ and $\mathbf{A}_{Xe}$ therefore represent, respectively, 3 per cent fluorine 2*s*- and 5 per cent xenon 5*s*-character in the orbital of the unpaired electron. Even these figures may be overestimates in view of the possibility of inner *s*-shell polarization, so that it can safely be asserted that the *s*-character of the orbital of the unpaired electron is very small.

A similar calculation can be carried out in order to estimate the valence *p*-character in the orbital of the odd electron. The experimental **B**-values are to be compared with theoretical values which are proportional to $\langle r^{-3}\rangle_{np}$. For fluorine 2*p* and xenon 5*p* the theoretical **B**-values, equal to $[2g\beta\gamma/5h]\langle r^{-3}\rangle$

were found to be 1515 mc. and 1052 mc., respectively. The experimental **B**-values then correspond, respectively, to 47 per cent fluorine $2p$- and 36 per cent xenon $5p$-character in the orbital of the unpaired electron.

It is possible that $4d$ or $5d$ atomic orbitals affect the experimentally determined $\mathbf{B}_{Xe}$, but in the absence of such wave functions it is not possible to estimate their contribution. It would appear, however, that a satisfactory description of the orbital of the unpaired electron in XeF is possible without invoking appreciable d contribution.

In Figure 3 a schematic energy-level diagram for the XeF molecule is shown. The molecular orbitals are constructed from fluorine $2p$ and xenon $5p$ atomic orbitals, and it will be seen that the odd electron in XeF must occupy the antibonding σ_2 orbital of the molecule. Spin-orbit interaction involving promotion of an electron of reversed spin from $\pi_{1,2}$ or $\pi_{3,4}$ into σ_2 (dotted arrows) would give rise to a g-value higher than 2.0023 (freespin) when the magnetic field is perpendicular to the Xe-F bond. This prediction is in accordance with the experimental results, for $g_\perp$ is equal to 2.125, and so $\Delta g_\perp$ equals $+0.123$.

It has been shown [12, 13], in the case of the "isoelectronic" molecule F_2^-, that these spin-orbit interactions should also give rise to a smaller, but negative shift in $g_\parallel$, and this is indeed observed ($\Delta g_\parallel = -0.028$).

REFERENCES

1. C. K. Jen, in A. M. Bass and H. P. Broide (eds.), *Formation and Trapping of Free Radicals*, Chap. 7, p. 213. New York: Academic Press, 1960.
2. J. E. Wertz, *Chem. Revs.* **55**, 829 (1955).
3. A. Carrington, *Quart. Revs.* **17**, 67 (1963).
4. H. H. Claassen, H. Selig, and J. G. Malm, *J. Am. Chem. Soc.* **84**, 3593 (1962).
5. F. P. Lossing, private communication.
6. L. D. Calvert, private communication.
7. By permission of the American Institute of Physics.
8. A. Horsfield, J. R. Morton, and D. H. Whiffen, *Mol. Phys.* **4**, 475 (1961).
9. M. W. Hanna and L. J. Altman, *J. Chem. Phys.* **36**, 1788 (1962).
10. E. Clementi, C. C. J. Roothaan, and M. Yoshimine, *Phys. Rev.* **127**, 1618 (1962).
11. D. F. Mayers, private communication.
12. T. G. Castner and W. Kanzig, *J. Phys. Chem. Solids* **3**, 178 (1957).
13. T. Inui, S. Harasawa, and Y. Obata, *J. Phys. Soc.* (Japan) **11**, 612 (1956).

NUCLEAR MAGNETIC RESONANCE STUDIES OF XENON FLUORIDES

J. C. HINDMAN AND A. SVIRMICKAS
Argonne National Laboratory

ABSTRACT

Shielding values of 310, 482, and 612 parts per million relative to gaseous fluorine have been measured for the solid xenon hexafluoride, tetrafluoride, and difluoride, respectively. The shielding of XeF_6 shows little change in going from the vapor to the solid phase. XeF_4O, with a shielding constant of 326 p.p.m. in the liquid phase, shows a doublet structure due to Xe^{129}-F^{19} coupling. The coupling constant is 1086 cycles per second. In hydrofluoric acid, XeF_6 undergoes exchange with the solvent. XeF_4 appears to dissolve without appreciable reaction with the solvent. A Xe^{129}-F^{19} coupling constant of 3860 cycles per second is found for this compound in HF media. Although XeF_2 undergoes exchange with the solvent at room temperature, the rate is sufficiently decreased on lowering the temperature to allow measurement of the Xe^{129}-F^{19} coupling. The observed coupling constant is 5600 cycles per second. Failure to observe any Xe^{131}-F^{19} coupling is consistent with the molecular symmetries of the solid compounds. The interpretation of the shielding data in terms of electron densities on the fluorine is discussed. Possible interpretations of the coupling constants in terms of bond configurations are also considered.

The isolation of the fluorine compounds [1–4] of the rare gas, xenon, has stimulated widespread interest among inorganic chemists. As a part of the general effort to characterize the properties of these compounds, a study of the fluorine resonance behavior was initiated.

As pointed out by Saika and Slichter [5] the p-electron density is the dominant factor determining the chemical shift in binary fluorides. As a consequence, the shielding changes in such compounds can be correlated with the degree of covalency of the binding and the electronegativities of the constituents [6]. In a more comprehensive examination of the factors affecting the shielding, Karplus and Das [7] have derived relationships between the shielding and localized bond properties, including hybridization, ionic character, and double bonding, in terms of gross orbital populations.

Of particular interest is the comparison between the fluorine resonance behavior of the xenon compounds and the interhalogens. The obvious possible relationships between the bonding in the interhalogens and xenon compounds have been pointed out [8, 9]. In particular, it has been suggested that the bonding only involves σ p orbitals on the xenon and fluorine. This would represent a particularly simple case, and a direct correlation between the electron densities of the fluorine and xenon and the chemical shifts would be expected.

Additional information about the nature of the structures of the xenon compounds can be obtained from the examination of the spin-spin coupling. Since xenon has two nuclei of spins $\frac{1}{2}$ and $\frac{3}{2}$ in abundances of 26.24 and 21.24 per cent, respectively, one can anticipate that a singlet, doublet, and quartet structure will be found in the fluorine resonance of molecules of appropriate symmetry. In molecules of lower symmetry (e.g., linear and square planar) with appreciable field gradients at the xenon nucleus, one would anticipate disappearance of the quartet structure because of quadrupole relaxation of the Xe^{131}. Non-equivalence of fluorines should be reflected in the appearance of second-order splittings as found in the interhalogens [6, 10, 11]. The relative magnitude of the coupling constants can also yield interesting information. If the bonding only involves p orbitals on the xenon and fluorine, then the coupling constants should decrease with increasing ionic character of the bond [10].

EXPERIMENTAL

Chemical-shift measurements were made at 40 and 56.4 mc. using a Varian 4300B Spectrometer with flux stabilizer. To avoid the troublesome problem of magnet cycling when measuring large chemical shifts, the Varian 4311 RF unit was modified so that the output of a Schomandl ND5 Variable Frequency Synthesizer (Schomandl K. G., Munich) could be fed into the crystal oscillator section of the 4311 unit. Provision has been made for driving the 100-cycle and kilocycle sections by motor for scanning purposes. For most measurements both the chemical shifts and line widths were measured at fixed field and variable frequency. Exceptions are noted in the text.

The samples were contained in fluorothene tubes. Because of the asymmetry of the sample tubes and because many were equipped with valves to permit removal of the samples, no spinning was employed. The probe-sample temperature was controlled, using the Varian Variable Temperature Accessory. Temperatures at the sample location were measured by substitution of a thermocouple probe. Where feasible, an external KF (14.68 M) reference was used. For shielding calculations this reference was assigned a $\sigma = 548.2$, relative to $\sigma_{F_2} = 0$ [6]. The shielding constants are given in units of 10^{-6} and correspond to the convention that positive shielding constants correspond to resonance at higher fields than fluorine gas. No susceptibility corrections have been made.

RESULTS

Chemical shifts in solid and liquid fluorides.—The chemical shifts of the solid halides were measured relative to the KF reference solution, using dispersion-mode signals and a lock-in amplifier. The derived shielding values are given in Table 1. Approximate line widths for the solid fluorides are also given in Table 1. These line widths were derived from the derivatives of the dispersion-mode signals. No attempt has been made to derive second moments or intermolecular parameters from the line widths of the solids, since preliminary analysis indicated the lines are not gaussian. Indeed, for the XeF_6 solid there is evidence of structure in the resonance. A more detailed description of the F^{19} chemical shift and resonance line widths in XeF_6 as a function of temperature is shown in Figure 1. For comparison the line widths of the Xe^{129} resonance in XeF_6 in the solid and liquid states are also given in Table 1.

Examination of these data shows a progressive decrease in the fluorine-shielding as the formal charge on the xenon increases. This behavior is similar to that of the interhalogens and presumably reflects, in part, an increase in electronegativity of the central atom as the oxidation number increases. A more detailed consideration of the possible relationships between the chemical shifts of xenon fluoride and those of other binary fluorides will be given later.

TABLE 1

SHIELDING CONSTANTS OF F^{19} IN XENON FLUORIDES AT 25° C.

$\sigma_{F_2} = 0$

Compound	$\sigma \times 10^6$	Approximate Line Widths (gauss)
XeF_2 (s)	612	6.0
XeF_4 (s)	482	3.8
XeF_6 (s)	310	1.9
(l)	309	0.1
(s)	....	1.4*
(l)	....	0.14*
XeF_4O (s)	(328)	....
(l)	326†	0.02

* Xe^{129} resonance line widths.

† Xe^{129}-F^{19} coupling observed. A = 1086 c/s^{-1}. Measured by high-resolution side-band technique.

It is noted that for XeF_6 and XeF_4O, there is, within our experimental error, no change in the shielding in going from the solid to the liquid phases, suggesting little contribution from crystal forces. Observations of a vapor-phase sample of XeF_6 at 100° indicate a small shift of approximately 5 p.p.m. to high field. The value for the shielding of the solid XeF_4O is bracketed, since it was an estimated value obtained from the chemical shift observed on warming a sample frozen at dry-ice temperature. Figure 1 indicates a progressive increase in molecular motion in the XeF_6 lattice from 0° C. up to the melting point of 46° C.

No multiplet structure was found for either the Xe^{129} or F^{19} resonances in liquid XeF_6. The observation of a single broadened line suggests moderately rapid exchange.

XeF_4O has been reported to be a molecule with a planar arrangement of four equivalent fluorines with an O-Xe-F bond angle near 90° [Chernick, Claassen, Malm, and Plurien, this volume, p. 294]. The triplet structure observed, consisting of a center line and a symmetric doublet, is consistent with this proposed symmetry. The doublet corresponds to the coupling of the Xe^{129} to equivalent fluorines. The failure to observe the coupling between the Xe^{131} and F^{19} spins is ascribed to the quadrupole relaxation of the Xe^{131} nucleus in the asymmetric electric environment of the molecule. The observed relative intensities of the

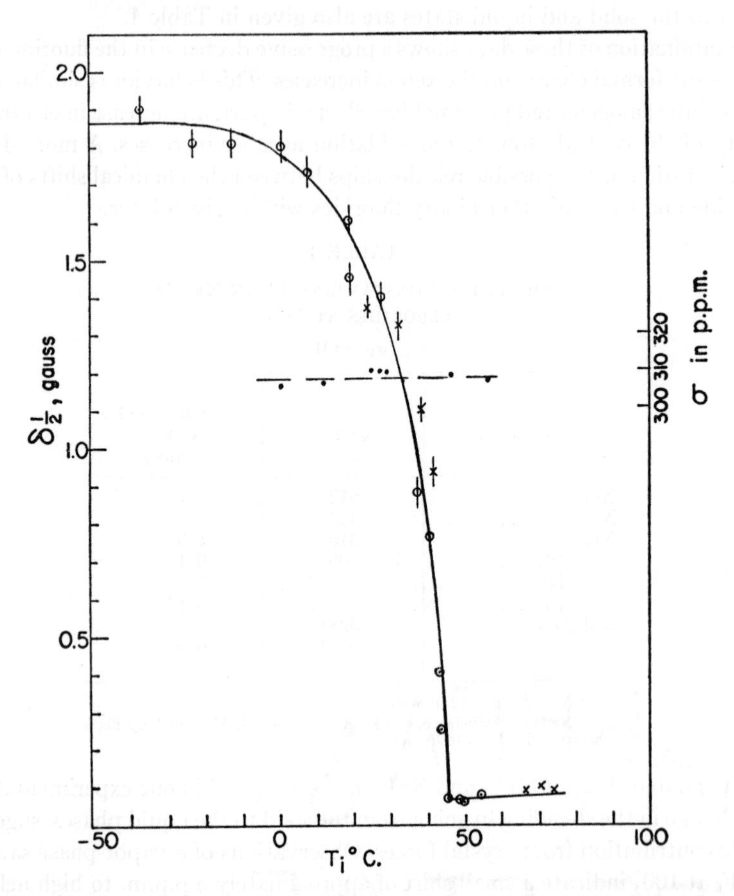

FIG. 1.—Effect of temperature on the F^{19} resonance line width and shielding of XeF_6

doublet lines to the center line is in accord with the isotopic ratio of the Xe^{129} and the above interpretation.

Chemical shifts in HF solutions.—XeF_6 dissolved in anhydrous HF appears to undergo exchange with the solvent. Only a single, broadened F^{19} resonance is observed. The chemical shift varies with the fraction of XeF_6 present. See Table 2. The half width, $\delta_{1/2}$, is increased by about 40 per cent on lowering the temperature from 25° C. to −13° C. Attempts to obtain separate XeF_6 and HF resonances by cooling the sample to dry-ice temperature were not successful.

Efforts to find a solvent in which XeF_6 does not undergo reaction have thus far been unsuccessful.

XeF_4 has a limited solubility in anhydrous HF. It dissolves without undergoing rapid exchange with the solvent, as shown by the fact that a separate resonance is observed at a lower field than the HF resonance. The nuclear magnetic resonance behavior is in accord with the observation that XeF_4 solutions in HF are non-conducting [12], indicating dissolution of XeF_4 without appreciable dissociation. A possible slow exchange is indicated by the observation that both the Xe_4-F^{19} and HF-F^{19} resonances are broader than in a pure HF solution (Table 2) and that the XeF_4-F^{19} line narrows with decreasing temperature. As shown in Table 2, a center line is observed at $\sigma = 456$ p.p.m. and a doublet

TABLE 2

CHEMICAL SHIFTS AND LINE WIDTHS OF FLUORINE RESONANCES IN XENON FLUORIDE–HYDROFLUORIC ACID SOLUTIONS

$\sigma_{F_2} = 0, \quad T = 25°$ C.

Chemical Species	Approximate Concentration (m/1000 g solvent)	σ (p.p.m.)	Approximate $\delta_{1/2}$ (gauss × 10^3)	Xe^{129}-F^{19} Coupling Constants (c/s)
HF	....	627	0.6	....
XeF_6	2.5	540	25	....
	8.4	473	....	....
XeF_4	0.25	452	8	....
	0.25 (saturated)	456	....	3860
HF(XeF_4)	....	....	2	....
XeF_2	1.0	629*	5	5600
HF(XeF_2)	....	....	5	....

* From data at −19.5° C. At this temperature doublet and center XeF_2-F resonance are clearly distinguishable from HF resonance.

with an *A* value of 3860 c/s. The quartet that could arise from the Xe^{131}-F^{19} coupling is not observed. This indicates an asymmetric environment for the xenon nucleus and would be consistent with the preservation of the square-planar structure of the molecule in the solvent. Again, the relative intensities of the center line and the satellites are in agreement with this explanation. The difference in chemical shift between the compound in the crystal and in solution could reasonably be attributed to crystal-field contributions in the solid.

At 25° C., the shielding constants of the XeF_2 and HF appear to be the same since only a single, intense resonance at the HF position and two weak satellites are observed. (See Table 2.) On decreasing the temperature, a separation of the fluorine resonance in the xenon compound and the HF is obtained and the satellites become more clearly defined as shown in Figure 2. Coincident with the amplitude increase of the satellites is a decrease in their half widths. These observations, coupled with the observation that the HF-F resonance is appreciably broadened at 25° C., shows exchange between XeF_2 and HF. Qualitatively, the exchange rate for XeF_2 solutions is more rapid than for the XeF_4 solutions. At −19.5° C., the xenon fluoride resonance is displaced 2.3 p.p.m. to high field from the HF resonance. Partial precipitation of the difluoride occurs at this temperature. The observations suggest that XeF_2, like XeF_4, exists as a largely undissociated species under these conditions. Again the resonance results agree with the observation that solutions of xenon difluoride in HF are not highly conducting. The electric asymmetry of the XeF_2 structure would again account for the failure to observe any Xe^{131}-F^{19} coupling.

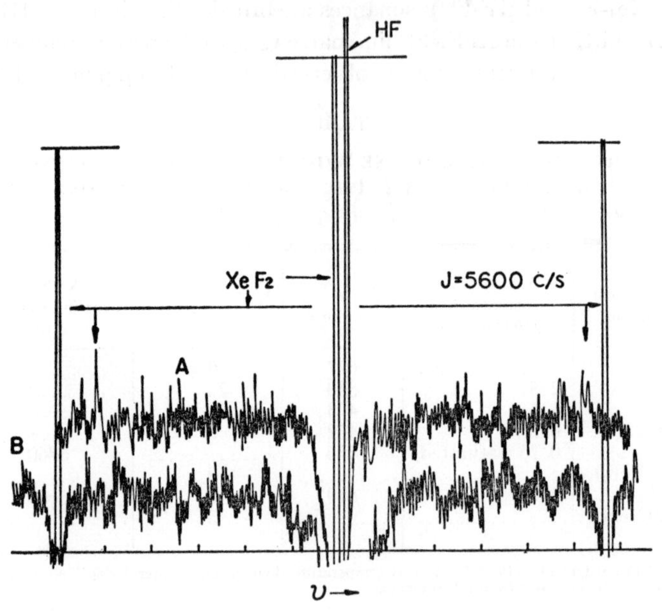

FIG. 2.—Effect of temperature on the fluorine resonance in XeF_2–HF solution: *A* at 25° C. and *B* at −19.5° C. Samples were run at different sweep rates for clarity.

DISCUSSION

As a starting point for the discussion, a correlation diagram has been prepared relating the chemical shielding to the neutral metal-atom electronegativity for a number of binary fluorides [13] (Fig. 3). Examination of these data indicates a much larger apparent change in xenon electronegativity with change in formal oxidation number than for the other fluorides. Precisely the same kind of effect would be noted for a similar comparison with the shielding data for the interhalogens. (See Fig. 4.) Such behavior could be interpreted as indicating a significant change in the bond properties of the different xenon fluorides.

At this point we shall consider how the observed shielding data might be correlated, depending on the assumptions made about the nature of the bonding in these species. Because of the inherent difficulties in direct application of Ramsey's theory of the chemical shift [14], it is necessary at present to adopt a semi-

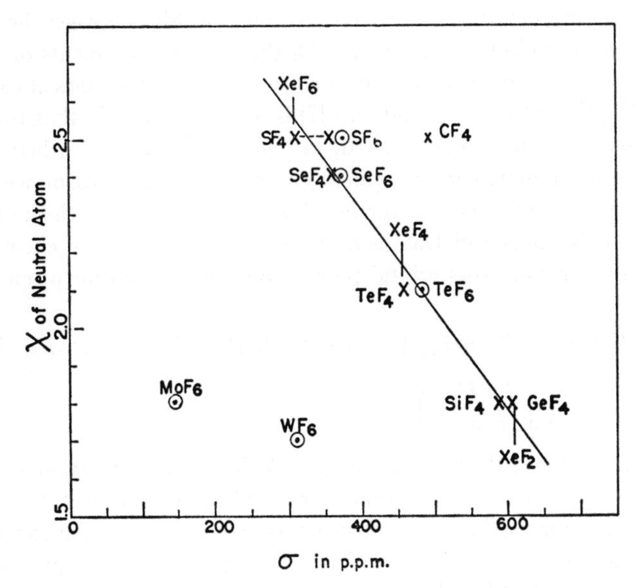

FIG. 3.—Plot of Pauling neutral atom electronegativity versus fluorine chemical shielding for some binary fluorides.

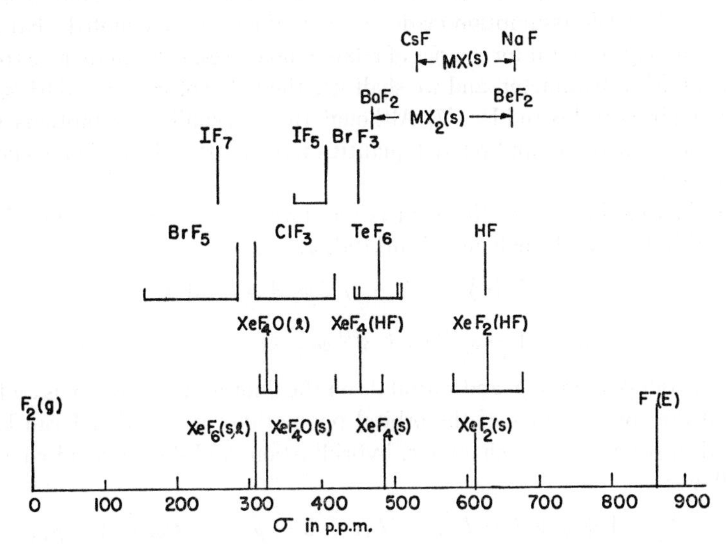

FIG. 4.—Shielding values for a number of binary fluorides. Multiplet structures for the xenon fluorides and TeF_6 are indicated by short vertical lines connected to the center resonance. The long and short interconnected lines for the ClF_3, BrF_5, and IF_5 represent nonequivalent fluorines. For these, the relative fluorine abundance is indicated by the length of the vertical line.

empirical approach in attempting to derive relationships between the shielding and the molecular electronic structure. Of the various treatments of the paramagnetic contribution to the chemical shift suitable for numerical calculation [5, 7, 15–17], that of Karplus and Das [7] is most complete in that the various possible contributions are considered, rather than the problem being treated simply in terms of ionic versus covalent character. For this reason we shall use their treatment as the basis for further discussion. Starting with the theory of Ramsey [14], Karplus and Das have derived relations for the chemical shift expressed in terms of gross orbital populations on the fluorine atom. In their notation

$$\sigma^{(2)} = \sigma^{(0)} [(P_{xx} + P_{yy} + P_{zz}) - \tfrac{1}{2} (P_{xx}P_{yy} + P_{yy}P_{zz} + P_{zz}P_{xx})] \qquad (1)$$

and

$$\sigma^{(0)} = -\left(\frac{2\, e^2 \hbar^2}{3 \Delta E m^2 c^2}\right) \langle r^{1/3} \rangle ,$$

where the P_{ii} indicate the gross orbital populations. Determination of the value of $\sigma^{(0)}$ involves an average excitation energy, ΔE, and the expectation value for the r^{-3} operator for the fluorine $2p$ electron. One of the principal uncertainties in any semi-quantitative application of this theory involves the question of the appropriate value of ΔE to be used. Various suggestions have been made [15–17]. Since, as Karplus [18] has pointed out, ΔE is not related in a simple manner to the excitation energies of the electronic states of the molecule, one is obliged to accept the basic assumption made by the various authors quoted: that ΔE is approximately constant for a series of related molecules. $\sigma^{(0)}$ can then be treated as an empirical parameter, and we shall use the value of $\sigma^{(0)} = -8.63 \times 10^{-4}$ estimated by Karplus and Das [7]. Although this and similar assumptions would affect the value of $\sigma^{(2)}$ and related quantities, qualitative deduction would not be affected.

Considering the z axis as the bond axis and writing the wave functions for the bond orbital, ψ_z, and the lone-pair orbital, ψ_{lp}, as

$$\psi_z = (1 + \tfrac{1}{2})^{1/2} [(1 - s)^{1/2} \phi_z + s^{1/2} \phi_s] + \cdots$$

and

$$\psi_{lp} = (1 - s)^{1/2} \phi_s + s^{1/2} \phi_z ,$$

where ϕ_z and ϕ_s are the p_z and s orbitals on the fluorine atom, Karplus and Das [7] find the contributions of the orbital populations of equation 1 can be expressed in terms of ionic character, hybridization, and double-bond character as follows:

$$P_{zz} = 1 + s + I - Is \; ; \qquad P_{xx} = 2 - \rho_x , \qquad P_{yy} = 2 - \rho_y ,$$

where s represents the extent of (sp) hybridization of the fluorine orbital, I the ionic character of the bond orbital, and ρ_x and ρ_y represent the double-bond contributions. In terms of these parameters equation 1 becomes

$$\sigma^{(2)} = \sigma^{(0)} \{ [1 - s - I + \tfrac{1}{2}(\rho_x + \rho_y)] + \tfrac{1}{2}[2Is + (\rho_x + \rho_y)(I - s - Is) - \rho_x \rho_y] \} . \qquad (2)$$

Equation 2 can be simplified. We might consider that s for fluorine should be small and can be neglected. Since there is no evidence from the fluorine resonance data for the existence of non-equivalent fluorines, and since the available structure data [19] indicate the molecules are symmetric, the bonds can be considered cylindrically symmetric about the bond axis and $\rho_x = \rho_y = \rho$. We then obtain the approximate result:

$$\sigma^{(2)} = \sigma^{(0)} \left(1 - I + \rho + \rho I - \frac{\rho^2}{2}\right). \tag{3}$$

Alternatively, if we neglect the possibility of double-bond character, we can write

$$\sigma^{(2)} = \sigma^{(0)} \left[(1 - s - I + Is)\right]. \tag{4}$$

TABLE 3

CALCULATION OF DEGREE OF IONICITY AND RELATED PROPERTIES OF XENON FLUORIDES AND INTERHALOGENS

Compound	I	P_{zz}(F)	q_F	q_{Xe}	χ_M
XeF_2 (s).....	0.71	1.71	−0.7	+1.4	2.65
(HF)...	.73	1.73	− .7	+1.4	2.65
XeF_4 (s).....	.56	1.56	− .55	+2.2	2.85
(HF)...	.53	1.53	− .55	+2.1	2.85
XeF_6 (s, l)....	.36	1.36	− .4	+2.4	3.05
XeF_4O (s, l)....	0.36	1.36	−0.4	+2.4	3.05
ClF_3 (lb)....	0.36	1.36	−0.35	....	3.05
(sb)....	.49	1.49	− .5	....	2.92
BrF_5 (lb)....	.33	1.35	− .35	....	3.13
(sb)....	.18	1.18	− .2	....	3.38
IF_5 (lb)....	.47	1.47	− .5	....	....
(sb)....	.42	1.42	−0.4	....	2.96
BrF_3 (av.)...	.52	....	....	....	2.91
IF_7 (av.)...	0.31	....	....	....	3.16

If we adopt the view that the only bonding is σ-bonding involving the appropriate p orbitals on the xenon and fluorine [8, 9], equations 3 and 4 reduce to the equation of Saika and Slichter [5], and the chemical-shift data can be related to the ionic character of the bond by the simple equation

$$\sigma^{(2)} = \sigma^{(0)} (1 - I). \tag{5}$$

Values of I from equation 5 are given in Table 3, together with derived values, P_{zz}(F), and the net charges, q_F and q_{Xe}, on the fluorine and xenon atoms. Also included in Table 3 are apparent electronegativity values, χ, for xenon in these compounds obtained from the I values and the I versus χ relations of Dailey and Townes [20]. Their data have been used in preference to other I versus χ relationships, since they are based on quadrupole-coupling data and hence are most directly related to shielding data.

It can be noted immediately from Table 3 that the calculated charges on the fluorine in the XeF_2 and XeF_4 molecules are greater than the $-0.5e$ required if a single xenon $5p$ orbital is to form a colinear σ bond to two fluorine atoms. These data would therefore be consistent with the simple molecular-orbital picture of the binding proposed for these molecules [9; see also Jortner, Rice, and Wilson, this volume, p. 358]. The charge distributions are in agreement with those calculated by Jortner, Rice, and Wilson [this volume, p. 374], using the model. It is of interest to consider how the lower fluorine formal charges calculated for the XeF_4O and XeF_6 molecules can be reconciled with the proposed bonding scheme. One way to do this, which does not disturb the σ-bond framework and does not require formation of Xe *spd* hybrids, is to introduce a small amount of π-character into the bonds via the $F_{p\pi}$ orbitals, utilizing the Xe $5d$ orbitals [21]. The possibility of π-character in the bonds is also suggested by consideration of the shielding equations in relation to, first, the electronegativity of neutral xenon (2.45) obtained from a plot of the apparent electronegativities in Table 3 as a function of the xenon formal charge and, second, the electronegativity (2.2) derived from the ionization potential [9]. It can be noted that the introduction of π-character to the extent of $\rho = 0.07$ would be sufficient to raise the population of the fluorine-bonding orbital, $P_{zz}(F)$, to $1.5e$ for the XeF_6 and XeF_4O.

The interhalogen data deserve comment, since the binding scheme for the xenon compounds is based on that proposed for these species [22–24]. For the particular compounds listed in Table 3, the long bonds, designated (lb), are those associated with the linear p-p bonds and those which should have fluorines with a formal charge exceeding $-0.5e$. Except for IF_5, the data indicate significantly lower fluorine charge densities in these bonds. ClF_3 would appear to be particularly anomalous, since for this compound the formal charge ascribed for the short bond would exceed that for the long bond. At this point we should note that the principle experimental evidence for the interhalogen-bonding scheme derives from quadrupole resonance data [25] for compounds not containing fluorine. Since fluorine has a greater tendency to form π bonds than either bromine or chlorine [cf. reference 7 and 21], it is reasonable to assume that the BrF_5 shielding could be reconciled to the bonding scheme in the manner discussed for the XeF_6 and XeF_4O. ClF_3 would appear to merit further consideration, since the data indicate marked charge migrations in this molecule. It might be noted in this connection that the F-F coupling constant [11] for ClF_3 $A \cong 400$ c/s, is considerably larger than for BrF_5 and IF_5 [10], $A = 76$ and 84 c/s, respectively, also suggesting extensive deviation from a simple σ-p bonding scheme.

For additional information about the nature of the bonds we turn to a consideration of the observed spin-coupling constants. The various contributions to the spin-coupling as given in the general theory of Ramsey [26] can be written as

$$A^{NN'} = A_{1a}{}^{NN'} + A_{1b}{}^{NN'} + A_2{}^{NN'} + A_3{}^{NN'} ,$$

where the subscripts refer to the separate coupling contributions corresponding to the terms in Ramsey's Hamiltonian. For directly bonded atoms the principal contributions to the spin-coupling arise from the terms $A_2^{NN'}$ and $A_3^{NN'}$, representing, respectively, contributions from dipolar interactions between nuclear moments and electrons in non-*s* orbitals and the Fermi contact interaction between *s* electrons and nuclear spins. For atoms bonded to hydrogen, the Fermi contact term, $A_3^{NN'}$, predominates, and the coupling is proportional to the fractional *s*-character of the bonding orbitals [27–29]. On the other hand, spin-coupling constants of fluorine directly bonded to an atom, *M*, have been interpreted on the hypothesis that the coupling is determined primarily by the fractional *p*-character of the atomic orbital of *M* in the *M*-F bond [10]. (The fluorine-bonding orbital is assumed to be pure *p*.) In this case, $A_2^{NN'}$ should vary in an approximately linear manner with the degree of ionicity of the bond and with the square of the coefficient of the *p* wave function in the LCAO hybrid orbital of *M* [10, 30]. It is immediately apparent that the predicted variation in coupling is opposite to that observed. Extensive studies on C^{13}-F^{19} coupling constants [31, 32] suggest other possibilities. It appears that the C^{13}-F^{19} direct coupling constant is increased by decreasing the C^+-F^- polarity, decreasing the *s*-character of the carbon orbital, or increasing the C-F double-bonding. The dominant factor appears to be the π contribution. Now in discussing the shielding data, we have noted that only for XeF_6 and XeF_4O was it necessary to suggest π contributions in order to appropriately adjust the electron density in the σ bonds for the proposed bonding scheme. Assuming the relative signs of the separate contributions remain the same as in the carbon-fluorine case, XeF_6 and XeF_4O would again be expected to exhibit the largest values for the coupling constant.

It must be noted, however, that the shielding data do not exclude significant π contributions for XeF_2 and XeF_4. An interesting point arises in connection with the formation of the $p\pi$-$d\pi$ type of bonds discussed earlier. For XeF_2, with the σ bonds on the *z* axis it is possible to have an interaction of the two π orbitals of a given fluorine atom with the d_{xz} and d_{yz} orbitals of the xenon, forming what might be considered a partial triple bond [21*b*] of a highly polar nature. Since the number of *d* orbitals of xenon of proper symmetry for π-bond formation is two for XeF_2, three for XeF_4, and three for XeF_6 or XeF_4O, part of the observed trend in the coupling constants may be accounted for if it is assumed that these π interactions contribute significantly to the coupling. This suggestion must be considered as highly speculative at present. It is hoped that the detailed calculations being undertaken will enable us to evaluate more quantitatively the various possible contributions.

Acknowledgments.—The authors wish to thank their colleagues in the Chemistry Division, particularly Drs. J. G. Malm, C. L. Chernick, and H. H. Hyman, for the samples used in these experiments. The authors also wish to acknowledge the assistance of Mrs. Myrna Heinen in making some of the measurements.

REFERENCES

1. H. H. Claassen, H. Selig, and J. G. Malm, *J. Am. Chem. Soc.* **84,** 3593 (1962).
2. J. L. Weeks, C. L. Chernick, and M. S. Matheson, *ibid.*, in press.
3. D. F. Smith, *J. Chem. Phys.* **38,** 270 (1963).
4. J. G. Malm, I. Sheft, and C. L. Chernick, *J. Am. Chem. Soc.* **85,** 110 (1963).
5. A. Saika and C. P. Slichter, *J. Chem. Phys.* **23,** 26 (1954).
6. H. S. Gutowsky and C. J. Hoffman, *ibid.* **19,** 1259 (1951).
7. M. Karplus and T. P. Das, *ibid.* **34,** 1683 (1961).
8. K. S. Pitzer, *Science* **139,** 414 (1963).
9. R. E. Rundle, *J. Am. Chem. Soc.* **85,** 112 (1963).
10. H. S. Gutowsky, D. W. McCall, and C. P. Slichter, *J. Chem. Phys.* **21,** 279 (1953).
11. E. L. Muetterties and W. D. Phillips, *J. Am. Chem. Soc.* **79,** 322 (1957).
12. H. H. Hyman, private communication.
13. From shielding data summarized by (*a*) E. L. Muetterties and W. D. Phillips, *J. Am. Chem. Soc.* **81,** 1084 (1959); and (*b*) Wang Yi-Ch'iu, *Soviet Physics, Doklady* **6,** 39 (1961), in English translation.
14. N. F. Ramsey, *Phys. Rev.* **77,** 567 (1950); **78,** 699 (1950).
15. T. Kanda, *J. Phys. Soc. (Japan)* **10,** 85 (1955).
16. K. Yosida and T. Moriya, *ibid.* **11,** 33 (1956).
17. J. Kondo and J. Yamashita, *J. Phys. Chem. Solids* **10,** 245 (1959).
18. M. Karplus, *J. Chem. Phys.* **33,** 941 (1960).
19. (*a*) S. Siegel and E. Gebert, *J. Am. Chem. Soc.* **85,** 240 (1963); (*b*) H. A. Levy and P. A. Agron, *ibid.* 241 (1963); (*c*) D. H. Templeton, A. Zalkin, J. D. Forrester, and S. M. Williamson, *ibid.* 242 (1963).
20. B. P. Dailey and C. H. Townes, *J. Chem. Phys.* **23,** 118 (1955).
21. For discussions of the possibility of the formation of such bonds as a result of the contraction of the *nd* orbitals of the central atom under the influence of highly electronegative ligands see (*a*) D. P. Craig *et al.*, *J. Chem. Soc.* 332 (1954); (*b*) H. H. Jaffé and M. Orchin, *Theory and Application of Ultraviolet Spectroscopy*, Chap. 17. New York: John Wiley & Sons, 1962.
22. R. J. Hach and R. E. Rundle, *J. Am. Chem. Soc.* **73,** 4321 (1951).
23. G. C. Pimentel, *J. Chem. Phys.* **19,** 446 (1951).
24. E. E. Havinga and E. H. Wiebenga, *Rev. trav. chim., des Pays-Bas* **78,** 724 (1959).
25. C. D. Cornwell and R. S. Yamasaki, *J. Chem. Phys.* **27,** 1060 (1957); *ibid.* **30,** 1265 (1959).
26. N. F. Ramsey, *Phys. Rev.* **91,** 303 (1953).
27. M. Karplus and D. M. Grant, *Proc. Natl. Acad. Sci.* **45,** 1269 (1959).
28. N. Muller and D. E. Pritchard, *J. Chem. Phys.* **31,** 768, 1471 (1959).
29. J. N. Shoolery, *J. Chem. Phys.* **31,** 1427 (1959).
30. K. Yosida and T. Moriya, *J. Phys. Soc. (Japan)* **11,** 33 (1956).
31. N. Muller and D. T. Carr, *J. Chem. Phys.* **67,** 112 (1963).
32. J. Bacon and R. J. Gillespie, *ibid.* **38,** 781 (1963).

HIGH-RESOLUTION MAGNETIC RESONANCE OF XENON COMPOUNDS

THOMAS H. BROWN, E. B. WHIPPLE, AND PETER H. VERDIER
Union Carbide Research Institute

The xenon fluorides are in several respects ideally suited to study by nuclear magnetic resonance. The 100 per cent abundant isotope F^{19} (spin $= \frac{1}{2}$) is one of the most favorable nuclei from the standpoint of experimental simplicity. Among the variety of naturally occurring isotopes of xenon, Xe^{129} (spin $= \frac{1}{2}$) and Xe^{131} (spin $= \frac{3}{2}$) are each about 25 per cent abundant and can yield complementary information. Only magnetic terms appear in the spin Hamiltonians of the fluorides of Xe^{129}, while electric quadrupole interactions in the fluorides of Xe^{131} can give additional information about the strength and symmetry of the electrostatic fields at the xenon nucleus. Both broad-line and high-resolution techniques are applicable, although only the latter are considered here [1, 2].

We have studied the magnetic resonance spectra of XeF_2, XeF_4, $XeOF_4$, and XeF_6, dissolved in anhydrous hydrofluoric acid. At room temperature, solutions of XeF_4 and $XeOF_4$ show resolved fluorine spectra, while those of XeF_2 and XeF_6 show a single fluorine resonance, indicating chemical exchange with the solvent. On cooling, a resolved fluorine spectrum is obtained from XeF_2. The resolved fluorine spectra all consist of symmetrical three-line patterns plus the strong solvent line, illustrated by the XeF_2 spectrum at 56.442 mc., given in Figure 1. (The very broad resonance from the Teflon container also occurs in the region of interest but presents no complication other than to give a sloping base line.) Qualitatively similar spectra are obtained at 15.000 mc. The separation of the weak outer lines is found to be independent of the magnetic field, while the separation of the center of the three-line pattern from the HF^{19} resonance is proportional to magnetic field. As discussed later, application of a second radio-frequency field at one of the Xe^{129} resonance frequencies perturbs the satellite F^{19} lines, demonstrating conclusively that they result from the bonding of fluorine to Xe^{129}. The single center resonance therefore includes the contribution from fluorine attached to the spinless isotopes of xenon. If the molecules were of tetrahedral or higher symmetry, an additional four-line spectrum would be expected from F^{19} coupled to Xe^{131}. In the absence of such symmetry, quadrupole relaxation of the Xe^{131} spin states may collapse the four transitions into the center line. Relative-intensity measurements of the three observed lines confirm, to within 1 per cent, that the center lines of all the spectra do include the fluorine bonded to Xe^{131}. A quadrupole relaxation rate fast on the time scale of the spin-spin coupling constant is therefore implied, setting lower limits of the order of 10 mc. on the quadrupole coupling constants, e^2qQ/h, in these compounds. The estimate of e^2qQ/h of 2700 mc. obtained from studies of the Mössbauer effect on XeF_4 [see G. J. Perlow, C. E. Johnson, and M. R. Perlow, this volume, p. 282] indicates that the NMR value is of little quantitative significance. Finally, one infers from the simplicity of the spectra that the

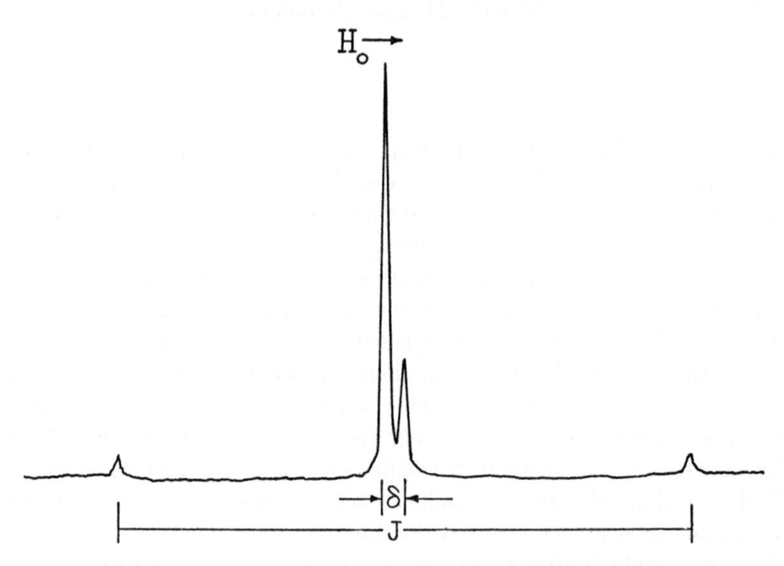

FIG. 1.—The F^{19} high-resolution magnetic resonance spectrum of XeF_2 at $-20°$ C. and $\nu = 56.422$ mc. The F^{19} chemical shift from HF^{19} is represented by δ, the F^{19}–Xe^{129} spin-spin coupling constant by J.

fluorine atoms are, on the time average of this experiment, magnetically equivalent. (The usual time scale of a nuclear magnetic resonance experiment is long, compared with the time scale of, for example, an electron-diffraction study. The case of PF_5 is a relevant example. On the short time scale of an electron-diffraction experiment, the structure of PF_5 is a trigonal bipyramid [3]. The results of a high-resolution magnetic resonance experiment indicate that all of the fluorine atoms are equivalent [8].) The F^{19} chemical shifts and F^{19}-Xe^{129} coupling constants obtained from the three resolved spectra are given in Table 1.

The Xe^{129} resonance spectra, whose observation might confirm the presence of xenon and show the number of equivalent fluorines in the molecules, are weak and therefore difficult to observe. In addition, they lack a convenient reference for the xenon shifts. Both problems are avoided in the double-resonance "tickling" experiment [4], in which one observes the effect on the F^{19} spectrum of a small radio-frequency field swept through each line in the Xe^{129} spectrum. As each Xe^{129} transition is irradiated, a slight splitting occurs in the F^{19} transitions that involve an energy level in common with the irradiated Xe^{129} transition [4]. Figure 2 shows the possible transitions in the triplet nuclear spin state resulting from the two equivalent F^{19} nuclei in XeF_2 [4]. It can be seen, for example, that

TABLE 1

F^{19} AND XE^{129} CHEMICAL SHIFTS AND F^{19}-XE^{129} COUPLING CONSTANTS IN XEF_2, XEF_4, AND $XEOF_4$

	$\delta_{F^{19}}$ (p.p.m.)*	$\delta_{Xe^{129}}$ (p.p.m.)*	$J_{F^{19}-Xe^{129}}$ (c/s)
XeF_2......	+ 3.3	−3930	5690
XeF_4......	−175	−5785	3864
$XeOF_4$.....	−292	−5511	1163

* These chemical shifts are defined as $10^6 \times (H_x - H_r)/H_r$, where the reference for F^{19} is HF^{19} and for xenon is a nominal value for xenon gas.

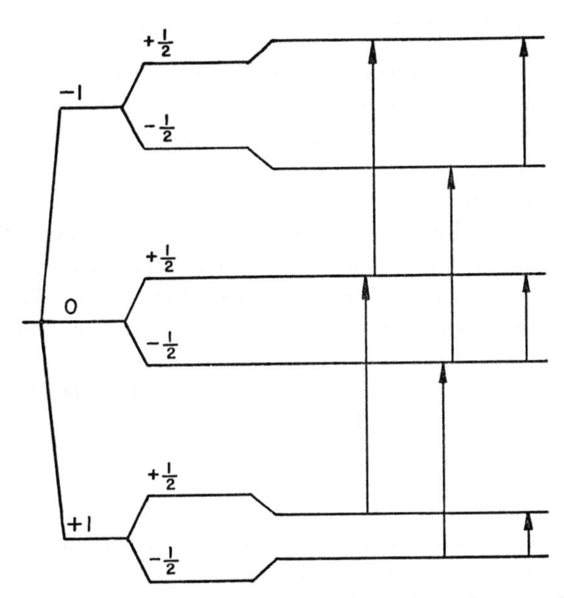

FIG. 2.—The allowed transitions in the triplet nuclear spin state of XeF_2

both components of the low-field F^{19} resonance will be split by irradiating the center xenon resonance, splitting the F^{19} line into a simple doublet, while only one of the two degenerate low-field components will be split by irradiating either of the two outer xenon frequencies, producing 1:2:1 triplets. The same considerations apply to $Xe^{129}F_4$ and $Xe^{129}OF_4$, although here the situation is complicated by the existence of both quintet and triplet states. The final result for these molecules is that 1:14:1, 1:2:1, and 3:2:3 triplets result from irradiation at the outer, intermediate, and central xenon frequencies, respectively.

Results of double-irradiation experiments on XeF_2 are shown in Figure 3; similar results for XeF_4 and $XeOF_4$ have been published previously [2]. The odd-line symmetry of the double-resonance spectra confirms the presence of even numbers of fluorines in the molecules; furthermore, irradiation with the central xenon frequency can produce a doublet only with a difluoride; and a triplet whose center is less intense than its side bands, only with a tetrafluoride.

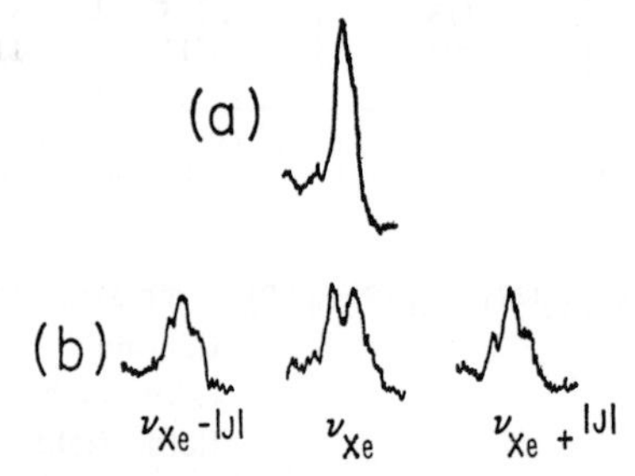

FIG. 3.—The low-field F^{19} transition in XeF_2 (HF solution at −20° C.): (*a*) unperturbed; (*b*) left to right, with simultaneous irradiation of the Xe^{129} transitions.

The separation between successive irradiating frequencies is equal to the coupling constant, and corresponding effects are observed in both F^{19} lines provided that, in field-swept spectrometers, the irradiating frequency is offset by the amount $J\gamma_{Xe}/\gamma_F$.

A reference point for the xenon chemical shifts is provided in the double-resonance experiments by the HF used to calibrate the F^{19} spectrum. In a typical experiment the magnetic field is swept to a value where, for example, the low-field component of the $Xe^{129}F^{19}$ doublet is observed at a fixed transmitter frequency, ν_o, and the tickling frequency, ν_{Xe}, of the center xenon transition is measured in this field. The HF resonance is observed in the same field at a sideband frequency, $\nu_s = \nu_o \pm \nu_m$, generated by audio modulation of the RF carrier. The magnetic field is determined from

$$H_o = \frac{\nu_s}{\gamma_F(1 - \sigma_{HF})},$$

which requires a calculated value, 413.7×10^{-6} [5], for the shielding in HF. The xenon shielding is then computed from

$$1 - \sigma_{Xe} = \frac{\nu_{Xe}}{\nu_s}\frac{\gamma_F}{\gamma_{Xe}}(1 - \sigma_{HF}),$$

using tabulated values for the magnetogyric ratios or magnetic moments of F^{19} and Xe^{129} [6]. The results are subject to any absolute errors in the HF shielding calculation and the magnetic-moment determinations, although these are of minor consequence in comparisons of relative shielding (chemical shifts) of xenon in its compounds. One can, with somewhat less accuracy, compare with the diamagnetic shielding in atomic xenon and predict chemical shifts which are, in principle, directly measurable. These are calculated from

$$10^{-6}\delta_{Xe} = \frac{1-\sigma_R}{1-\sigma_{HF}}\frac{\gamma_{Xe}}{\gamma_F}\frac{\nu_s}{\nu_{Xe}} - 1\,,$$

using $\sigma_R = 5590 \times 10^{-6}$ for atomic xenon ($Z = 54$) [7], and are listed in Table 1 above.

In addition to the qualitative information obtained from the resolved high-resolution spectra of the xenon fluorides, it is relevant to consider briefly the more quantitative aspects of the problem. It is clear that *accurate* molecular wave functions for the xenon fluorides must be able to predict the F^{19}-Xe^{129} coupling constants and F^{19} and Xe^{129} chemical shifts, as well as the qualitative changes found in going from one xenon compound to another. For the present we must settle for more approximate interpretations.

The chemical shifts of F^{19} in the xenon fluorides may be compared with the F^{19} shifts observed for some of the interhalogens, such as IF_7, IF_5, BrF_5, BrF_3, etc. [8]. In a number of simple fluorides, the F^{19} chemical shifts have been related to the ionic character of the fluorine bond [9]. The end-points of this scale are F_2, which of course is completely covalent, and HF, which is highly ionic. The Xe-F bond in XeF_2 would then be about as ionic as the H-F bond in hydrogen fluoride. This interpretation on the basis of ionic character is, however, probably not unique. It has been pointed out that changes in hybridization might produce effects similar to those produced by changes in ionic character [9]. A detailed treatment has been given for the case of F^{19} shifts in a number of fluorobenzenes [10], where an expression is obtained for the F^{19} shielding in terms of hybridization and double bonding, as well as ionic character. Such a detailed treatment may be needed in the present case.

The magnitudes of the F^{19}-Xe^{129} spin-spin coupling constants might be expected to shed further information on the bonding in the xenon fluorides [11]. The theories of spin-spin coupling constants, which have been used so successfully in the case of proton-proton coupling constants in organic molecules [12, 13], are not directly applicable to the case of heavier nuclei. Though some attempts have been made to evaluate the additional terms which are needed [12, 14], the present state of the theory does not allow detailed evaluations to be made. Even empirical correlations between spin-spin coupling constants and chemical shifts, when they do exist, yield a relationship in the direction opposite to that found for the xenon fluorides [15]. Thus, until detailed theories of spin-spin coupling become available, the F^{19}-Xe^{129} coupling constants must simply remain as interesting results. In this respect, the F^{19}-Xe^{129} coupling constants are very large. The value of 5690 cycles per second (c/s) obtained for XeF_2 is, relatively speaking, one of the largest coupling constants ever observed.

The present state of understanding of the chemical shifts of larger nuclei is equally unclear. Chemical shifts at least as large as those reported here for Xe^{129} have been observed, for example, in compounds of Pb^{207} [16]. Few attempts have been made to understand these in detail. From the three Xe^{129} chemical shifts available, one can conclude that the second-order (paramagnetic) contribution to the shielding is large. For the particular case of XeF_4, this contribution is larger even than the accepted value of 5590 p.p.m. for the diamagnetic shielding in atomic xenon [7]. A simple model for the xenon fluorides, which is an extension of the model used for the F^{19} chemical shifts in fluorobenzenes [10], does indeed predict large values for the second-order contribution to the shielding of Xe^{129} [17]. However, since the differences in shieldings for the different xenon compounds depend upon such quantities as ionic character, excitation energies, and radial-distribution functions, more refined calculations will be required before accurate values for the Xe^{129} shieldings can be calculated.

The unresolved, room-temperature spectra of XeF_2 and XeF_6 yield chemical information concerning the exchange of F^{19} between these fluorides and HF. A slow exchange with measurable activation energy is indicated by the emergence of an XeF_2 spectrum upon cooling, while no separate XeF_6 spectrum could be detected above temperatures approaching the freezing point of the solvent. The single F^{19} line observed in XeF_6 is shifted from an external reference by amounts proportional to the solute concentration, while the effect of solute on the proton resonance of the solvent is not appreciable. The fluorine resonance line shape closely approximates a single Lorentzian curve, whose width increases with the solute concentration, approaching a limiting value of about 1200 c/s. Saturation measurements show that T_2 is less than T_1 in the solution. Line-width measurements at several concentrations show no significant change between magnetic fields of 14.1×10^3 and 3.75×10^3 gauss. These facts collectively indicate a very rapid chemical exchange of fluorine between XeF_6 and HF; from the concentration-dependent shift of the F^{19} resonance, the average lifetime of an F^{19} atom on XeF_6 appears to be less than 10 microseconds.

In principle, the line shapes can yield quantitative information about the exchange rates and mechanism(s). However, one is uncertain about the role of $Xe^{131}F_6$, which would have a four-line spectrum in the slow-exchange limit, provided the molecule has octahedral symmetry.

Xenon tetrafluoride was obtained from Peninsular Chem-research. The other compounds reported here were kindly supplied by D. F. Smith of the Oak Ridge Gaseous Diffusion Plant. The samples were studied in 5- and 10-mm. Teflon tubes, in a standard Varian High-Resolution Spectrometer. The double-irradiation experiments were carried out initially with an improvised coupling network and Varian Variable-Frequency RF unit, and later, more precisely with an NMR Specialties Company SD-60 Heteronuclear Spin Decoupler.

REFERENCES

1. Two broad-line NMR studies have been reported for solid XeF_4: S. MARIČIĆ and Z. VEKSLI, *Croat. Chem. Acta* **34,** 189 (1962); R. BLINC, P. PODNAR, J. SLIVNIK, and B. VOLAVŠEK, *Physics Letters* **4,** 124 (1963). The F^{19} line is sufficiently narrow at room temperature that the latter group has been able to obtain an approximate value for the F^{19} chemical shift in XeF_4 (448 ± 20 p.p.m. upfield from F_2). This value is consistent with the high-resolution value given here.
2. Some of the results of the high-resolution magnetic-resonance studies have been given previously: T. H. Brown, E. B. WHIPPLE, and P. H. VERDIER, *Science* **140,** 178 (1963); T. H. BROWN, E. B. WHIPPLE, and P. H. VERDIER, *J. Chem. Phys.* **38,** 3029 (1963).
3. L. O. BROCKWAY and J. Y. BEACH, *J. Am. Chem. Soc.* **60,** 1863 (1938).
4. R. FREEMAN and D. H. WHIFFEN, *Proc. Phys. Soc.* **79,** 794 (1962); R. FREEMAN and W. A. ANDERSON, *J. Chem. Phys.* **37,** 2053 (1962); A. L. BLOOM and J. N. SHOOLERY, *Phys. Rev.* **97,** 1261 (1955).
5. C. W. KERN and M. KARPLUS, private communication. Their value is based upon a σ^d computed theoretically and a σ^p determined from measurement of the spin-rotational constant of HF by M. R. BAKER, H. M. NELSON, J. A. LEAVITT, and N. F. RAMSEY, *Phys. Rev.* **121,** 807 (1961).
6. N. F. RAMSEY, *Molecular Beams.* London: Oxford University Press, 1956.
7. W. C. DICKINSON, *Phys. Rev.* **80,** 563 (1950).
8. H. S. GUTOWSKY and C. J. HOFFMAN, *J. Chem. Phys.* **19,** 1259 (1951).
9. A. SAIKA and C. P. SLICHTER, *ibid.* **22,** 26 (1954).
10. M. KARPLUS and T. P. DAS, *ibid.* **34,** 1683 (1961).
11. N. F. RAMSEY and E. M. PURCELL, *Phys. Rev.* **85,** 143 (1952); N. F. RAMSEY, *ibid.* **91,** 303 (1953).
12. H. M. MCCONNELL, *J. Chem. Phys.* **24,** 460 (1956).
13. M. KARPLUS, D. H. ANDERSON, T. C. FARRAR, and H. S. GUTOWSKY, *ibid.* **27,** 597 (1957), and subsequent references.
14. J. A. POPLE, *Mol. Phys.* **1,** 216 (1958).
15. See, for example, J. A. POPLE, W. G. SCHNEIDER, and H. J. BERNSTEIN, *High-Resolution Nuclear Magnetic Resonance,* p. 329. New York: McGraw-Hill Book Co., 1959.
16. L. H. PIETTE and H. E. WEAVER, *J. Chem. Phys.* **28,** 735 (1958).
17. D. LAZDINS, C. W. KERN, and M. KARPLUS, private communication.

ANISOTROPY OF FLUORINE CHEMICAL SHIFT IN SOLID XENON TETRAFLUORIDE

R. BLINC, P. PODNAR, J. SLIVNIK, AND B. VOLAVŠEK
Nuclear Institute Jožef Stefan

S. MARIČIĆ AND Z. VEKSLI
Rudjer Bošković Institute

Static magnetic-susceptibility measurements [1] confirmed the expected diamagnetism of XeF_4, from room temperature down to 77° K. In the same temperature range, the fluorine magnetic resonance line shapes are asymmetric, undergoing a change of width as depicted in Figure 1 (for 29 mc/s). We wish to report here on these measurements and discuss the fluorine chemical shift [2] and its anisotropy, with the aim of shedding some light on the nature of the xenon-fluorine bond.

The paramagnetic fluorine shift has been measured in XeF_4 dissolved in HF and in solid XeF_4 relative to HF as an external standard [2]. Its field dependence at room temperature is shown in Figure 2. Expressed in p.p.m., this isotropic shift (σ_o) is 448 $\pm$ 30 with respect to F_2 or -415 with respect to the ideal F^- ion (using the σ_o value for F_2 of Burns, Agron, and Levy [5]). The F^{19} magnetic resonance of XeF_4 is thus closer to the value for HF (625 p.p.m., relative to F_2) than to the value for F_2, which indicates that the Xe-F bond has a substantial ionic character.

A similar conclusion may be reached independently by discussing the anisotropy of the chemical-shift tensor. This parameter can be obtained in two ways:

a) It can be obtained directly by plotting the experimentally determined second moments *vs.* the squares of the applied magnetic field. The slope of the straight line is related to the difference ($\Delta\sigma$) between the principal values of the shift tensor parallel ($\sigma_{\parallel}$) and perpendicular ($\sigma_{\perp}$) and to the bond direction (for bonds of axial symmetry) [3]. Figure 3 shows such a plot of second moments obtained at sufficiently low temperatures to freeze in any molecular motion. The two points at $H_o^2 = 0$ are calculated second moments due to pure nuclear dipole-dipole interactions. The intermolecular second moment was calculated from the X-ray crystal-structure data [4] and added to each intramolecular term obtained (Xe-F = 1.93 A., according to Templeton *et al.* [4], and Xe-F = 1.95 A., according to a neutron-diffraction study [5]). From Figure 3 one obtains $|\sigma_{\parallel} - \sigma_{\perp}| = \Delta\sigma = 7.6 \times 10^{-4}$.

b) It can be obtained indirectly by using Lebedev's [6] nomograms and three characteristic parameters of low-temperature asymmetric lines. Using this method, we estimate $\Delta\sigma = 7.9 \times 10^{-4}$, which is in good agreement with the former result.

Karplus and Das [7] were able to calculate the isotropic shift for a covalently bonded fluorine atom, to which molecular fluorine approximates. Their value,

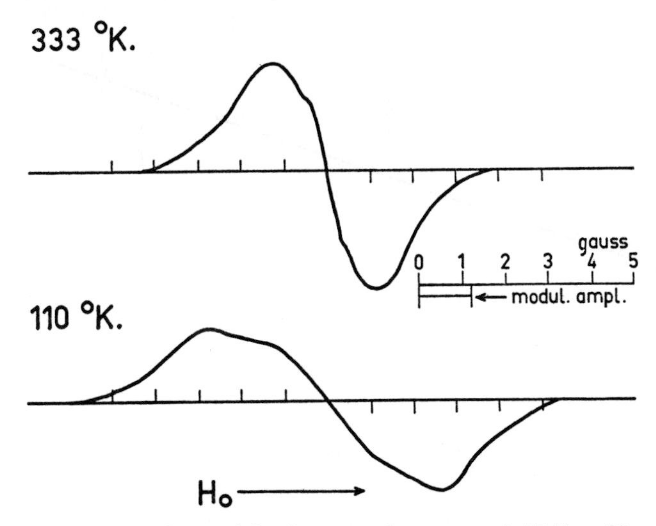

FIG. 1.—The line shapes of fluorine magnetic resonance in XeF_4 at 29 mc/s

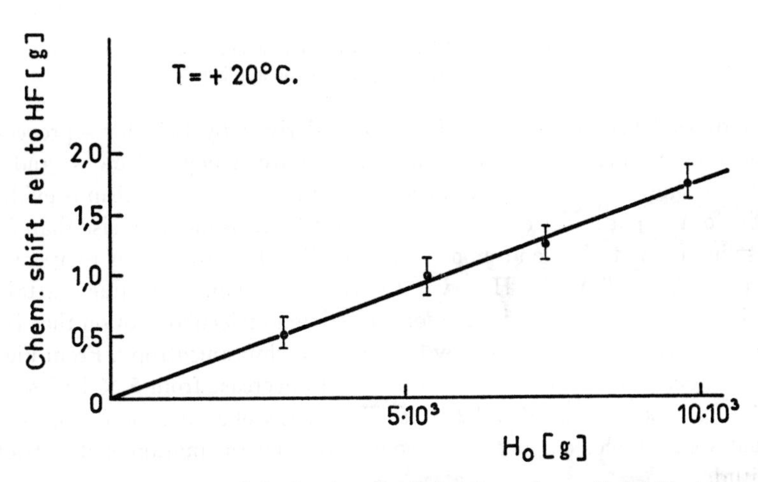

FIG. 2.—The dependence of the fluorine chemical shift on the applied field at room temperature.

relative to the F^- ion, is $\sigma_o = -8.63 \times 10^{-4}$. Since $\frac{2}{3}\Delta\sigma = \sigma_o$ (see Andrew and Tunstall [3]), we obtained $\sigma_o = -5.2 \times 10^{-4}$ for the Xe-F bond (with the shift anisotropy $\Delta\sigma_{\text{mean}} = 7.75 \times 10^{-4}$), which is in relatively good agreement with the value obtained from the chemical-shift data: $\sigma_o = -4.15 \times 10^{-4}$. This is significantly smaller than the covalent value, so that we may confidently conclude that the Xe–F bond is not devoid of ionic character, hybridization, and double-bonding. It requires more detailed analysis [3] to elucidate the partial contribution of each of the three listed possibilities.

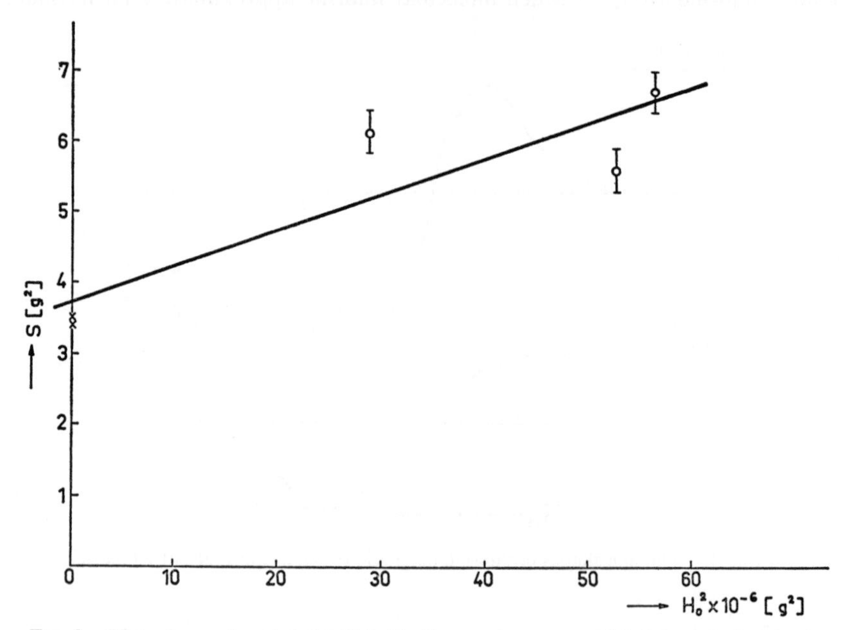

FIG. 3.—Plot of experimental "rigid-lattice" second moments (*circles*) and those calculated from crystal structure data (*crosses*) *vs.* square of the applied field.

The dependence of $\Delta H_{\text{anis}} = |H_{\parallel} - H_{\perp}|$ (derived by Lebedev's procedure [6]) on temperature was measured at constant frequency (29 mc/s) and the results are shown in Figure 4. A molecular reorientation mechanism is evident, and the location of the critical temperature (250° K.) is more precise than for a corresponding plot of second-moment values. The data above this temperature are only "effective" ΔH_{anis}. However, the fact that a non-zero value is attained at higher temperatures is evidence for a restricted molecular motion that is insufficient to average out completely the chemical-shift anisotropy. From Figure 4 the motional correlation time is estimated to decrease from 5×10^{-5} sec. to 2×10^{-7} sec. between 220° and 273° K. The energy of activation for this effect is about 4 kcal/mole, supporting the conclusion that the motion is of restricted amplitude.

The main difficulty in relating the NMR data of XeF_4 to the dimensions of this molecule is the anisotropy contribution to the "rigid-lattice" second moment. More detailed experimental evidence is required in order to obtain precisely the pure nuclear second moment by extrapolating $S = f(H_o^2)$ to $H_o^2 = 0$, independently of the crystal-structure data. However, there seems to be a good deal of agreement between the crystal-structure second moment of 3.45 G^2 and the NMR value. The latter is given by the intersect on the ordinate in Figure 3, 3.73 G^2. From the low-temperature $\Delta H_{\text{anis}} = 5.7 \pm 0.2\ G$ at 29 mc/s, one can work out (by Lebedev's nomograms [6]) the line width of the component lines in

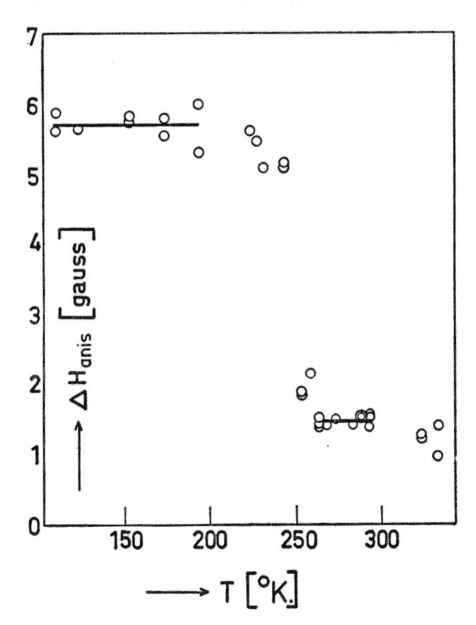

FIG. 4.—The anisotropy of fluorine (NMR) chemical shift in XeF_4 at 29 mc/s *vs.* temperature.

an anisotropic spectrum. This width is attributed to the pure nuclear dipole-dipole broadening, from which the corresponding "rigid-lattice" second moment is estimated to be 3.2 G^2 in reasonable agreement with the values in Figure 3.

Lebedev's [6] treatment was originally devised for an isotropic, motionless distribution of anisotropic *g*-factors in electron paramagnetic resonance. However, the agreement we reach here between estimates using Lebedev's nomograms and other, more direct determinations is encouraging. The use of this simple method may greatly facilitate the very promising application of fluorine chemical-shift anisotropy to molecular- and crystal-structure studies.

Acknowledgment.—We wish to thank Professor Andrew for having sent us his manuscript on fluorine chemical-shift anisotropy.

REFERENCES

1. S. Maričić, J. Slivnik, Z. Veksli, and B. Volavšek, *Croat. Chem. Acta* **35**, 77 (1963), in press.
2. R. Blinc, P. Podnar, J. Slivnik, and B. Volavšek, *Physics Letters* **4**, 124 (1963).
3. E. R. Andrew and D. P. Tunstall, *Proc. Phys. Soc.*, in press.
4. D. H. Templeton, A. Zalkin, J. D. Forrester, and S. M. Williamson, *J. Am. Chem. Soc.* **85**, 242 (1963). See also J. A. Ibers and W. C. Hamilton, *Science* **139**, 106 (1963).
5. J. H. Burns, P. A. Agron, and H. A. Levy, *Science* **139**, 1208 (1963).
6. J. S. Lebedev, *Zhur. strukt. khim.* **4**, 22 (1963).
7. M. Karplus and T. P. Das, *J. Chem. Phys.* **34**, 1683 (1961).

HYDROGEN FLUORIDE SOLUTIONS CONTAINING XENON DIFLUORIDE, XENON TETRAFLUORIDE, AND XENON HEXAFLUORIDE

H. H. Hyman and L. A. Quarterman
Argonne National Laboratory

Abstract

Xenon tetrafluoride is sparingly soluble in anhydrous hydrogen fluoride; xenon difluoride and xenon hexafluoride are very soluble. Xenon hexafluoride ionizes extensively; the others do not.

Shortly after xenon tetrafluoride became available, its solubility in anhydrous hydrogen fluoride was investigated. It was soon established that xenon tetrafluoride (XeF_4) was appreciably soluble in anhydrous hydrogen fluoride, that the electrical conductivity was not appreciably different from that of the pure solvent, and that the Raman spectrum and NMR spectrum could be best interpreted in terms of dissolution without reaction, ionization, or unusual interaction.

The chemical reactions of a solution of xenon tetrafluoride in hydrogen fluoride were those of an active fluorinating reagent. Organic samples such as benzene reacted rapidly. Metals such as platinum and molybdenum were converted to fluorides [1].

When xenon difluoride and xenon hexafluoride became available, similar studies were initiated. Both the difluoride and hexafluoride of xenon proved to be very soluble at room temperature in anhydrous hydrogen fluoride—much more soluble indeed than the xenon tetrafluoride.

Xenon difluoride (XeF_2), though highly soluble in hydrogen fluoride, appears to undergo no reaction or ionization in solution.

Xenon hexafluoride (XeF_6), on the other hand, is very soluble in anhydrous hydrogen fluoride; furthermore, the solutions are electrically conducting. Solutions containing XeF_6 are yellow liquids, comparable in color to liquid XeF_6. This color disappears on cooling.

NMR spectra are consistent with extensive ionization. The Raman spectrum of the colored solution is rather hard to observe, but a single peak has been found. While this may be most simply interpreted as the fundamental of an octahedrally symmetrical XeF_6 molecule, extensive ionization in more dilute solution is certainly not inconsistent with this observation.

Indeed, while any alternative symmetry would yield a more complicated pattern of Raman bands than the octahedron, the observations are not good enough to exclude this possibility.

EXPERIMENTAL

All xenon compounds were prepared at Argonne National Laboratory, as described elsewhere in this volume [C. L. Chernick, p. 35]. All hydrogen fluoride was redistilled commercial material with a conductivity less than 10^{-4} ohm^{-1} cm.$^{-1}$ at 0° C. [2]. Solutions were prepared by weight on a vacuum line using all polychlorotrifluoroethylene (Kel-F) equipment as described elsewhere [2]. Solubilities were determined by preparing solutions of appropriate concentration by

TABLE 1

SOLUBILITY OF XeF_2, XeF_4, AND XeF_6 IN ANHYDROUS HF

Compound	Temperature (T° C.)	Solubility (moles/1000 gm)
XeF_2..........	− 2.0	6.38
	12.25	7.82
	29.95	9.88
XeF_4..........	20	0.18
	27	0.26
	40	0.44
	60	0.73
XeF_6..........	15.8	3.16
	21.7	6.06
	28.5	11.2
	30.25	19.45

weight and observing the temperature at which the last crystal disappeared. Repeated freezing and thawing of the same sample did not affect the observed temperature.

The solubility data are summarized in Table 1 and in Figure 1, where the log of the solubility is plotted against the reciprocal of the absolute temperature. The heats of solution estimated from the lines drawn in Figure 1 are 2.5, 6.7, and 18 kcal/mole for XeF_2, XeF_4, and XeF_6, respectively.

The electrical conductivities found for solutions of XeF_6 in hydrogen fluoride are summarized in Table 2. For complete ionization at 0° C., a molar conductivity of about 300 would be reasonable.

Raman spectra were recorded with a Cary Model 81 Photoelectric Recording Spectra Photometer. The Raman tubes used were the most recent version of similar Kel-F tubes described elsewhere [2]. They were fabricated from 10-in. lengths of $\frac{3}{4}$-in. tubing 0.030 in. thick. Appropriate fittings at each end of the tube permitted the attachment of the sapphire window and a Kel-F valve. Only sapphire and Kel-F were in contact with the solutions.

Unfortunately, these tubes require an excess of 50 ml. of solution, and for reagents in short supply it has not always been possible to prepare hydrogen fluoride solutions containing enough of the sample to give a good spectrum.

An additional problem was encountered in manipulating xenon hexafluoride

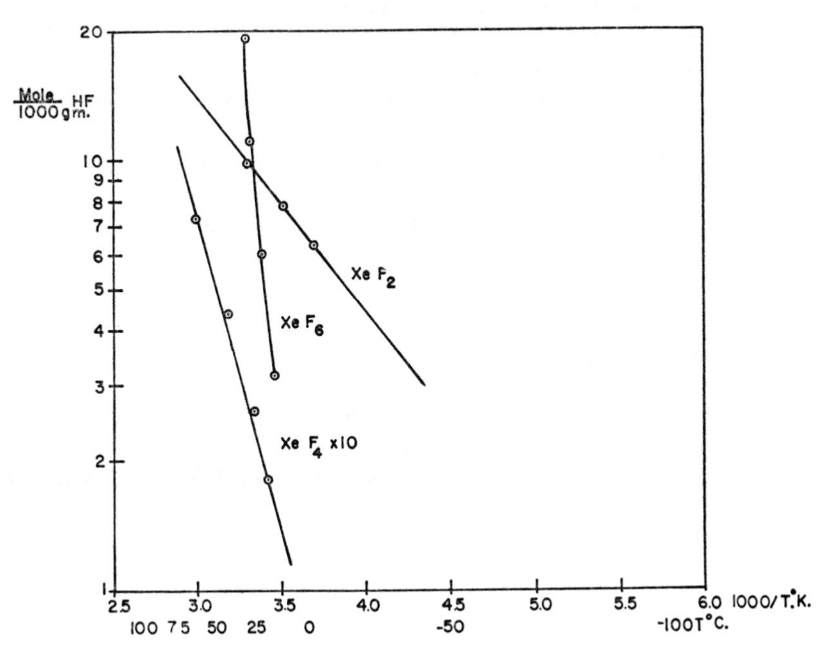

FIG. 1.—Solubility of XeF_2, XeF_4, and XeF_6 in anhydrous HF

TABLE 2

ELECTRICAL CONDUCTIVITY OF XeF_6 IN HF AT 0° C.

Concentration (mole/l)	Series*	Specific Conductivity (ohm^{-1} $cm.^{-1}$)	Molar Conductivity (ohm^{-1} $cm.^{2}$)
0.02_4.....	B	3.53×10^{-3}	147
$.06_7$.....	B	8.44	126
.09......	A	13.4	150
$.13_5$.....	B	15.0	110
$.16_3$.....	B	16.7	102
$.17_5$.....	A	19.2	110
.24.....	B	23.0	96
.49......	B	33.6	69
0.75......	B	44.9	60

* Independent serial dilutions of different batches.

solutions. The tube walls are thin, and flaws occur in the material from which they are fabricated. Cells containing solutions of xenon hexafluoride in hydrogen fluoride developed a leak during use in every case, and it was not possible to employ optimum Raman scans. Xenon hexafluoride solutions in hydrogen fluoride have been stored for moderately long periods in heavy-walled Kel-F containers, but it was often possible to find evidence of substantial cracks in the exposed plastic when the solutions were removed.

While somewhat similar observations were made with xenon tetrafluoride, it is not yet known whether this would be true of pure samples, since many xenon tetrafluoride samples handled at the beginning of this program contained xenon hexafluoride.

The Raman displacement found for the principal band of xenon tetrafluoride in hydrogen fluoride solution is 550–553 cm.$^{-1}$. The intensity for a 0.1 molar solution corresponds to approximately 1 per cent of that for the 460 cm.$^{-1}$ band of carbon tetrachloride in a similar cell.

The xenon hexafluoride absorption band was found at about 620 cm.$^{-1}$. The very low intensity observed for this band is undoubtedly due to light absorption by the yellow solution.

REFERENCES

1. C. L. Chernick *et al.*, *Science* **138,** 136 (1962).
2. H. H. Hyman, L. A. Quarterman, M. Kilpatrick, and J. J. Katz, *J. Phys. Chem.* **65,** 123 (1961).

THE MÖSSBAUER EFFECT IN CHEMICAL COMPOUNDS OF Xe^{129}

G. J. PERLOW, C. E. JOHNSON,* AND M. R. PERLOW
Argonne National Laboratory

The Mössbauer effect [1] can be a useful tool for testing chemical structures. Wherever in the list of stable nuclides a low-lying nuclear state can be excited (for example, by a previous beta decay), some fraction of the gamma rays emitted in de-excitation will come off without nuclear recoil. Such a γ-ray has the proper energy to be absorbed in the inverse process and to be detected in a simple transmission experiment. The resonant character of the absorption makes for high cross-section. The resonance is scanned by using the Doppler effect to alter the γ-ray energy. In the present experiments, speeds of a few centimeters per second were adequate. The relevance of the effect to chemical structure arises chiefly from three circumstances: (1) In either source or absorber, or in both, the atomic environment of the nucleus may cause a hyperfine splitting of ground and/or excited state. In the compounds of Xe^{129} we are interested in the electric quadrupole interaction between the field gradient at the nucleus attributable to xenon orbitals and the quadrupole moment of the excited state. (2) If the ground and excited states of the nucleus do not have the same mean-square charge radius, the γ-ray energy is altered by that part of the electronic charge cloud which is contained within the nuclear volume. This so-called isomer effect is sensitive to the *s*-electron density at the nucleus. (3) The fraction of the radiation which is emitted or absorbed without recoil is dependent on the binding of the atom in the crystal, on the γ-ray energy, and on the temperature of the experiment. We shall not make use of this property further except to point out that it is necessary to make all measurements at liquid-helium temperature in order to get sufficient effect.

Our measurements were made with the 40.0-kev. $\frac{3}{2}^+ \rightarrow \frac{1}{2}^+$ transition in Xe^{129}, which follows the β decay of I^{129} (1.6×10^7 *y*). The half life of the state is known to be 0.7 ± 0.3 nsec. from delayed-coincidence measurements [2]. For a source we used 0.5 gm. of NaI^{129}, while the absorbers were successively: the hydroquinone clathrate containing 28 per cent Xe by weight; sodium perxenate (nominally $Na_4XeO_6 \cdot 2H_2O$); xenon tetrafluoride; and xenon difluoride. All

* C. E. Johnson is on leave from AERE Harwell.

xenon compounds were prepared by C. L. Chernick and J. G. Malm, using methods described elsewhere in this volume [e.g., pp. 35 and 167].

The apparatus is shown schematically in Figure 1. The source and absorber were immersed in liquid helium in the inner Dewar, and there was liquid nitrogen in the outer. The radiation emerged through helium, nitrogen, and the Dewar walls and was detected in the proportional counter below. The crank motion was communicated to the source (or in some cases to the absorber) by Mylar tubes. The 40-kev. line was selected in a single-channel pulse-height

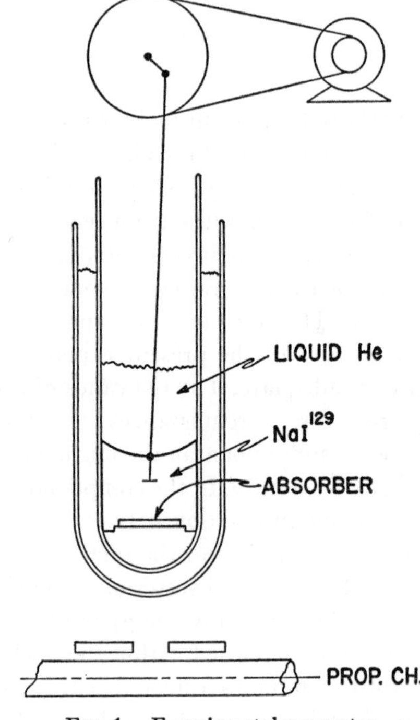

FIG. 1.—Experimental apparatus

analyzer. The data were registered in a 400-channel pulse-height analyzer used as a multiple scaler. An optical shutter served to synchronize the crank with the oscillator which advanced the channel number, so that each channel could be assigned a velocity.

We show representative spectra in Figure 2 and a summary of data in Table 1. The clathrate and perxenate show a single line unshifted from zero velocity within experimental error. Both XeF_4 and XeF_2 show large and equal quadrupole splittings. We observe no isomer shift for them either. All data have been fitted and errors determined by the method of least squares.

The single-line spectrum in clathrate and perxenate means that the field gradient eq ($\equiv \delta^2 V/\delta_z^2$) vanishes for both these substances and for the NaI^{129}

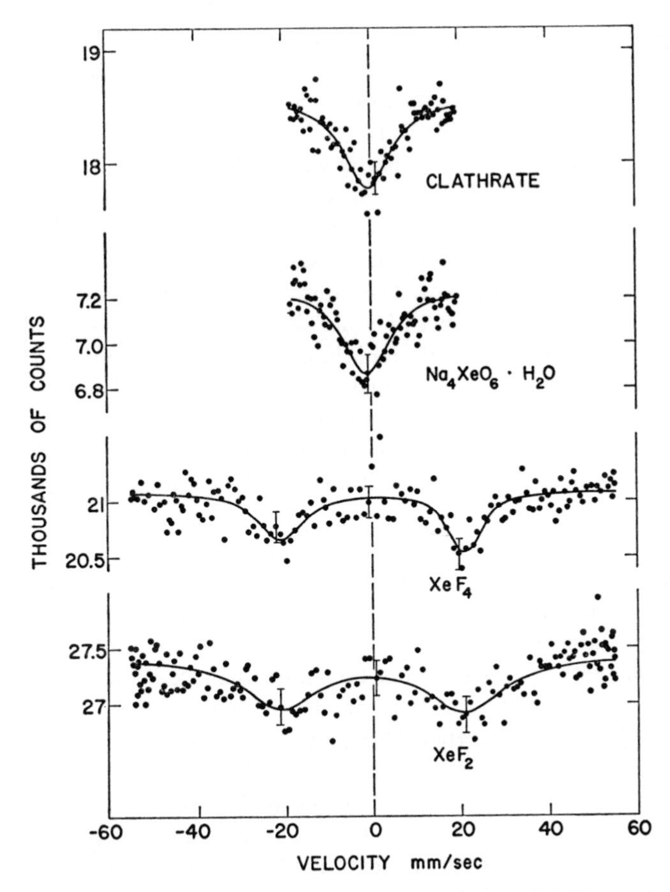

FIG. 2.—Velocity spectrum of xenon compounds using NaI^{129} source

TABLE 1

LEAST SQUARES FIT TO VELOCITY SPECTRA

Compound	Thickness (mg/cm² of Xe)	Line position (mm/sec)	Isomer shift (mm/sec)	Line width (mm/sec)	Dip (%)
Hydroquinone clathrate	23	− 0.2±0.3	−0.2±0.3	11.9±1.4	4.8±0.1
Xenate	24	− 0.3±0.4	− .3±0.4	12.6±1.5	5.4± .3
Line 1		21.1±0.7	....	9 ±2	2.6± .3
XeF_4	23	....	.1±1.1	....	....
Line 2		−20.8±0.9	....	13 ±3	2.0± .2
Line 1		20.9±1.2	....	21 ±4	1.8± .2
XeF_2	23	....	0.0±1.9	....	....
Line 2		−20.9±1.4	....	21 ±4	1.6±0.2

source. In the latter case, the β^- decay of I^- should lead predominantly [3] to excited Xe^0. The xenon atom finds itself in a cubic iodine site, and q necessarily vanishes. The clathrate environment does not necessarily have such high symmetry, as was pointed out by Ruby and collaborators [4], who studied the Mössbauer effect of Kr^{83} in a similar environment. However, the line width is greater for xenon than for krypton, the energy higher, and the expected effect on our absorption line is less than our error. From our measured line width we obtain the half life of the excited nuclear state as 0.58 ± 0.07 nsec., consistent with the electronic measurement. The vanishing of q for the xenate is consistent with the assignment of a valence of either six or eight as the chemical evidence suggests. This is easily seen for the ionic limit, where the loss of the $5p$ shell from the xenon leaves it in an orbital singlet state. We expect, however, that the valence of eight would result in reduction of the $5s$ density at the xenon nucleus and thus in an isomer shift. We have not found such a shift, but it is likely to be small for even-Z nuclei in any case. More refined measurements may cast light on this point in the future.

In order to analyze the results of the fluorides we need to know the quadrupole moment, Q, of the excited state in Xe^{129}. In general, such knowledge would come from a measurement of the quadrupole coupling e^2qQ in a case for which q can be calculated. We have, of course, the opposite problem, but, Xe^{131}, which for energetic reasons is less favorable than Xe^{129} for observing the Mössbauer effect, has the same low-lying levels as Xe^{129}, except that the order of the $\frac{3}{2}^+$ and $\frac{1}{2}^+$ levels is inverted. This has permitted measurement of the quadrupole moment of the $\frac{3}{2}^+$ state in Xe^{131} by optical spectroscopy [5]. We feel justified in using this value ($-0.12b$) for the analogous state in Xe^{129}.

Considering the tetrafluoride first, the large value of the splitting, $e^2qQ/2 = 41.9 \pm 1.1$ mm/sec (or 1360 mc/s), coupled with the small value of the moment requires extreme assumptions about the bonding. Following a suggestion of Gordon Goodman, we assume that the field gradient is to be attributed to a doubly-occupied p_z orbital, the charge distribution in the bonding xy plane having been stretched outward toward the fluorines so much that it does not appreciably contribute to the gradient. This is possible without assuming pure ionic bonding because of the rapid variation of q with radius. We obtain $q = -(8/5)\langle 1/r^3\rangle$. If we use the value $\langle (a_0/r)^3\rangle = 22.4$ calculated for the singly ionized $5p$ shell, we get $e^2qQ/2 = 15.7$ mm/sec (506 mc/sec). This is too small by a factor of 2.7. If we wish to take an ionic model seriously, then a correction has to be applied to $\langle 1/r^3\rangle$ because of the decrease in screening. The effect is not large, however, since the important contributions to the integral come from regions where the effective nuclear charge approaches z in any case. After making a correction [6] for the fourfold ionization the calculated splitting becomes about 24 mm/sec.

In the case of the linear XeF_2, the picture consistent with that adopted for XeF_4 is to ascribe the field gradient to two p_z holes, the z axis now being the bonding axis. The bonding electrons are now close to the fluorines as in XeF_4 and, similarly, contribute zero gradient. For this one obtains $q = +8/5 \langle 1/r^3 \rangle$, differing only in sign from XeF_4. Since we measure the same splitting in both fluorides, we must assign them the same value for $\langle 1/r^3 \rangle$. Unfortunately we cannot have the same correction for ionization in both compounds. Thus, in both XeF_2 and XeF_4, we find it difficult to find a large enough source of the field gradient.

Acknowledgments.—We wish to thank John Oyler for help with the measurements and Curtis Rockwood for some of the electronic design and to acknowledge illuminating discussions with Drs. Gordon Goodman, John Gabriel, Murray Peshkin, Harry Lipkin, and Malcolm Macfarlane.

REFERENCES

1. R. L. Mössbauer, *Z. Phys.* **151**, 124 (1958); *Naturwissenschaften* **45**, 538 (1958). These articles and a selection of later ones on the subject are reprinted in Hans Frauenfelder, *Mössbauer Effect.* New York: W. A. Benjamin, 1962. For later developments, see D. M. J. Compton and A. H. Schoen (eds.), *The Mössbauer Effect, Proceedings of the Second International Conference.* New York: John Wiley & Sons, 1962.
2. T. Alväger, B. Johansson, and W. Zuk, *Ark. Fysik* **14**, 373 (1958).
3. T. A. Carlson, A. H. Snell, F. Pleasanton, and C. H. Johnson, *IAEA Symposium on the Chemical Effects of Nuclear Transformations*, Prague, 1960. Vienna: International Atomic Energy Agency, 1961.
4. Y. Hazoni, P. Hillman, M. Pasternak, and S. Ruby, *Physics Letters* **2**, 337 (1962).
5. A. Bohr, J. Koch, and E. Rasmussen, *Ark. Fysik* **4**, 455 (1951).
6. C. H. Townes, *Encyclopedia of Physics*, p. 423. Berlin: Springer Verlag, 1958.

MÖSSBAUER EFFECT IN KRYPTON

Stanley L. Ruby
Westinghouse Electric Corporation

I would like to make it clear that we have nothing to report on krypton compounds; the main reason for my not being able to report on the Mössbauer effect in such compounds is that I do not have any such compounds. In fact, a major reason for my contribution to this volume is to suggest that people who have krypton compounds and no Mössbauer analyzer get together with people who have Mössbauer analyzers and no krypton compounds.

I want to spend a minute to disabuse the notion that a man who studies Mössbauer effect must spend his whole life on Fe^{57}. The essential parameters for a good nucleus are low energy and longish lifetime. Kr^{83} has an energy lower than that of Fe^{57}, and its lifetime is 1.47 times as long. It even has a reasonably long-lived parent and a satisfactorily high isotopic abundance, and in most regards it is equal, if not superior, to the notorious Fe^{57}. The decay scheme is shown in Figure 1 [1], and the goodly size of the experimental effect in Figure 2 [2]. The lower curve is obtained from a still absorber, and the upper curve from a vibrating one. Clearly then, one can investigate krypton compounds using the Mössbauer effect, if you can get the krypton compounds. Successful experiments with Xe^{129} are described in this volume, despite the fact that this nucleus is much less well suited than Kr^{83} for Mössbauer experiments. [Perlow, Johnson, and Perlow, p. 279.] Thus the solid state physics and chemistry of the noble gases are relatively well off in regard to suitable Mössbauer nuclear probes.

Mössbauer analysis can give information not available by other methods or at least the same information more easily. This is no place to review the many chemical phenomena which can affect the observed patterns; I wish merely to briefly describe two of the most important. For a good review of the work to date, I believe the article by Boyle and Hall is the most complete and authoritative available [3].

First consider the so-called isomeric shift. This is the result of the change in charge radius of the nucleus between the ground and excited states interacting with the electronic charge density at the nucleus. The result is a measurable change in the energy of the emitted and absorbed γ-ray which gives the difference in electron density at the nucleus. This is most directly useful in Sn^{119}, where the $(5s)^2$, $(5s)^1$, and $(5s)^0$ states show up quite clearly. Things are more complicated in Fe^{57}, where the change between Fe^{3+} and Fe^{2+} is largely that of one more $3d$ electron; since only s electrons have a finite density at $r = 0$, this simple analysis would lead one to expect no isomeric shift. Actually, the change in shielding affects the $3s$ and $4s$ functions enough to make even this "indirect" isomer shift easily visible—and useful. I think it will be amusing to look for this shift in, say, KrF_4 as compared to, say, solid Kr.

A second, useful observation is $f(T)$—the Debye-Waller factor as a function of temperature. This can also be observed by X-ray techniques. I'm sure they would do things more precisely. Essentially, this is a measurement of $\langle r^2 \rangle$ (ac-

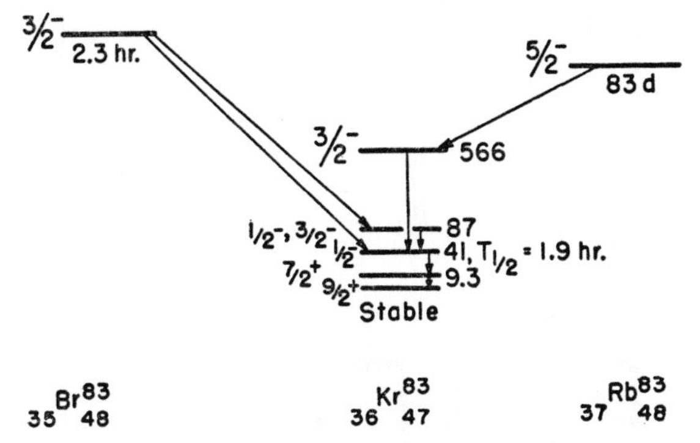

FIG. 1.—Simplified energy level for A = 83. (Redrawn from Nuclear Data Sheet NRC 59-1-77.) We report the half life of the 9.3 kev. level of Kr^{83} as 147 ± 4 nsec.

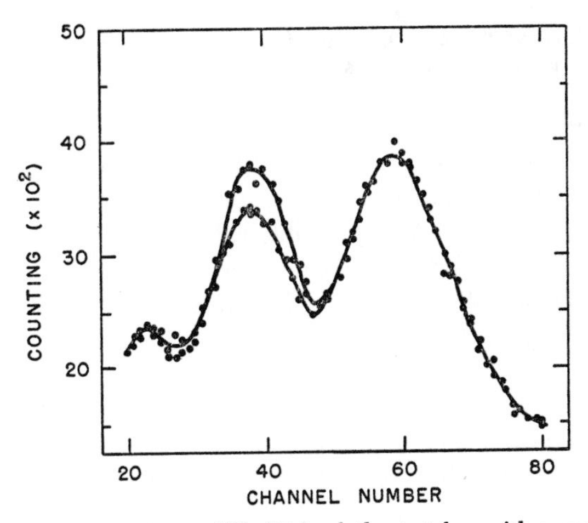

FIG. 2.—The γ-energy spectrum of Kr ^{63m} in clathrate taken with a proportional counter and zinc critical absorber with Mössbauer source and absorber at 107° K. The upper curve is for source vibrating, the lower for source fixed.

tually $f = \exp - (\kappa^2\langle r^2 \rangle)$ where r is measured along the direction of the γ-ray). But $\langle r^2 \rangle$ in turn is (sometimes) simply related to cell dimensions and lattice constants. In some cases, these results can be used to find the actual force constants between the Mössbauer atom and its neighbors. As an example of the usefulness of this, consider Figure 3; this shows $f(T)$ for Kr in β-quinol clathrate. Let us not now discuss exactly what a clathrate is but merely point out that this graph shows a fairly clear picture of a particle not harmonically bound, nor in a

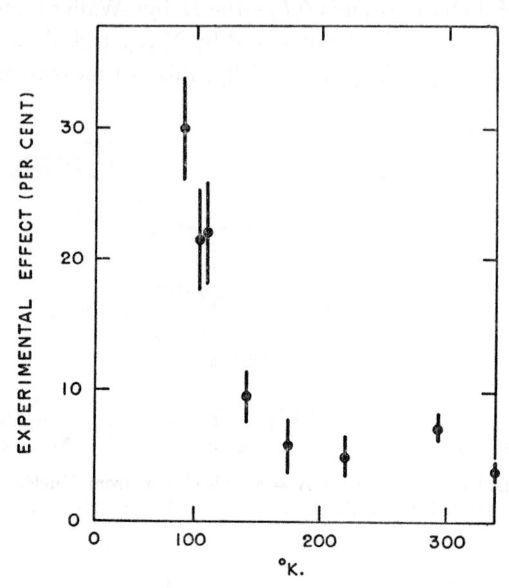

FIG. 3.—The observed "on-off" difference for the 9.3 kev. Kr^{83} γ-ray with clathrate source and absorber both at temperature shown.

box, but rather in between. The rapid drop in f as the temperature rises from 0° is typical of ordinary harmonic binding, but the rather constant plateau found later on is just what one would expect for a square well potential. Numerically, it appears that $\sqrt{\langle r^2 \rangle} \simeq 0.3$ A., a plausible value.

To conclude: please consider Mössbauer analysis as a way to get information on krypton compounds. We will pay the freight to Pittsburgh if you supply the crystals. If they are xenon, though, the address is Chicago.

REFERENCES

1. S. L. RUBY, Y. HAZONI, and M. PASTERNAK, *Phys. Rev.* **129**, 2, 826 (1963).
2. Y. HAZONI, P. HILLMAN, M. PASTERNAK, and S. L. RUBY, *Physics Letters* **2**, 7 (1962).
3. A. J. F. BOYLE and H. E. HALL, *Reports on Progress in Physics*, vol. 21. Institute of Physics and Physical Society, 1962.

VIBRATIONAL SPECTRA AND STRUCTURES OF XENON TETRAFLUORIDE AND XENON OXYTETRAFLUORIDE

H. H. CLAASSEN,* CEDRIC L. CHERNICK, AND JOHN G. MALM
Argonne National Laboratory

ABSTRACT

The infrared spectrum of XeF_4 vapor has strong bands at 123 cm.$^{-1}$, 291 cm.$^{-1}$, and 586 cm.$^{-1}$. The Raman spectrum of the solid has very intense peaks at 502 cm.$^{-1}$ and 543 cm.$^{-1}$ and a weaker one at 235 cm.$^{-1}$. These data show that the molecule is planar and of symmetry D_{4h}. The seven fundamental frequencies have been assigned as 543 (a_{1g}), 291 (a_{2u}), 235 (b_{1g}), 221 (b_{1u}), 502 (b_{2g}), 586 (e_u), and 123 (e_u)—the assignment of 221 cm.$^{-1}$ to b_{1u} being uncertain.

The infrared spectrum of $XeOF_4$ vapor has intense peaks at 288 cm.$^{-1}$, 362 cm.$^{-1}$, 578 cm.$^{-1}$, 609 cm.$^{-1}$, and 928.2 cm.$^{-1}$ and the Raman spectrum of the liquid has bands at 231 cm.$^{-1}$, 286 cm.$^{-1}$, 364 cm.$^{-1}$, 530 cm.$^{-1}$, 566 cm.$^{-1}$, and 918 cm.$^{-1}$. The spectra fit a C_{4v} model very well. The fundamentals determined are: 928.2 (a_1), 578 (a_1), 288 (a_1), 231 (b_1), 530 (b_2), 609 (e), and 362 (e). This leaves one (b_2) and one (e) fundamental undetermined. Very close correspondence between vibrations in the two molecules indicates that the O-Xe-F angle in $XeOF_4$ must be rather close to 90°.

INTRODUCTION

The preparation of XeF_4 has been described previously [1], and the results of a preliminary study of its vibrational spectra reported briefly. [2] The preparation of $XeOF_4$ is described by Chernick, Claassen, Malm, and Plurien [this volume, p. 106]. We report here the results of a more complete study of the Raman spectrum of solid XeF_4 and of the infrared spectrum of the vapor. For $XeOF_4$ we have studied the Raman spectrum of the liquid and the infrared spectrum of the vapor.

PURITY OF COMPOUNDS

The purity of the XeF_4 was checked by infrared analysis. The probable impurities are XeF_6 and XeF_2, which have absorption peaks at 610 cm.$^{-1}$ and 566 cm.$^{-1}$, respectively. The sample was found to contain small amounts of the more volatile XeF_6, but this was easily removed, since its vapor pressure is higher by a factor of 10. Pumping the equilibrium vapor rapidly out of the storage can several times removed the XeF_6, so that none of the 610 cm.$^{-1}$ absorption could be detected in the bulk of sample remaining. The $XeOF_4$ was prepared by reacting XeF_6 with SiO_2 [see p. 106]. The completion of the reaction was indicated by the

* The permanent address of Dr. Claassen is Wheaton College, Wheaton, Illinois.

disappearance of the yellow color. It was purified from materials of lower volatility, such as XeF_4, by sublimation at $-25°$ C. from the quartz bulb in which it had been prepared.

TECHNIQUES OF SPECTROSCOPY

The infrared spectra were obtained with a Beckman IR-7 spectrophotometer with CsI and NaCl prisms and Perkin-Elmer 421 and 301 spectrophotometers. The cells were made of nickel and were used with either AgCl or polyethylene windows. For $XeOF_4$, with a 30-mm. vapor pressure at room temperature, the usual 10-cm. path length was enough, but for XeF_4 a 60-cm. absorbing path was also used.

The Raman spectrum of XeF was obtained for the solid using the Cary 81 photoelectric instrument with the lens system designed for solids. The sample used was approximately 1 gm. that had grown to a single crystal in a sealed quartz tube. The $XeOF_4$ was studied as the liquid at room temperature, using a Pyrex Raman tube 7 mm. in outside diameter, filled to a length of 3 cm. Qualitative indications of polarization were obtained for the stronger bands of $XeOF_4$ by using Polaroid cylinders around the Raman tube.

RESULTS AND INTERPRETATION

XeF_4.—Figure 1 shows tracings of the regions of the infrared spectrum where bands were observed, and Figure 2 is a tracing of the Raman spectrum. Judging from their positions and intensities, the three infrared bands at 123 $cm.^{-1}$, 291 $cm.^{-1}$, and 586 $cm.^{-1}$ are probably fundamentals. Of the four bands observed in the Raman spectrum, the one at 442 $cm.^{-1}$ is the least intense and probably does not represent a fundamental. It may, in fact, represent a combination of 543 $cm.^{-1}$ excited by 4339 A. and 502 $cm.^{-1}$ excited by 4347 A. which would occur at apparent shifts of 442 $cm.^{-1}$ and 445 $cm.^{-1}$ from 4358 A.

In considering the information the spectral data furnish on the molecular symmetry, it must be noted that the Raman measurements are for the solid compound and the infrared ones are for the vapor. Some solid-vapor shifts in frequencies are to be expected, and the Raman spectrum might contain frequencies due to lattice modes.

From the infrared spectrum alone, one can conclude that there is high symmetry in the XeF_4 molecule. Only one band is observed in the region where bond stretching motions (500–700 $cm.^{-1}$) are expected. Of all the symmetries possible for the YZ_4-molecule, only for the T_d (tetrahedral) and D_{4h} (square-planar) would there be just one infrared-active bond-stretching fundamental. The infrared spectrum also allows the distinction to be made between these two symmetries, since a T_d molecule would have one bending mode that would be infrared active while a D_{4h} molecule would have two. As two are observed for XeF_4, the D_{4h} model is the preferred one. Strong support for this is provided in the Raman spectrum also.

The modes of atomic motions, symmetries, numbering, and spectral activity

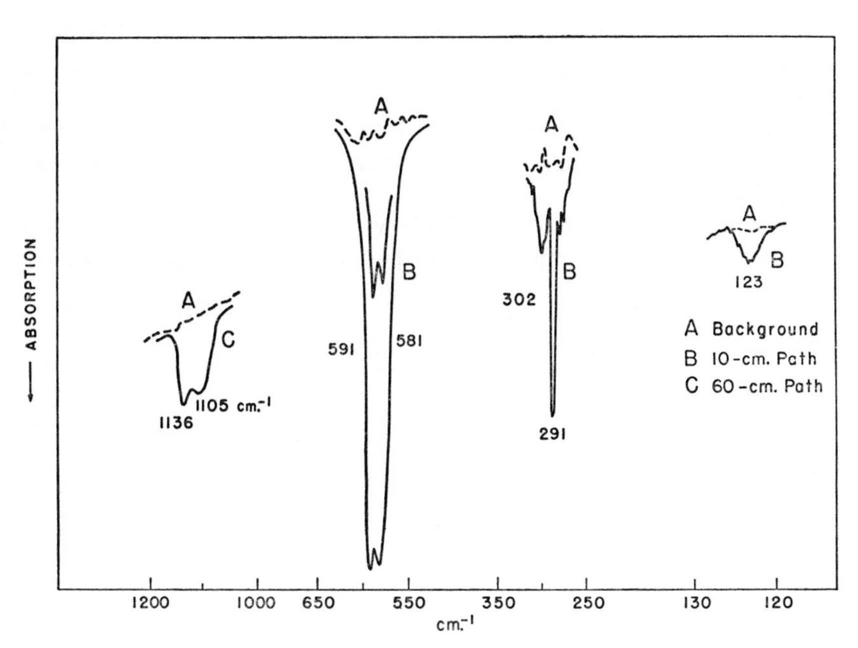

FIG. 1.—Infrared spectrum of XeF_4 vapor at approximately 3 mm. pressure

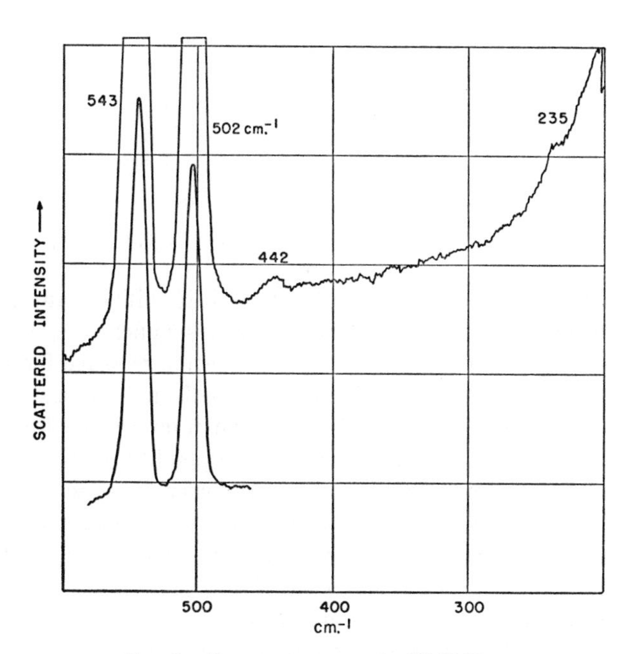

FIG. 2.—Raman spectrum of solid XeF_4

of the fundamental vibrations of a D_{4h} YZ_4 molecule are described in the left half of Figure 3. The assignment of ν_2 is definite from the band contours predicted for planar molecules by Gerhard and Dennison [3]. Only for the out-of-plane motion, ν_2, should there be a very intense Q-branch and this is observed at 291 cm.$^{-1}$. The other two infrared fundamentals are then assigned without ambiguity.

The Raman spectrum of the solid also fits very well. The two very intense bands at 543 cm.$^{-1}$ and 502 cm.$^{-1}$ must be due to the two stretching vibrations. Although

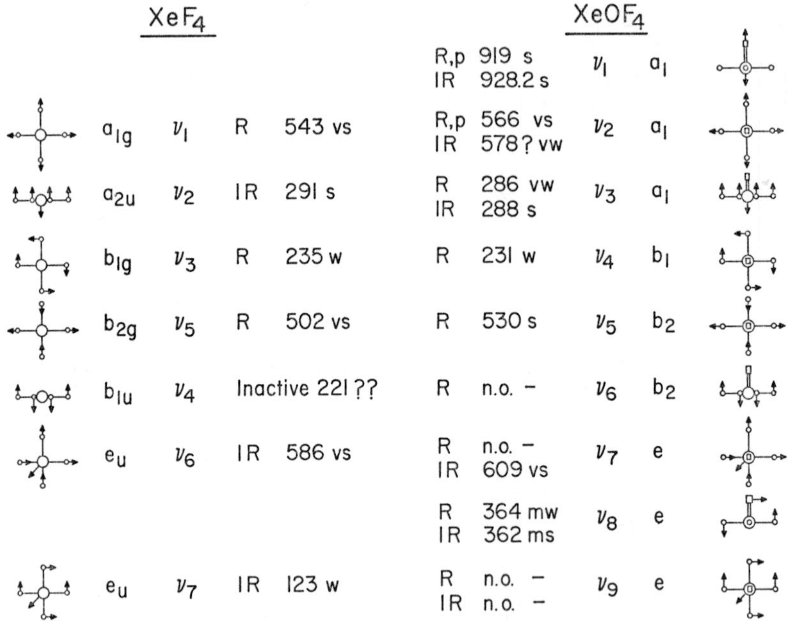

XeF_4				$XeOF_4$		
				R,p 919 s IR 928.2 s	ν_1	a_1
a_{1g}	ν_1	R	543 vs	R,p 566 vs IR 578? vw	ν_2	a_1
a_{2u}	ν_2	IR	291 s	R 286 vw IR 288 s	ν_3	a_1
b_{1g}	ν_3	R	235 w	R 231 w	ν_4	b_1
b_{2g}	ν_5	R	502 vs	R 530 s	ν_5	b_2
b_{1u}	ν_4	Inactive	221 ??	R n.o. –	ν_6	b_2
e_u	ν_6	IR	586 vs	R n.o. – IR 609 vs	ν_7	e
				R 364 mw IR 362 ms	ν_8	e
e_u	ν_7	IR	123 w	R n.o. – IR n.o. –	ν_9	e

FIG. 3.—Vibrational modes and assignments for XeF_4 and $XeOF_4$

polarization measurements could not be made, it is quite certain that the symmetric vibration is the higher one, because any significant repulsion between fluorines would almost require this. The observation reported for a solution in HF offers further support for this interpretation of the Raman spectrum of the solid XeF_4. H. H. Hyman and L. A. Quarterman observed a Raman band at 553 cm.$^{-1}$ [this volume, p. 278]. This must correspond to the 543 cm.$^{-1}$ band for the solid. The ν_5 band was not observed in the very dilute solution. This indicates that the higher frequency is a sharper band in the solution and is therefore the totally symmetric one. The 235 cm.$^{-1}$ Raman band is assigned to ν_3 without question. The band at 442 cm.$^{-1}$, if an actual frequency, must be an overtone of the inactive ν_4. This gives the value of 221 for ν_4, which, however, must be considered as quite doubtful and is therefore listed in Figure 3 with two question marks.

The infrared absorption peaks at 1105 cm.$^{-1}$ and 1136 cm.$^{-1}$ may be assigned as $\nu_5 + \nu_6 = 1088$ cm.$^{-1}$ and $\nu_1 + \nu_6 = 1129$ cm.$^{-1}$. The fit is satisfactory when account is taken of corrections needed due to vapor-to-solid shift of frequencies.

The one feature of the infrared spectrum that we do not understand is the doublet appearance of ν_6. This has been traced many times and the peaks reproducibly found at 581 cm.$^{-1}$ and 591 cm.$^{-1}$. A triplet band is expected, with all three peaks of about equal intensity and with a P-R separation of approximately 14 cm.$^{-1}$ [3]. The observed splitting is much too large to ascribe to isotopes of xenon and may be due to a Coriolis coupling between the doubly degenerate vibration and rotation.

The structure of XeF_4 solid has been obtained by X-ray diffraction [4, 5, 6] and by neutron diffraction [7]; [see also Part VI, this volume]. It has been found that the molecule is square planar within experimental error. Confirmation of the square-planar molecule in the vapor phase has been obtained by electron diffraction [Bohn, Katada, Martinez, and Bauer, this volume, p. 241].

Several theoretical discussions [8, 9] have stated that the square-planar model best fits the theory, and one of them [8] suggests that the molecule could possibly be distorted by Coulomb repulsion. Therefore, it seems interesting to question whether the vibrational data require an exactly planar molecule or whether the "ring" of fluorines might be slightly puckered. If the latter were true, the Raman-active ν_5 would be infrared active, but a slight distortion would, of course, result in a very weak infrared band. One can set a rough upper limit to the amount of possible puckering if one looks at the infrared spectrum in the region of 502 cm.$^{-1}$ and makes the plausible assumption that the rate of change of bond moment with stretching is approximately the same for ν_5 and ν_6. The result is that an upper limit can be set for deviation of the Xe-F bond from the plane of about 0.5 degrees or fluorine distances of 0.02 A. from the plane.

The *Q-R* separation of 11 ± 1 cm.$^{-1}$ in the 291 cm.$^{-1}$ band can be used to calculate a bond length. This gives 1.85 ± 0.2 A. for the Xe-F bond, in good agreement with the value of 1.94 A. for the vapor obtained from electron diffraction [this volume, p. 240]. Since the value of ν_4 is uncertain, we have not calculated thermodynamic functions.

$XeOF_4$.—Figure 4 is a tracing of the infrared spectrum of $XeOF_4$ vapor, and Figure 5 is a photograph of the Raman spectrum of the liquid. The intense absorption band at 928.2 cm.$^{-1}$ for the vapor and the peak at 919 cm.$^{-1}$ in the Raman of the liquid must represent a Xe = O stretching motion. The high value of this frequency rules out the possibility that the O is bonded to both Xe and one of the fluorines. Rather, the valence of the Xe is 6, and the most likely molecular symmetry, then, is C_{4v}, which is adequately verified by the spectra. The right half of Figure 3 above gives the spectral activity, our assignments, the species, and schematic indications of vibrational motions of $XeOF_4$, assuming C_{4v} symmetry. The XeF_4 part of the molecule is drawn in the figures as plane, although the symmetry does not require this. The two Raman bands with no

counterparts in the infrared must be assigned to b species and can be identified with ν_4 and ν_5 by comparison with corresponding motions in the XeF_4 molecule. The polarization scans indicated that both 919 cm.$^{-1}$ and 566 cm.$^{-1}$ are polarized and therefore must be assigned to a_1. The identical and distinctive shapes of the infrared bands at 928.2 cm.$^{-1}$ and 288 cm.$^{-1}$ indicate that they belong to the same species, so the 288 cm.$^{-1}$ is assigned also to a_1. The 566 (a_1) is the most intense Raman band and probably has its infrared counterpart at 578 cm.$^{-1}$, where its sharp Q-branch is seen on the side of an intense band. It is the weakest of the infrared fundamentals.

The two other strong infrared bands then must be e fundamentals. They are

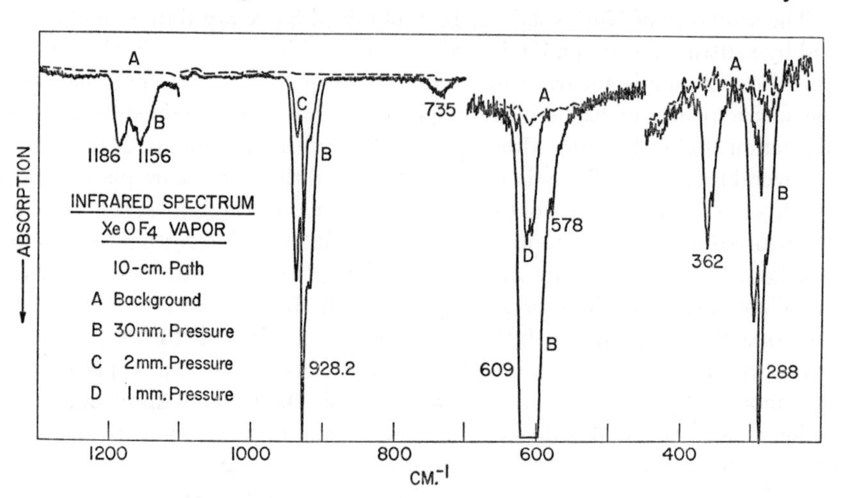

FIG. 4.—Infrared spectrum of $XeOF_4$ vapor

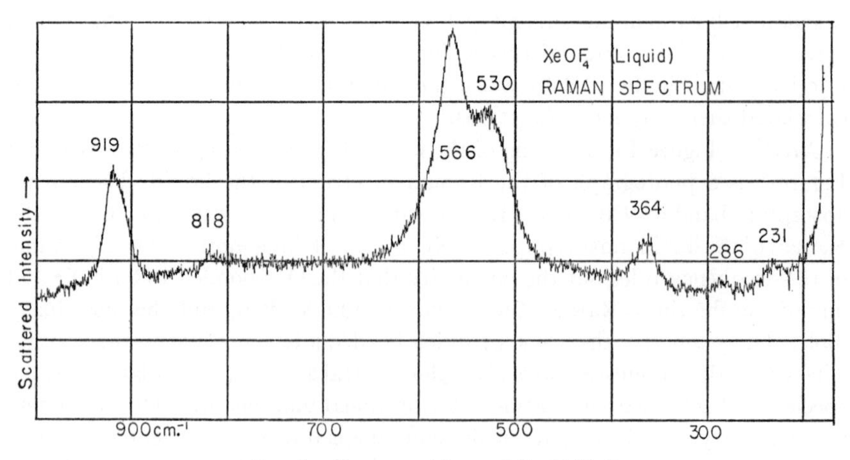

FIG. 5.—Raman spectrum of liquid XeO_4

assigned as ν_7 and ν_8, because ν_9 is expected to be lower in frequency. This leaves ν_6 and ν_9 undetermined. The former may be described as an "F_4 ring puckering" motion that is probably very weak in the Raman effect; the corresponding motion in XeF_4, ν_4, is inactive. A search was made for ν_9 between 100 and 200 cm.$^{-1}$ in the Raman spectrum. The scans are not shown in Figure 5 because no band was found. It may be expected to be very weak in the Raman, because the corresponding vibration in XeF_4 is active only in the infrared. When ν_9 has been obtained by extending the infrared study to lower frequencies, and the heat capacity of $XeOF_4$ has been measured, it should be possible to calculate an approximate value for ν_6.

Three weak bands in the infrared and one in the Raman may be assigned as combinations. The infrared bands at 1186 cm.$^{-1}$, 1156 cm.$^{-1}$, and 735 cm.$^{-1}$ are

TABLE 1

COMPARISON OF FORCE CONSTANTS

	MOLECULE		
	XeF_4	XeF_2	PuF_6
Bond length	1.94 A.	2.00 A.	1.972 A.
Stetching force constant	3.00 md/A	2.85 md/A	3.59 md/A
Interaction constant for perpendicular bonds	0.12	....	0.22
Interaction constant for opposite bonds	~0.06	0.11	−0.08

assigned as $\nu_2 + \nu_7 = 1187$ (E), $2 \times \nu_2 = 1156(A_1)$ and $2 \times \nu_8 = 724(A_1 + B_1 + B_2)$. The last one is also close to expected values for $\nu_7 + \nu_9$ $(A_1 + A_2 + B_1 + B_2)$ and $\nu_5 + \nu_6(A_1)$. The weak Raman band at 818 cm.$^{-1}$ is assigned as $\nu_3 + \nu_5 = 816(B_2)$.

COMPARISONS OF THE TWO MOLECULES

As is indicated in Figure 3, there is a close correspondence between the vibrations of XeF_4 and those of $XeOF_4$. Each vibration of XeF_4 has a corresponding one in $XeOF_4$, and the latter has two additional ones. There is only one pair, however, for which frequencies can be compared directly between the two molecules, because only for ν_3 in XeF_4 and ν_4 in $XeOF_4$ do we have comparable motions, and each is the only vibration in the species, so that there are no interactions with other vibrations. For this pair we have 235 cm.$^{-1}$ for solid XeF_4 and 231 cm.$^{-1}$ for liquid $XeOF_4$, i.e., essentially the same frequency. Intensity comparisons provide further confirmation for the close correspondence. Thus, ν_2 for $XeOF_4$ is very strong in the Raman and very weak in the infrared, and the corresponding motion in XeF_4 is allowed only in the Raman spectrum. In $XeOF_4$, ν_6, ν_7, and ν_9 are all allowed in the Raman effect but not observed, presumably because they are very weak. The three corresponding motions in XeF_4 are all forbidden in the Raman effect. These observations indicate that the XeF_4 part of the $XeOF_4$ has very nearly the same configuration as the XeF_4 molecule. Thus the O-Xe-F angle must be rather near 90°, although the C_{4v} symmetry does not require 90°, and repulsion between the oxygen and the fluorines could well cause a larger angle.

FORCE CONSTANTS

Preliminary force-constant calculations using a valence plus interaction-terms type of potential function, similar to that used by Claassen for hexafluorides [10], gave a value of 3.00 md/A for the bond-stretching constant and 0.12 md/A for the interaction constant between bonds at right angles (Table 1). The interaction constant for opposite bonds cannot be determined accurately, but is approximately 0.06 md/A. These may be compared with values given by Smith [11] for XeF_2 and with those for PuF_6 [10], a molecule that also has fluorine bonds at right angles and of comparable bond length.

Acknowledgments.—We are indebted to the Perkin-Elmer Corporation for the opportunity to use the 301 instrument at Norwalk, Connecticut, and to Charles Helms and Robert Anacreon for their help with the operation of that spectrophotometer.

REFERENCES

1. H. H. Claassen, H. Selig, and J. G. Malm, *J. Am. Chem. Soc.* **84**, 3593 (1962).
2. C. L. Chernick *et al.*, *Science* **138**, 136 (1962).
3. S. L. Gerhard and D. M. Dennison, *Phys. Rev.* **43**, 197 (1933).
4. J. A. Ibers and W. E. Hamilton, *Science* **139**, 106 (1963).
5. D. H. Templeton, A. Zalkin, J. D. Forrester, and S. M. Williamson, *J. Am. Chem. Soc.* **85**, 242 (1962).
6. S. Siegel and E. Gebert, *ibid.* 240 (1963).
7. J. H. Burns, P. A. Agron, and H. A. Levy, *Science* **139**, 1209 (1963).
8. R. E. Rundle, *ibid.* 112 (1963).
9. L. L. Lohr, Jr., and W. N. Lipscomb, *ibid.* 240 (1963).
10. H. H. Claassen, *J. Chem. Phys.* **30**, 968 (1959).
11. D. F. Smith, *ibid.* **38**, 270 (1963).

INFORMATION ON BONDING IN XENON COMPOUNDS FROM INFRARED SPECTRA

D. F. SMITH
Oak Ridge Gaseous Diffusion Plant

Infrared spectra have been obtained for XeF_2, XeF_4, XeF_6, $XeOF_4$, and XeO_3 samples. In each instance some information on the chemical bonding has been derived from the spectrum. The spectrum of XeF_2 showed the molecule to be linear and yielded an estimate of the Xe-F bond distance. The molecule is so simple that all of the vibrational force constants have been derived from the spectrum, in contrast to the more usual situation where there are more force constants than vibrational frequencies. These force constants have been interpreted as indicating covalent bonding in XeF_2.

Our experimental work on the infrared spectrum of XeF_4 has added nothing new to the work of Claassen. Even though there are more vibrational constants than vibrational frequencies, the stretching force constants can be derived from the spectrum initially reported by Claassen [1]. The same arguments used to support the existence of covalent bonding in XeF_2 can be applied to XeF_4. The infrared spectrum of XeF_6 contains only four or five remarkably broad bands, one of which occurs at twice the frequency of the other. If this is an overtone, then XeF_6 has the non-octahedral structure expected for covalent bonding, but not for other bonding schemes. The infrared spectrum, together with the Raman spectrum of $XeOF_4$, shows this molecule to have C_{4v} symmetry, consistent with covalent bonding. The Xe-O stretching constant clearly indicates double-bond character of the Xe-O bond. In XeO_3 the stretching force constant is smaller, but still indicative of double-bond character.

A tracing of the absorption by the intense asymmetric stretching vibration ν_3 of XeF_2 is shown in Figure 1. As can be seen, this band has no Q-branch. Since bent XY_2 molecules show the Q-branch predicted for them, the absence of a Q-branch indicates that XeF_2 is not bent, but is a linear molecule. A second weak band without a Q-branch is observed at 1070 cm.$^{-1}$. This band has been interpreted as the combination $\nu_1 + \nu_3$, where ν_1 is the symmetric stretching vibration frequency. No evidence for the absorption of ν_1 has been found. Absorption in the region of 513 cm.$^{-1}$ (1070 cm.$^{-1}$ − 557 cm.$^{-1}$) occurs only with such high pressures of XeF_2 that the wing of the 557 cm.$^{-1}$ can be held responsible. Since the symmetric vibration ν_1 is inactive in the infrared for a linear molecule, but active and normally observed for a bent molecule, this is further evidence for the linearity of XeF_2. This assignment of ν_1 has been verified by Begun of Oak Ridge National Laboratory, who observed the intense ν_1 at 496 cm.$^{-1}$ in solid XeF_2. Mason, of the University of Tennessee, has observed the bending vibration band ν_2 for XeF_2. It has the expected Q-branch at 213 cm.$^{-1}$.

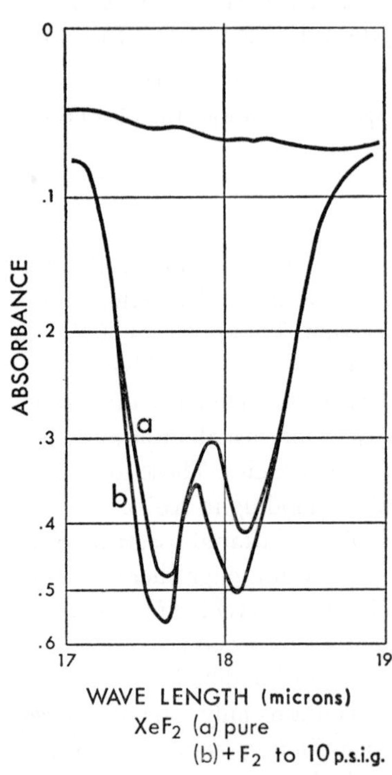

FIG. 1.—The infrared absorption spectrum of XeF_2

An estimate of the Xe-F bond length can be obtained from the separation of the peaks of the P and R-branches of ν_3. Herzberg [2] gives

$$\Delta\nu = \sqrt{\frac{8\,kTB}{hc}}, \tag{1}$$

where the rotational constant B is given by

$$B = \frac{h}{8\pi^2 c I}. \tag{2}$$

The moment of inertia, I, for XeF_2 is $2\,M_F\,r_{XeF}^2$. At 298° K. these can be combined to give

$$r_{XeF} = \frac{271.05}{\Delta\nu}, \tag{3}$$

where $\Delta\nu$ is in cm.$^{-1}$ and r_{XeF} in angstrom units. The effect of the vibrational isotope shift, about 0.5 cm.$^{-1}$ per mass unit, on the composite band is to broaden the peaks, but to leave the peak separation as for a single species, so that no correction need be made for this.

As can be seen in Figure 1, the observed absorption increased and the peak separation decreased when a non-absorbing gas was added. This is a pressure-broadening effect, and for equation 1 to apply, the XeF_2 lines should be essentially completely broadened. Tests with foreign gases at pressures up to 10 atm. showed that about the same value for the *P-R* peak separation was obtained for all pressures above 300 mm. Hg. The average value for $\Delta\nu$ obtained for different cell lengths and spectrometers is 14.2 ± 0.2 cm.$^{-1}$, corresponding to an Xe–F bond length of 1.9 A. This is in much better agreement with the diffraction values [3] than the initially reported value, 1.7 A., based on the 16 cm.$^{-1}$ separation observed with XeF_2 unbroadened [4]. It should be clearly understood that this method of obtaining bond lengths from *P-R* peak separations is less direct than those used for either resolved rotational structure or for diffraction studies. It is not only much less precise but may also be subject to unrecognized biases. The main uses were to yield an early estimate of the bond distance and to demonstrate that the bond distance in XeF_2 vapor is not drastically different from that in the solid.

The vibrational force constants have been obtained for XeF_2: f_r, the principal stretching force constant, 2.82 md/A; f_{rr}, the interaction force constant between the two stretching coordinates, 0.15 md/A; $f\delta/l^2$, the bending force constant, 0.19 md/A.

There are three kinds of bonding which might be considered for XeF_2: (*a*) the ionic model $F^-Xe^{++}F^-$, which involves six electrons in the xenon valence shell; (*b*) two one-electron bonds which might be represented by resonance between F^-Xe^+-F and F-Xe^+F^-, with each bond having 50 per cent covalent, 50 per cent ionic character—this model has eight electrons in the xenon valence shell; and (*c*) the covalent model F-Xe-F, which requires ten electrons in the xenon valence shell, with promotion to the 5*d* levels.

Each model can be linear. Thus, the observed configuration does not discriminate between them. Model (*a*) (the ionic model) would be expected to have a bond distance appreciably longer than 2.00 A. in view of the 2.345 A. bond length found for diatomic CsF [8] and the 2.26 A. bond length estimated for gaseous RbF. The pure ionic model can be ruled out on this basis. If the covalent radius for xenon is taken to be about 0.05 A. shorter than it is for iodine, the radius sum 1.95 A. should apply to the covalent model (*c*). Model (*b*), by analogy with ClF_3 and BrF_3, should have a bond length 0.1 A. longer, about 2.05 A. Thus a choice between models (*b*) and (*c*) cannot be made on the basis of bond length.

The interpretation of the stretching force constants in terms of chemical bonding is based on direct analogy with ICl_2^-. The bonding in ICl_2^- has been convincingly established by Cornwell and Yamasaki [5]. They have measured the chlorine nuclear quadrupole coupling in solids containing the ICl_2^- ion as 38.4 mc/sec, as contrasted with a value of 74.4 mc. for solid iodine monochloride, ICl. They concluded that the I-Cl bonds in ICl_2^- have a rather large proportion of ionic character corresponding to the presence of a net negative charge on each chlorine atom of roughly half that of an electron. It is not necessary for the iodine to provide one independent orbital for each bonded atom. A single *p* orbital can serve to form colinear bonds to two atoms. The ICl_2^- can be regarded as a resonance hybrid of the structures,

$$Cl^- \; I\text{–}Cl \quad \text{and} \quad Cl\text{–}I \; Cl^- \tag{4}$$

if the dash is taken to represent a bond. This is just the model that might be expected to correspond to XeF_2.

The infrared and Raman spectra of ICl_2^- have been found by Person *et al.* [6], to lead to stretching force constants which show the effects of this kind of bonding quite strikingly. First the principal stretching force constant has a low value, 1.00 md/A, as compared with 2.38 md/A for the ICl force constant, on the one hand, and a value of about 1 md/A for the completely ionic CsCl, on the other. In addition, the interaction constant is large, 0.36 md/A, over a third of the value of the principal force constant. This large and positive interaction constant is interpreted as a direct consequence of the bonding in ICl_2^- as represented by structures shown above (4). When one bond is stretched, the resonant structure in which this bond is ionic is stabilized, and the other bond takes on a more covalent character. This effect in ICl_2^- is much more prominent than in CO_2. A parallel argument was used by Thompson and Linnett [7] to account for the sign of the small interaction term of CO_2.

The direct relationship between a large f_{rr}/f_r ratio and the bonding of the kind pertaining in ICl_2^- seems so obvious that this large ratio is taken as a distinguishing mark for this kind of bonding.

Indeed, Cornwell and Yamasaki have used arguments nearly identical with those used for ICl_2^- to explain their measurement of the chlorine nuclear quadrupole in solids containing the ICl_4^- ion to show that the ICl_4^- ion can be regarded as a resonance hybrid of such structures as

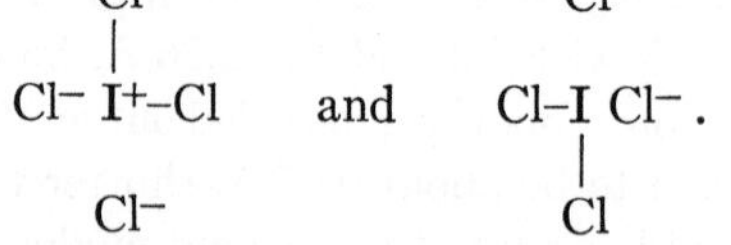

$$\begin{array}{c} Cl \\ | \\ Cl^- \; I^+\text{–}Cl \\ \\ Cl^- \end{array} \quad \text{and} \quad \begin{array}{c} Cl^- \\ \\ Cl\text{–}I \; Cl^- . \\ | \\ Cl \end{array} \tag{5}$$

Person *et al.* find, as a consequence of this kind of bonding, a low stretching constant, 1.25 md/A; a large interaction constant for bonds 180° apart, 0.33 md/A; and a small interaction constant for bonds 90° apart, 0.08 md/A. There is no Xe-F stretching force constant which can be taken as a standard with which to compare the value f_{rr} = 2.82 md/A for XeF_2, as the value f_r = 2.38

md/A for ICl was taken as a standard for the ICl_2^- and ICl_4^- ions. Most stretching force constants for a stretching vibration of the type f_{Y-F} are within 20 per cent of the value 4.55 md/A found for F_2 in which the bonding must be covalent. The exceptions fall into three classes: higher values encountered when Y is a light atom, very low values when the bond is clearly an ionic bond, such as in CsF for which the force constant is about 1 md/A, or moderately low values in molecules where the bonding must involve either structures such as in ICl_2^- or appreciable *d* orbital hybridization. The example which seems to come closest to XeF_2 is ClF_3. This is a T-shaped molecule with the stretching constant for the unique bond 4.1 md/A, while the principal stretching constant for the other two bonds is 2.89 md/A. The chlorine nuclear quadrupole coupling in ClF_3 can be simply explained on the basis of the same kind of covalent-ionic resonance used

TABLE 1

Ion or Molecule	Principal Stretching Constant (f_r md/A)	Interaction Stretching Constants		Ratio ($f_{rr'}/f_r$)
		Bonds at 90° (f_{rr} md/A)	Bonds at 180° ($f_{rr'}$ md/A)	
ICl_2^-	1.0	....	0.36	0.36
ICl_4^-	1.25	0.08	.33	.26
ClF_3	2.89*	....	.23	.08
XeF_2	2.82	....	.15	.05
XeF_4	3.02	.12	.04	.014
SF_6	4.99	.30	.54	.11
SF_4	2.08*	0.43	1.13	0.54

* Constant applies to longer bonds only.

for ICl_2^-. The quadrupole coupling in ClF_3 is not decisive, as was the case with ICl_2^-, since an explanation based on a model that has bonding scarcely more ionic than ClF is equally successful. The analogy of XeF_2 and ClF_3 with ICl_2^- fails when the interaction constant for opposite bonds is considered. This is shown in Table 1, where one additional entry is made for SF_6, which has a larger principal stretching constant and another entry for SF_4 [8], which has two long nearly coaxial bonds characterized by a low force constant and a large interaction constant.

For XeF_4 the principal stretching constant is 3.01 md/A, and the stretching interaction constant for bonds at 180° is only .04 md/A. The analogy to ICl_4^- fails when the interaction constants are considered. Two alternatives have been proposed to account for the failure of the analogies. The first was previously stated by the writer [4] as indicating that the failures occurred because the bonding in XeF_2 and XeF_4 is different than the bonding in ICl_2^- and hence is covalent. Another explanation was offered by Pimentel and Spratley [9]. They pointed out that, although the formal charge implication in ICl_2^- is that the excess charge of the ion is distributed on the chlorine atoms while the iodine atom has a charge close to zero, the case of XeF_2 contrasts since the central atom must have significant positive formal charge to balance the negative charge (placed) on the terminal atoms. This difference can be expected to strengthen the bond somewhat because of electrostatic attraction, raising f_r. At the same time, they stated, the interaction constants should be reduced, because an asymmetric displacement is no longer favored over a symmetric displacement. In the symmetric mode, the terminal atom repulsions that tend to raise the energy are counteracted by the central atom positive charge. Their explanation retains the bonding scheme proposed by Pimentel, which accounts so handsomely for the bonding in the bifluoride and polyhalide ions.

Fortunately, an example exists in which the bonding is the same as in the ICl_2^- ion while the charge distribution is the same as proposed for XeF_2. In the ICl_4^- ion, the colinear Cl I Cl group should involve the same orbitals as in ICl_2^-; but because of the negative charge on the other two Cl atoms, the iodine atom is no longer neutral, but must have significant formal charge to balance the negative charge (placed) on these other chlorine atoms. This charge distribution, differing as it does from the ICl_2^- case, may account for the larger principal stretching constant, 1.25 md/A, as compared with 1.00 md/A in ICl_2^-, and for the reduction of the value of f_{rr} for the opposite bond interaction force constant from 0.36 to 0.33 md/A. A large f_{rr}/f_r ratio, however, remains a distinguishing mark of this kind of bonding despite the different charge distribution.

The explanation of the force constants of XeF_2 and XeF_4 in terms of covalent bonding requires that separate orbitals for each bond must be provided by the xenon so that 5*d* orbitals are involved. For XeF_2, these may be the sp^3d hybrid orbitals with colinear hybrid orbitals, each half *p*-, half *d*-character assigned to the bonding electrons, while the unshared pair are assigned to the sp^2 orbitals normal to the bonds. In XeF_4 the xenon orbitals are sp^3d hybrids directed toward the corners of an octahedron. Ideally equivalent orbitals, $\frac{1}{6}\ s$, $\frac{1}{2}\ p$, $\frac{1}{3}\ d$, are filled by the bonding electrons as well as the unshared pairs. The hybrid orbitals in XeF_4 are then different from those in XeF_2, and the shorter bond length found in XeF_4 [10] is accounted for by the presence of some *s*-character. It is not insisted that the ICl_2^- kind of bonding plays no part in XeF_2 but rather that its contribution is not so overwhelming as in ICl_2^-.

It should be pointed out here that the explanation of the stability of the xenon fluorides and the accurate prediction of the instability of the xenon oxides which was presented by Pitzer [11] involved a discussion of the similarity of the bonding of XeF_2, XeF_4, and ClF_3 to the bonding in the ICl_2^- ion. Only the similarity of bonding in XeF_2, XeF_4, and ClF_3 is necessary for the validity of the explanation; and the similarity to, or distinction from, the bonding in ICl_2^- does not enter the calculation.

The infrared spectrum of XeF_6 vapor is shown in Figure 2. The bands all have unusual band contours and abnormally great breadth. The Raman spectrum of

XeF_6 solid has been observed by Begun of the Oak Ridge National Laboratory. He has found Raman lines at 655 cm.$^{-1}$ (intensity 10) 635 cm.$^{-1}$ (intensity 8) and 582 cm.$^{-1}$ (intensity 4), which are all in the region where the Xe-F stretching vibrations occur. No Raman lines or infrared bands have yet been recognized at the lower frequencies where the bending vibrations are expected. There does not appear to be enough detail in the spectrum to permit assignment of a structure. The octahedral model, however, has only two Raman lines in the Xe-F stretching region, so that it does appear to be excluded. Since there are no

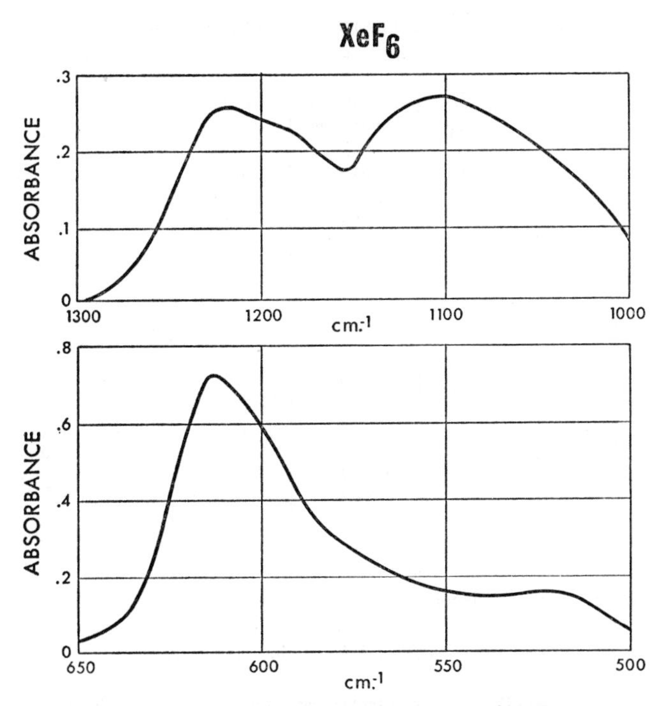

FIG. 2.—The infrared absorption spectrum of XeF_6

obvious coincidences of infrared bands and Raman lines, the existence of a center of symmetry cannot be ruled out, and the molecular symmetry may be as high as D_{4h}. The 1225 cm.$^{-1}$ infrared band can be accounted for either as 612 cm.$^{-1}$ + 582 cm.$^{-1}$ or as 2 × 612 cm.$^{-1}$. The first choice would result from the 582 cm.$^{-1}$ being reduced 31 cm.$^{-1}$ in frequency in going from the vapor to the solid phase. The second choice would exclude all models with a center of symmetry.

The evidence is thus that XeF_6 does not have the octahedral configuration that would result if the bonding in XeF_6 were the same as in the polyhalide ions.

Table 2 lists the fundamental vibration frequencies of $XeOF_4$ as determined from the Raman spectrum of the liquid observed by Begun and from the infrared spectrum of the gas. The intensity and polarizations of the Raman lines are given. The similarity of the vibration spectrum of BrF_5 [12] is so close that no difficulty was experienced in making the assignments listed in Table 2 on the basis of a structure with C_{4v} symmetry. The polarizations and band shapes can also be accounted for on the basis of this C_{4v} symmetry. It seems clear that the fluorine atoms are at the corner of a square, while the xenon and oxygen atoms are on a line passing through the center of the square and normal to it. The

TABLE 2

FUNDAMENTAL VIBRATION FREQUENCIES OF $XeOF_4$

Designation	Species	Raman (liquid) (cm.$^{-1}$)	Intensity and Polarization	I.R. (gas) (cm.$^{-1}$)	Intensity	Description
ν_1	A_1	920	20 P	926	s	ν(Xe-O)
ν_2	A_1	567	100 P	576	m	ν(Xe-F)symmetric
ν_3	A_1	285	2 P	294†	s	δ(F-Xe-F) symmetric out of plane
ν_4	B_1	527	40 D	....	....	ν(Xe-F) antisymmetric
ν_5	B_1	....	not observed	....	....	δ(F-Xe-F) antisymmetric out of plane
ν_6*	B_2	233	6 D	....	....	δ(F-Xe-F) symmetric in plane
ν_7	E_u	....	....	608	vvs	ν(Xe-F) antisymmetric
ν_8	E_u	365	15 D	361	s	δ(F-Xe-O)
ν_9	E_u	161	3 D	....	....	δ(F-Xe-F) antisymmetric in plane

P—polarized; D—largely depolarized; ν—stretching vib.; δ—bending vibration.

Vibrations of symmetry species B_1 and B_2 are inactive in the infrared for C_{4v} molecular symmetry.

* Following discussion with H. H. Claassen after the conference the assignment of the 233 cm.$^{-1}$ band has been changed from ν_5 as reported verbally to ν_6.

† The 294 vibration as observed appears to be associated with the presence of HF impurity. The value 288 is assigned to this vibration by Claassen, Chernick, and Malm [this volume, p. 290].

infrared band corresponding to the vibration ν_2 is less intense than the ν_1 band, just as in the BrF_5 spectrum. This can be taken as evidence that the Xe-F vibrational component in the direction of the symmetry axis is small. The structure may not differ much from the BrF_5 structure and can be taken to be a square pyramid with the oxygen atom at the apex and the xenon atom close to the plane of the fluorine atoms.

The force constants have been calculated, assuming the idealized 90° angles. The force-constant analysis given by Jones and Stephenson for BrF_5 required only minor modification for application to $XeOF_4$. The principal stretching force constants are k_{XeF} 3.20 md/A, and k_{XeO} 7.10 md/A. The interaction force constant for the Xe-F coordinates at right angles is $k_{dd} = 0.12$ md/A, and for bonds at 180° $k_{dd'} = 0.15$ md/A. The large value of the xenon-oxygen force constant corresponds to a large amount of double-bond character.

Xenon trioxide is too dangerously explosive to consider an attempt at obtaining the Raman spectrum. The infrared spectrum can be recorded with only a thin XeO_3 film on thin polyethylene. No damage has occurred when the very small amount of XeO_3 in the film has exploded. Absorption bands have been observed in the 800 cm.$^{-1}$ and 300 cm.$^{-1}$ region, and fundamental vibrations have been assigned on the basis of analogy with ClO_3^-, BrO_3^-, and IO_3^-, as ν_1 (A) 770 cm.$^{-1}$, ν_2 (A) 311 cm.$^{-1}$, ν_3 (E) 820 cm.$^{-1}$, and ν_4 (E) 298 cm.$^{-1}$. The force constants have been calculated using the 103° bond angle reported by Templeton *et al.* [13]. The principal stretching force constant is 5.66 md/A, about as in the halide trioxide ions, and is large enough to indicate considerable double-bond character to the Xe-O bond here, as might have been expected from the 1.76 A. Xe-O distance reported by Templeton *et al.*

REFERENCES

1. C. L. Chernick *et al.*, *Science* **138**, 136 (1962).
2. G. Herzberg, *Infrared and Raman Spectra of Polyatomic Molecules*, p. 391. Princeton, New Jersey: D. Van Nostrand Company, 1945.
3. H. A. Levy and P. A. Agron, *J. Am. Chem. Soc.* **85**, 250 (1963).
4. D. F. Smith, *J. Chem. Phys.* **38**, 276 (1963).
5. C. D. Cornwell and R. S. Yamasaki, *ibid.* **27**, 1060 (1957).
6. W. B. Person *et al.*, *ibid.* **35**, 908 (1961).
7. H. W. Thompson and J. W. Linnett, *J. Chem. Soc.* 1384 (1937).
8. K. Venkateswarlu and M. G. Krishnapillai, *Optics and Spectroscopy* **11**, 26 (1961).
9. G. C. Pimentel and R. D. Spratley, *J. Am. Chem. Soc.* **85**, 826 (1963).
10. J. H. Burns, P. A. Agron, and H. A. Levy, *Science* **138**, 1208 (1963).
11. K. S. Pitzer, *ibid.* **139**, 414 (1963).
12. C. V. Stephenson and E. A. Jones, *J. Chem. Phys.* **20**, 1830 (1952).
13. D. H. Templeton, A. Zalkin, J. D. Forrester, and S. M. Williamson, *J. Am. Chem. Soc.* **85**, 817 (1963).

RÉSUMÉ ON VIBRATIONAL SPECTRA OF XENON COMPOUNDS

H. H. Claassen
Argonne National Laboratory

The structures of XeF_2 and XeF_4—linear symmetric and square planar, respectively—are unambiguously indicated by the vibrational spectra, and the assignment of observed bands is straightforward.

Both the infrared and Raman spectra for $XeOF_4$ were studied at Oak Ridge and at Argonne at about the same time. The essential agreement between the two groups of investigators in regard to both frequencies and assignments is gratifying. The molecular symmetry is C_{4v}, and both groups came to the conclusion that the O-Xe-F angle must be near 90°. Future theoretical treatments of this molecule should attempt to explain why the XeF_4 molecule changes so very little in shape and frequencies of vibration when an oxygen atom is attached on one side of the plane.

For XeF_6 the problem is much more difficult. The infrared spectrum has been studied at Argonne and at Ford, while at Oak Ridge both the infrared and Raman spectra have been observed, the latter for the solid. Because of difficulties with impurities and with reactions with window materials, the spectra are incomplete and uncertain, and the structure of the molecule is debatable. The fifteen other hexafluoride molecules that have been studied have all been found to have the symmetry of a regular octahedron, O_h [1].

Smith tentatively concluded that XeF_6 is of lower symmetry than O_h because Begun observed three Raman bands in the bond-stretching region and because he considered the 520 $cm.^{-1}$ infrared band as a genuine XeF_6 band. The O_h model allows only two Raman-active and one infrared-active fundamentals in the bond-stretching region. Claassen noted in discussion that he had evidence for a non-constant intensity ratio for the 612 $cm.^{-1}$ and the 520 $cm.^{-1}$ bands, and Weinstock *et al.* showed a trace they ascribed to XeF_6 with an intense 610 $cm.^{-1}$ peak but no 520 $cm.^{-1}$ peak. Unfortunately, their trace does not extend to 928 $cm.^{-1}$, so that one could confirm that it was definitely not $XeOF_4$, which is known to absorb at 610 $cm.^{-1}$ without absorbing at 520 $cm.^{-1}$. As for the three observed Raman bands, it is possible that they represent only two vibrational modes, one of which interacts with lattice interactions so as to split the Raman band.

The two infrared combination bands observed for XeF_6 both at Oak Ridge and at Argonne at 1100 cm.$^{-1}$ and 1230 cm.$^{-1}$ are indicative of O_h symmetry, since, for this model, there are only two allowed binary combination bands for the bond-stretching vibrations. For the symmetrical hexafluorides, both characteristically show up as reasonably intense bands. From these combination bands for XeF_6, assuming O_h symmetry, one gets values of 620 cm.$^{-1}$ and 490 cm.$^{-1}$ for the Raman-active fundamentals, but these are not in agreement with the 655 cm.$^{-1}$, 635 cm.$^{-1}$, and 582 cm.$^{-1}$ Raman frequencies observed at Oak Ridge. In HF solution, however, a 620 cm.$^{-1}$ band has been observed and tentatively assigned to a symmetrical stretching vibration [Hyman and Quarterman, this volume, p. 278].

One has to admit that the structure of the XeF_6 molecule is not known as yet. To the writer the evidence seems to favor the O_h model, but more work is needed. Most urgent, perhaps, is repeating some of the experimental work using different windows for infrared cells and observing the Raman spectrum for a liquid or gaseous phase. The decision about the symmetry of XeF_6 is especially important because of the bearing it will have on theories of Xe-F bonds (see Introduction, Part IX, p. 315].

REFERENCE

1. B. Weinstock, H. H. Claassen, and C. L. Chernick, *J. Chem. Phys.* **38**, 1470 (1963).

PART 8 PHYSIOLOGICAL PROPERTIES OF NOBLE-GAS COMPOUNDS

The physiological properties of xenon compounds are potentially exceedingly interesting, and chemists working with these reagents will want to know their toxicological properties. The first information of this nature was reported at the conference and is included in this volume.

METABOLIC AND TOXICOLOGIC STUDIES OF WATER-SOLUBLE XENON COMPOUNDS

A. J. Finkel, C. E. Miller, and J. J. Katz
Argonne National Laboratory

Abstract

Some observations have been made of the metabolic and toxicologic effects on mice of injections of aqueous solutions of sodium xenate. Sodium xenate has been found to be moderately toxic when injected intravenously, and the median lethal dose lies between 15 and 30 mg/kg. The administered sodium xenate appears to be rapidly reduced to xenon gas, and by the use of radio-xenon and sensitive detection equipment, it has been shown that the bulk of the xenon gas leaves the body within minutes after injection.

The recent discovery of xenon-fluorine compounds and the even more surprising finding that water-soluble xenon compounds exist have raised the question of the biological properties of the latter substances. In an area of research in which so many startling discoveries have already been made, the possibility that unusual behavior may be evoked in living organisms by these unusual compounds deserves exploration. The compounds that we used in some exploratory experiments were obtained from John G. Malm, who has expressed keen interest in this work and whose generous co-operation enabled us to carry out these studies.

The first observations that we made were concerned with the gross toxicologic properties which the water-soluble xenates exhibited when injected into mice. Sodium xenate was dissolved in 0.2 M phosphate buffer to provide solutions of different concentrations and pH. (We shall use the term "xenate" to refer to the mixture of water-soluble xenon compounds employed here. The composition and structure of these substances are discussed elsewhere in this volume [Part V, pp. 155–90].) Two main types of solution were used, one with pH $\sim$ 10 and the other in the pH range 6–8. It is now recognized that chemistry of the xenates is strongly pH dependent, and the nature of the solutions appears to play an important role in biological systems as well.

For intravenous administration the solutions were injected into a lateral tail vein of adult female CF-1 mice in amounts ranging from 0.05 to 0.30 ml. The results, which are summarized in Table 1, indicate that the toxic level for injected sodium xenate at essentially neutral pH lies between 0.5 and 1.0 mg. per mouse. This range is roughly equivalent to a median lethal dose of 15 to 30 mg. per kg. Comparable values based on a smaller series of mice injected with strongly alkaline solutions lie between 1.2 and 1.9 mg. per mouse, or 36 to 57 mg. per kg. The more alkaline solutions, despite their higher oxidizing power, are apparently less toxic. Moreover, death resulting from injections of the alkaline solutions occurred after a longer interval than it did after injections of comparable amounts of a neutral solution.

The early deaths were preceded by marked convulsions and tetanic contractions, while those that occurred later were characterized by a deepening comatose state. No unusual gross pathology was seen at post-mortem examination except for an unusual darkness of the blood. This condition may have been

TABLE 1

TOXICITY OF WATER-SOLUBLE XENATES INJECTED INTRAVENOUSLY INTO MICE

Amount Injected (mg/mouse)	Proportion Dead	Remarks
pH 6–8		
0.2	0/3	No ill effects
0.4	0/5	No ill effects
0.5	0/7	One mouse initially catatonic (2 min.)
1.0	6/6	Died in 8, 157, 207, 208, >360, >360 min.
1.5	5/5	Died in 4 to 48 hours
2.2	6/6	Died in 4, 5, 16, 38, 49, 77 min.
4.3	5/5	Died in 3, 6, 55, 59, 79 min.
pH∼10		
0.48	0/2	No ill effects
1.20	0/1	No ill effects
1.90	2/3	Died between 3 and 19 hours
3.90	2/3	Died between 3 and 19 hours

a consequence of the formation of methemoglobin by oxidation, and it requires further investigation.

Because we had access to highly radioactive samples of sodium xenate labeled with Xe^{133}, we were able to study the retention of xenon when it was administered to mice as a water-soluble compound. The mice were measured for Xe^{133} content by means of a sodium iodide scintillation detector that fed impulses into two multiscalers. Each mouse was put into a thin-walled cellulose nitrate tube through which air was passed at a rate of about 200 ml. per minute to carry off the exhaled xenon. The axis of the tube was placed parallel to, and 7 cm. below, the lower face of the NaI(T1) crystal, which is a right cylinder 29.5 cm. in diameter and 10 cm. high. Since the mouse occupied a roughly cylindrical volume 2 cm. in diameter and 6 cm. in length, its movements within the confining tube produced essentially no changes in geometry when it was under the center of the large crystal.

This counting system was originally designed for gamma-ray spectrometry of low levels of radioactivity in man, and it is located in a low background enclosure [1]. This equipment is more elaborate than is necessary for the study of mice, but it was readily available, and it was adapted to this work by provision of a tray for holding the tube containing the mouse in place and by the addition of adequate air flow past the mouse. The pulses from the crystal that were in the energy range from 50 to 110 kev. were counted simultaneously with two TMC Model 404 multichannel analyzers employed as multiscalers. One of these counted at 1-second intervals and the other at 0.1-minute intervals. The background count with this detector system and iron-room enclosure was 6 counts per second (c.p.s.), and it could be ignored in the early phases of each run where the peak counts ranged up to 1200 c.p.s.

For intravenous injection, the mouse was placed in the tube under the crystal detector, the multiscalers were turned on, and the injection was started and was usually completed within two to three seconds. The mouse tail was then stuffed into the tube, which was sealed with a cork within fifteen seconds. For intraperitoneal administration, the mouse was placed in the tube immediately after completion of the injection, and again the system was sealed within fifteen seconds. Each mouse was injected with 0.1 ml. of a solution containing 5 mg. per ml. sodium xenate in 0.2 M phosphate buffer.

The results of a typical intravenous injection are shown in Figure 1. Xenon leaves the body very rapidly, presumably after reduction of the xenate to xenon gas. The peak value is reduced to 50 per cent within twenty seconds and to 20 per cent in seventy-five seconds; thereafter the content of xenon declines more slowly.

By contrast, intraperitoneally injected sodium xenate leaves the body more slowly (Fig. 2), so that a value half that of the peak is reached in six minutes. The data used in Figures 1 and 2 are plotted on a semi-log grid in Figure 3. Here, during the first five minutes, the data for intraperitoneally administered xenon appear to fall along a straight line. When the data for longer periods of time are plotted on an arithmetic grid (Fig. 4), the differences between modes of administration are less pronounced, and they disappear within 150 minutes, at which time all of the values are very close to background. A semi-log plot of these curves, given in Figure 5, suggests that at least for the intraperitoneally administered material there is a final slow component that lies along a straight line. The earlier portions of both curves appear to be non-linear when plotted on this scale.

One mouse that had been injected intraperitoneally in an early experiment died because its nose was forced into the tube outlet by a stronger air stream than was used subsequently. Death of the animal was discovered when the radio-xenon values did not fall during counting, and actually they remained constant for the next forty-eight hours despite some decomposition of the animal.

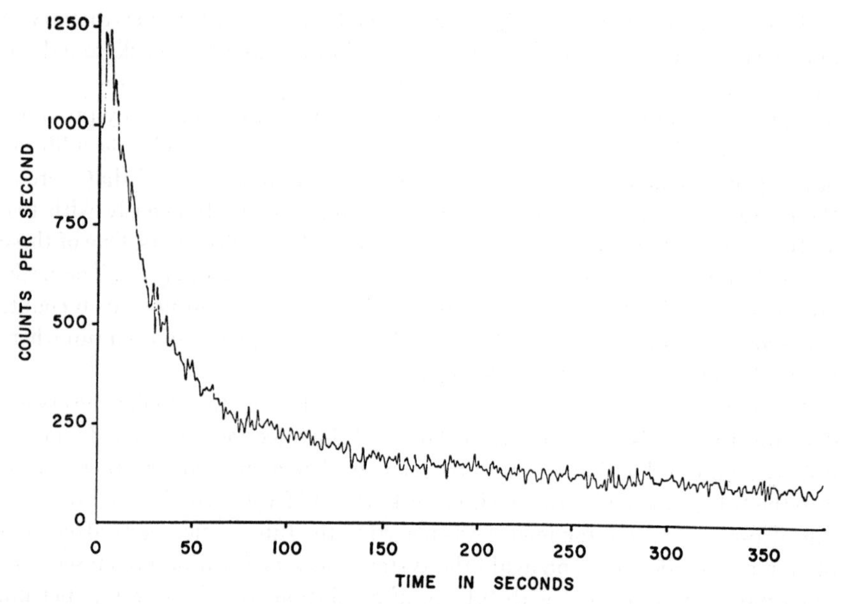

FIG. 1.—Retention of radio-xenon (Xe^{133}) by a mouse after intravenous injection of sodium xenate.

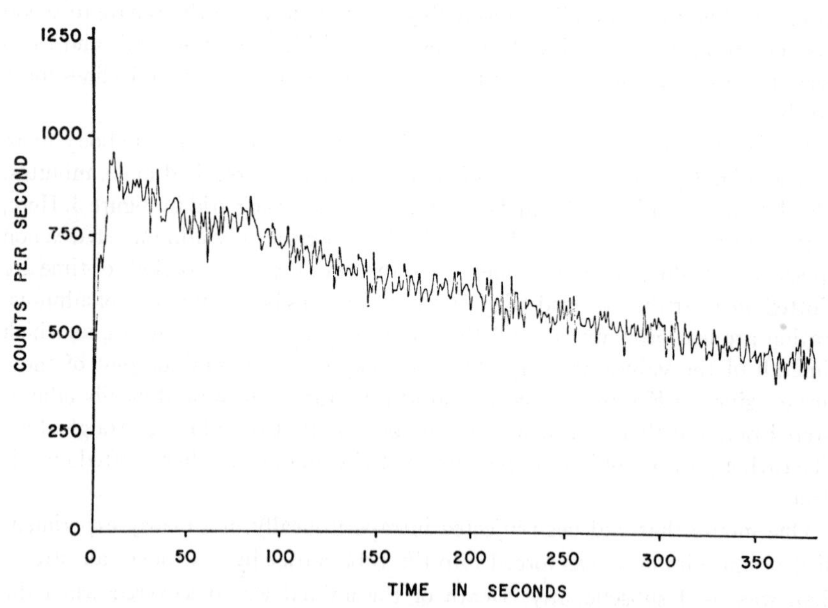

FIG. 2.—Retention of radio-xenon (Xe^{133}) by a mouse after intraperitoneal injection of sodium xenate.

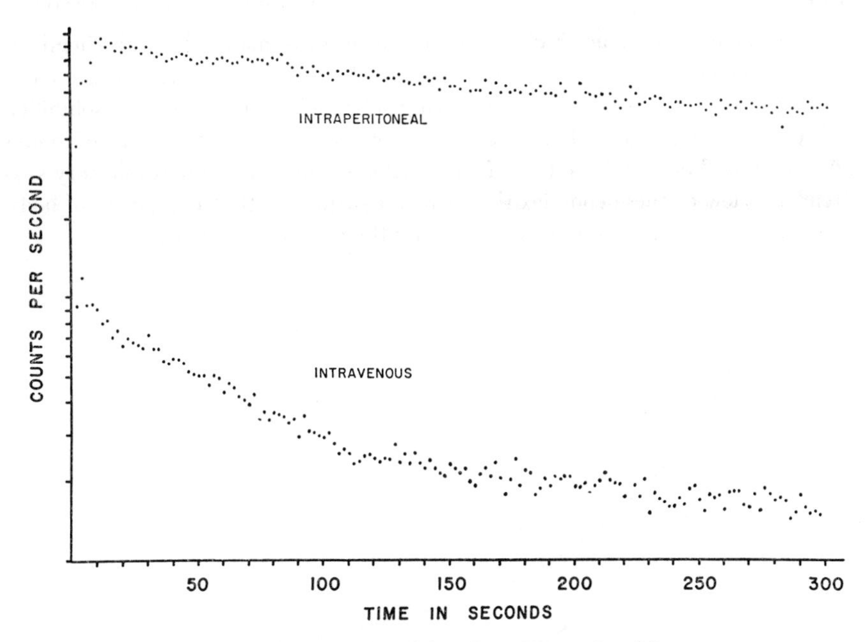

FIG. 3.—Semi-log plot of data from Figures 1 and 2

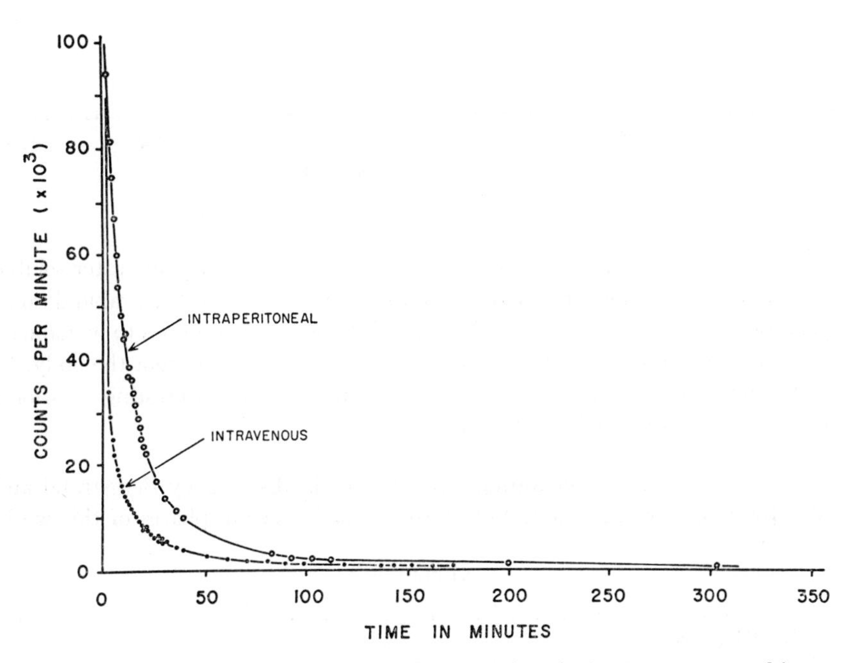

FIG. 4.—Long-term retention by mice of radio-xenon (Xe^{133}) after intravenous and intraperitoneal injections of sodium xenate.

The retention of xenon by the mouse is a result of many physical, chemical, and physiological factors. Among these are the instability of the xenate compound in the presence of organic reducing substances, the differential solubility of xenon in water and body fats, the transport of xenate and the liberated xenon in the body fluids, and the rate of elimination of the gas by the respiratory system. In view of these complexities, it is not surprising that the curves of body content cannot be expressed by simple mathematical equations.

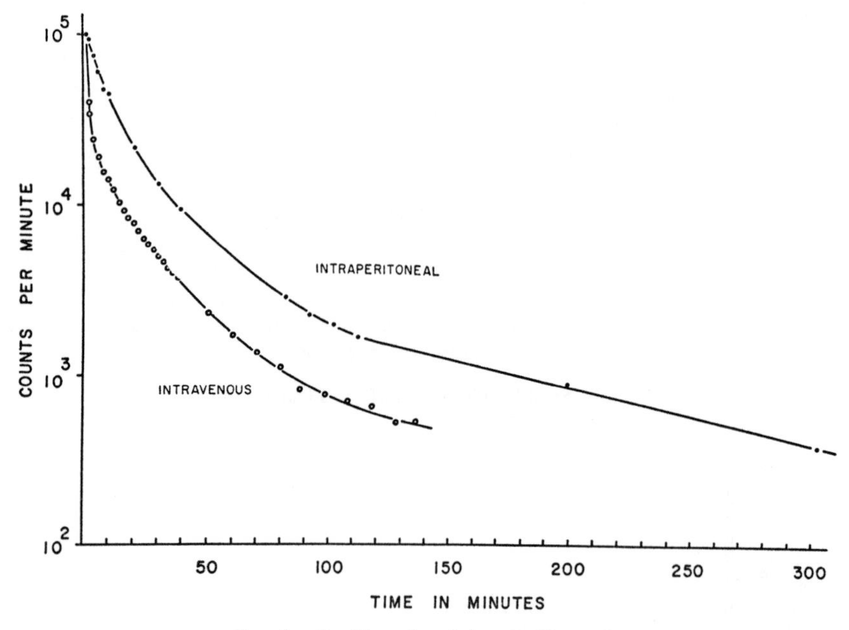

Fig. 5.—Semi-log plot of data in Figure 4

The outstanding features of these studies on the biology of water-soluble xenates are (*a*) the moderate toxicity of these substances, (*b*) their rapid decomposition in the body, (*c*) the speed with which the xenate appears to be reduced to xenon gas, and (*d*) the very rapid elimination of this gas from the body. In addition to these results, the work to date has raised many interesting questions, which warrant further investigation.

Acknowledgments.—The authors wish to thank Miss Nancy Bergstrand and Mr. John B. Corcoran for technical assistance in certain phases of this work.

REFERENCE

1. C. E. Miller, *Progress in Nuclear Energy*, Series VII. *Medical Sciences*, vol. 2, pp. 87–104 (1959).

PART 9 THEORETICAL STUDIES OF NOBLE-GAS COMPOUNDS

Edited and with an introduction by G. L. Goodman

The primarily theoretical contributions to the conference are collected in this last section of the proceedings (as they were relegated to the final hours of the meeting itself). Of course, most theoretical chemists must feel at least slight chagrin in offering explanations, rather than predictions, of the relatively easy formation of fluorides of xenon. The chemists who seem to have come closest to predicting generally the stability of such compounds are von Antropoff,* Oddo,† Pauling,‡ and Pimentel.§

The theoretical contributions in this volume are, for the most part, rather detailed descriptions of the electronic structures for compounds of the noble gases. Some of the papers emphasize the electronic state giving the lowest energy for a molecule, in order to indicate the role of electronic structure in dictating the molecular geometry, or more specifically the equilibrium nuclear configuration. The papers by Allen, Michels, Gillespie, and Lohr and Lipscomb seem to fall into this category. Allen feels most strongly the need of a reconciliation between the classification of "inert" gases in the periodic table and the demonstrated ability of some of them to form simple molecules.

The other group of articles are more empirical in that they treat equilibrium nuclear configurations as experimental data. The aim of the papers by Boudreaux, Hinze and Pitzer, and Jortner, Wilson, and Rice is to extract an understanding of the electronic structure of a molecule by studying the amount of

* A. von Antropoff, *Z. Angew. Chem.* **37**, 217 (1924).

† G. Oddo, *Gazz. chem. ital.* **63**, 38 (1933).

‡ L. Pauling, *J. Am. Chem. Soc.* **55**, 1895 (1933).

§ G. C. Pimentel, *J. Chem. Phys.* **19**, 446, (1951).

energy required to raise the molecule from its lowest energy level to some of its electronically excited levels.

The following are notes of explanation about those versions of the contributions that are published here.

After the meeting Dr. Michels submitted some comments in writing. (There was no time for discussion after each paper in this section was read or at the end of the meeting itself.) Those of Michels' comments that discuss recent *ab initio* molecular calculations are printed here.

The energy level diagram for XeF_6 presented in this volume by Boudreaux yields $^1A_{1g}$ as the symmetry of the electronic state with lowest energy for this molecule under the assumption that its equilibrium nuclear configuration has octahedral symmetry. This statement must be contrasted with that statement based on the same assumption and made at the time of the meeting, that an orbitally degenerate level of this molecule is at lowest energy. The revised version of Dr. Boudreaux's work is in agreement on this point with the related work of Lohr and Lipscomb.

Except for those cases noted above, the contributions printed in this section of the proceedings correspond rather closely to those made orally at the meeting.

G. L. G.

THEORY OF BINDING IN INERT-GAS MOLECULES

LELAND C. ALLEN
Princeton University

ABSTRACT

A many-electron theory for binding in inert-gas molecules is formulated and discussed. The importance of instantaneous electron-electron correlation in specifying the relationship between the periodic table and molecular formation is pointed out. A survey of other theoretical efforts and their relation to this work is also included.

INTRODUCTION

The discovery of inert-gas molecules presents chemistry with three theoretical challenges.

The first is to understand and explain why inert-gas compounds should be formed at all. How does one reconcile the well-established free-atom properties represented by the periodic table with the advent of molecular formation? The achievement of a satisfactory answer to this question along with the prediction of other properties is the goal which creates a special interest in the noble-gas compounds.

The second challenge is to justify and rationalize the geometries (particularly the square-planar structure of XeF_4) and other properties of the observed species at their equilibrium separation. It turns out that there are several routes to a semi-quantitative justification of the equilibrium geometries, and these are treated below in the section on related work.

The third challenge is to formulate quantitatively based theoretical predictions about the properties of inert-gas molecules before these are found experimentally. There have been relatively few opportunities in the history of chemistry, in contrast to that of physics, for theory to play a part as important as that of experiment in the development of new chemistry. The recent and rapid rise of noble-gas chemistry and the unusual relationship between the periodic table and the molecular structure in these compounds thus offer a unique theoretical opportunity.

MANY-ELECTRON FORMULATION

Since xenon difluoride illustrates most of the electronic structure problems which arise, we shall work out the formulation for this molecule. We shall also consider the molecule as being formed by two fluorine atoms approaching along a line from opposite sides of the xenon atom and always maintaining equal separation between each other and the xenon atom. This enables one to examine the bonding mechanism directly and avoid simultaneous consideration of the activation energy.* Because we are interested in the connection between the periodic table and molecular binding, we shall initially consider a large internuclear separation. As the atoms come together they are subject to van der Waals attraction. This force is very small—though greater between two xenon atoms than between a fluorine and a xenon atom—and does not lead to directional bonding. It can be ignored for our considerations. When the atoms first enter the overlap region, it is most appropriate to set up the wave function within the framework of the valence-bond method. An adequate model can probably be set up initially using p orbitals only. Three types of evidence may be cited to support this view for XeF_2 and XeF_4. First, experimental evidence may be cited, such as the quadrupole coupling-constant measurements on ICl_2^- made by Cornwell and Yamasaki [1], NMR experiments [Hindman and Svirmickas, p. 251; Brown, Whipple and Verdier, p. 263, this volume], and Mössbauer experiments reported by Perlow *et al.* [p. 279]. Theoretical interpretations of these experiments can be made in each case, using p orbitals either solely or as the dominant contribution, although this involves some rather crude approximations. The second argument arises from the fact that the difference in the one-electron energies for the outer s and p electrons is 10.3 ev. in fluorine and 11.3 ev. [2] in xenon, implying a relatively small hybridization of these orbitals. The spectroscopic excitation energy of xenon to the $5d$ level is 9.9 ev., a third type of evidence, implies, at least for a first approximation, that d mixing is not required. This point is discussed further in the section on related work. It is also a reasonable first approximation to treat the molecule as a four-electron problem. This excludes π-bonding and closed-shell repulsion between fluorine and xenon. These effects will both be smaller than those arising from the σ-bonding and will tend to cancel each other.

Within these limits, the neutral valence-bond state for XeF and XeF_2 is symbolically represented in Figure 1. The wave function for XeF_2 is a linear combination of 4×4 determinants formed from two singly occupied fluorine $2p$ orbitals and a doubly occupied xenon $5p$ orbital. Multiplicative coefficients in the linear combination of determinants are chosen to give a singlet spin function. The most satisfactory one-electron orbitals for our purposes are free-atom Hartree-Fock functions. A Hartree-Fock fluorine $2p$ orbital is available from the literature [3], and a $5p$ radial function has been interpolated from other available Hartree-Fock solutions. The many-electron wave function that we have specified is built from atomic basis orbitals mounted on their nuclear

* In the gas-phase reaction with heat, the reactants are Xe + F_2 rather than F + Xe + F. It would be interesting to carry through the quantum mechanical calculation for Xe + F_2 and seek to predict the activation energy.

centers, thereby generating a non-orthogonal basis set which must be retained. Lack of orthogonality considerably complicates the numerical details, but the formalism for carrying it through is straightforward [4]. Because we have selected only four of the seventy-two electrons present in XeF_2, an effective static potential must be constructed to represent the other sixty-eight. The simplest scheme is to use effective nuclear charges. These charges are chosen to reproduce the observed ionization potentials of 17.42 ev. and 12.13 ev. for fluorine and xenon, respectively. The resultant effective charges which reproduce the ionization potentials with our free atomic orbitals are slightly greater than unity for both fluorine and xenon. This produces a superficial lack of neutrality for the molecule as a whole. However, the separated atoms are defined by the same atomic orbitals and same effective nuclear potentials; thus our model of the molecular electronic structure is not invalidated. A more complete, a priori method for obtaining an

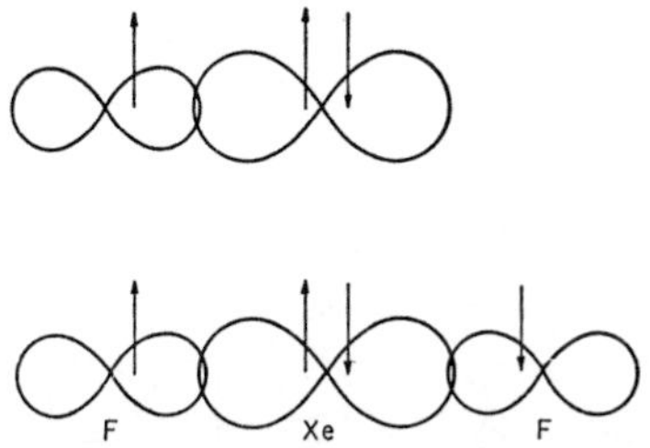

F Xe F

FIG. 1.—Valence-bond state for XeF and XeF_2, using *p*-orbitals only

effective potential is to construct that potential which results from a superposition of the atomic charge density of all other electrons. The modification to the expected from this distributed potential is being estimated.

Although numerical calculations for this wave function and others to be described below are not complete, it seems clear, both from approximate numerical estimates and qualitative chemical arguments based on saturation of valence, that neither the diatomic radical nor xenon difluoride is bound in the standard, neutral valence-bond approximation of Figure 1. The central conceptual element necessary to establish the relationship between the periodic table and molecule formation is the effect of instantaneous electron-electron correlation. This is present, of course, as a relatively small perturbation in all many-electron systems, but, in most other molecular problems, a simple uncorrelated, one-electron orbital picture (following along the usual valence-bond or molecular-orbital lines) is adequate for qualitative understanding of the process of bond creation. In inert-gas compounds it is the *sine qua non* for molecular formation. Historically, it seems evident that the lack of a simple, qualitative, chemical textbook description of correlation effects is responsible for the failure to intensively investigate the possibility of noble-gas compounds during the last forty-five years. Part of this correlation—that which may be termed "in-out"—can be represented by a double set of p orbitals on the xenon center. The radial function of one p orbital is slightly drawn in and the other slightly extended with respect to the unperturbed p orbital. This allows for "spin correlated," "split," or "open-shell" orbitals and lowers the total energy by approximately 1 ev. [5]. Figure 2 symbolically shows these spin-correlated radial functions as two different sizes of ordinary p spherical harmonic diagrams. Each of the four drawings of Figure 2 corresponds to one of the four determinants required to produce an eigenfunction of S^2:

$$\Psi = |F\alpha(1)Xe'\beta(2)Xe\alpha(3)F\beta(4)| - |F\beta(1)Xe'\beta(2)Xe\alpha(3)F\alpha(4)| - |F\alpha(1)Xe'\alpha(2)Xe\beta(3)F\beta(4)| + |F\beta(1)Xe'\alpha(2)Xe\beta(3)F\alpha(4)| .$$

A heavy cross in the region between an F and Xe in Figure 2 indicates that bonding between these atoms is discouraged by the Pauli principle, while the

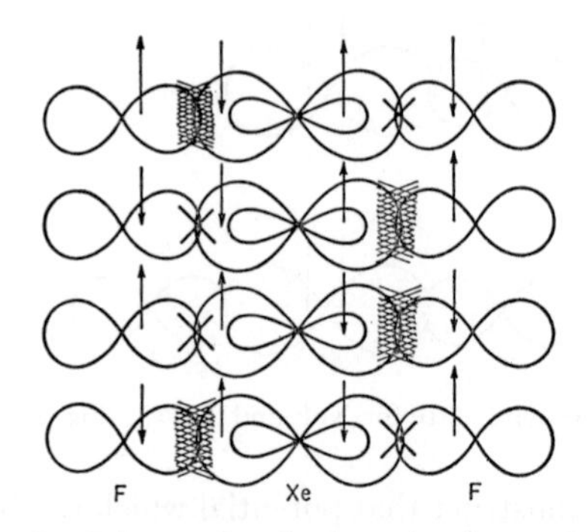

FIG. 2.—Spin-correlated valence-bond state for XeF_2

dense cross-hatching indicates that spin polarization provides an opportunity for binding. Examination of the four diagrams together shows a type of resonance between the determinants and implies the potentiality for lowering the energy. Until numerical calculations have been completed it is not possible to say with confidence whether or not this degree of correlation will lead to binding. Recent numerical results by Michels [6; see also this volume, p. 329] indicate that it may be sufficient.

Another part of the correlation—that which may be termed "left-right"—comes into the valence-bond method by inclusion of ionic states.* The charge-transfer notation for the possible ionic states is shown in Figure 3. Since xenon

* The chemical concept of charge transfer can be described with an effective one-electron Hamiltonian as well as with the many-electron Hamiltonian we are employing. However, mixing of neutral and ionic states implies the concept of a correlated wave function. In transferring a quantity of charge from a noble-gas atom to another atom, one must ask what is the spin distribution of the transferred charge. If it is one-half α and one-half β, as it must be in a one-electron theory, no binding can result. To accomplish a transition from the complete spin and space symmetry of an inert-gas atom to molecular formation, one must set up an appropriately spin-symmetrized many-electron wave function.

cannot accept an electron, no states containing Xe^- are included. The drawings in Figure 3 only show the orbitals and atomic sites to be employed. Each state of Figure 3 represents a combination of several determinants and spin distributions, e.g., {F Xe F} represents Figure 2. Each of the states of Figure 3 is added to the wave function with an arbitrary multiplicative coefficient, which is then determined by energy minimization. It is also important to note that as splitting in the spin-correlated orbitals is formally enlarged there is a smooth transition from one state into another. Just as the neutral valence-bond state is unsatisfactory by itself, so each individual ionic state is incorrect because it requires an excitation energy from the neutral state and separates into ions at infinity. The most important ionic state at large separations is $\{F^- Xe^+ F + F Xe^+ F^-\}$ and the cross term in the energy expression between this state and

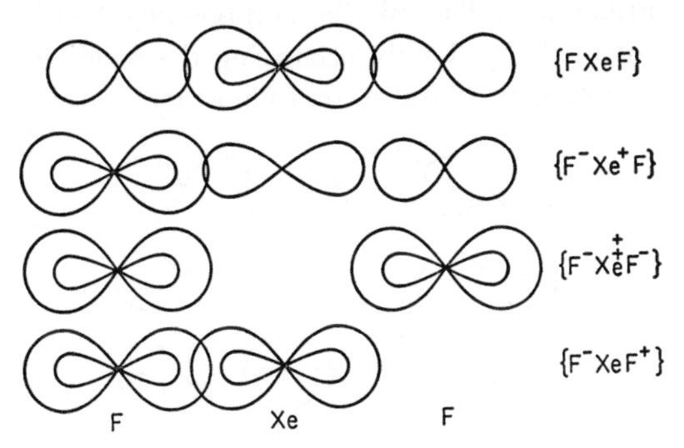

FIG. 3.—Ionic states for XeF_2

{F Xe F} brings in left-right correlation. Both left-right correlation and the in-out correlation introduced by spin-correlated orbitals cause a slight amount of ionic character to be mixed into the wave function at large separation, and this produces the onset of molecular binding. Our model lacks any manifestation of the tendency for two electrons to be on opposite sides of the internuclear axis. This angular correlation may be introduced by extending the basis set to include *d*- and *s*-like atomic orbitals on the xenon and fluorine centers. However, general experience indicates that a sufficient part of the correlation is already included to obtain binding and the considerable increase of labor required to include angular terms justifies their omission initially.* There is still another aspect of the correlation problem which must be considered: as a fluorine atom approaches a xenon atom, they will polarize each other destroying the free-atom form of the basis orbitals. Michels finds this effect to be significant in his calculations on the $^2\pi$ and $^2\Sigma$ states of HeH [6]. Some of this polarization is implicitly included in the spin-polarized orbitals of Figure 1. The ionic states which we

* As noted in the text, inclusion of *s* and *d* orbitals as a principal ingredient in the binding mechanism for XeF_2 and XeF_4 does not seem to be required for a first approximation, but orbitals of these symmetries will still make an appreciable contribution to the correlation energy.

have included are also quite effective in representing polarization, and thus it is probably unnecessary to further complicate the essential simplicity of our model by introducing distorted basis orbitals.*

As the internuclear separation is decreased, a large transfer of charge occurs because of the large electronegativity of fluorine. At the equilibrium separation, experimental evidence indicates a major contribution from $\{F^- Xe^+ F + F Xe^+ F^-\}$. A similar conclusion follows from the molecular-orbital considerations discussed below.

A many-electron formulation for XeF_2 can proceed as well from the molecular-orbital viewpoint as from the valence-bond approach. Within the same limits assumed in the valence-bond method, the molecular orbitals are displayed in Figure 4. The top and bottom drawings have ungerade symmetry, one bonding and the other antibonding. The middle, non-bonding MO is gerade. A single 4×4 determinant made by doubly occupying the bonding and non-bonding orbitals forms the molecular-orbital state. In most molecular problems this state

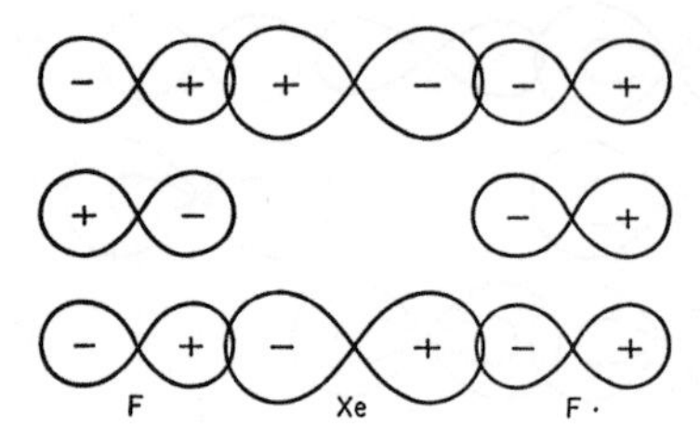

FIG. 4.—Molecular orbitals for XeF_2, using *p*-orbitals only: bonding (*top*), non-bonding (*middle*), and antibonding (*bottom*).

yields a fairly good representation of the molecule at its equilibrium separation, and there is no reason to expect any different behavior for XeF_2. The highly favorable potential environment for charge transfer onto the fluorine sites in the bonding orbital will result in a relatively large delocalization of charge and give approximately the same qualitative weight to ionic character as in the valence-bond approach. At infinity the molecular-orbital state for XeF_2 separates into $2F^- + Xe^{+2}$ and thus does not provide a connection with the periodic table. This can be corrected by adding other determinants to the wave function. These are made by taking some of the electrons out of either the bonding or non-bonding orbitals and putting them into the antibonding MO. In effect, the partial occupancy of the antibonding orbital at large separation cancels the overemphasis on ionic character present in the molecular-orbital state. It is well known that both the molecular-orbital state plus configuration interaction or the valence-bond method, if carried to the same approximation, give identical results. Calculations for both schemes are being performed simultaneously in this

* Polarization distortion can also be partly expressed by hybridizing the atomic orbitals, but we have already given reasons for excluding hybridization effects in our first-order approximate model.

laboratory. In general it is harder to assign a qualitative chemical significance to the configuration interaction terms in the molecular-orbital framework than to the valence-bond states.

PREDICTION OF OTHER SPECIES

At present only a semi-quantitative search has been made for new species, and this has been further limited to linear molecules containing a single inert-gas atom. We predict that XeO_2, XeN_2, and XeF are electronically possible. It was reported that Morton and Falconer [this volume, p. 245] have produced a stable XeF radical in an XeF_4 matrix. Pitzer [7] has established criteria for the thermodynamic stability of inert-gas molecules by analogy to interhalogen compounds. His study indicates that the thermodynamic stability is predominantly dependent on the ionization potentials of the rare-gas atoms and less dependent on the detailed electronic structure of the molecules. Therefore, the known stability of XeO_3, the relatively tight binding of O_2, and the strong binding in N_2 undoubtedly preclude the observation of XeO_2 or XeN_2 under normal conditions. Indirect methods accompanied by trapping in a matrix will be required to observe these species.

RELATION TO OTHER WORK

One of the first tasks to which theory must address itself is justification of the observed linear structure for XeF_2 and the square-planar, rather than tetrahedral, structure for XeF_4. All of the theoretical descriptions that have been proposed are capable of rationalizing these structures. The simplest scheme is merely to consider the six outer *p* electrons in a free noble-gas atom. These electrons form a spherically symmetric charge distribution, but if we allow repulsion between them, they will arrange themselves octahedrally around a sphere with the atomic radius. With this electron arrangement, two fluorines may be attached in a linear structure and four with a square-planar structure [8]. This type of free atom–based argument is analogous to that employed to explain the tetrahedral bonding of carbon. It is also interesting to note that the simple addition of standard atomic radii predicts a value for the XeF bond distance midway between the observed distance of 1.95 A. for XeF_4 and 2.00 A. for XeF_2, roughly suggesting that the bonding structure is not grossly different than in interhalide molecules [9]:

$$1.33 + 0.64 = 1.97 \text{ A.}$$

The conventional electron-pair theory was applied by Allen and Horrocks [9] to demonstrate a linear and square-planar structure for XeF_2 and XeF_4, respectively, and to predict the geometry of several potential xenon oxyfluoride molecules. This model was developed by Gillespie, and he has recently used it to predict the geometry and bond lengths of a great variety of rare-gas oxyfluorides [this volume, p. 333]. The widespread success of this model in justifying the structures found in inorganic chemistry and its extreme simplicity give it an elegance worthy of note. The number of electrons in the outer shell of the rare-gas atom, the valence of attached atoms, the tendency of electrons to form pair bonds, and simple electrostatics constitute the necessary information. This model predicts a non-octahedral symmetry for XeF_6. Although a crystallographic study has not yet been made, there is some experimental evidence to support this contention [D. F. Smith, this volume, p. 301]. Workers at the Oak Ridge National Laboratory [10, 11] have cited bond-lengths and force-constant data in an attempt to justify the conventional electron-pair model as opposed to an ionic model, $F^- Xe^{+2} F^-$, or to a resonance molecular-orbital picture made with p-orbitals:

$$F^- Xe^+\text{–}F$$

$$F\text{–}Xe^+ F^- .$$

Actually, none of these three simple models has a complete quantitative significance, and it is not really meaningful to make a selection from such models. Either a valence-bond or molecular-orbital approach can be made thoroughly quantitative and encompass all of the present experimental results that are well established [12].

The most popular theoretical model has been the simple molecular-orbital scheme built from p atomic orbitals only. Pimentel [13] appears to have been the first to qualitatively discuss inert-gas compounds with this model. The molecular orbitals employed are those shown in Figure 4. Rundle [14] and Pitzer [7] have also qualitatively used this model to rationalize the geometries and discuss the molecular charge distribution. A quantitative version of this model, based on the work of Mulliken [15, 16], can be worked out. This has been done by Lohr and Lipscomb [17], Waters and Gray [18], and also by us. [See also, in this volume: Boudreaux, p. 354; Jortner *et al.*, p. 358; Hinze and Pitzer, p. 340.] For XeF_2 the one-electron energies for the bonding and non-bonding orbitals are:

$$E_u = \frac{-I_{Xe}-I_F-4\beta S-\sqrt{[(-I_{Xe}-I_F-4\beta S)^2-4(1-2S^2)(I_{Xe}I_F-2\beta^2)]}}{2(1-2S^2)},$$

$$E_g = -I_F ,$$

and

$$\beta = -2S\sqrt{I_{Xe}I_F} ,$$

where S is the overlap integral between a $5p$ xenon orbital and a $2p$ fluorine orbital and I_{Xe} and I_F are the ionization potentials of xenon and fluorine, respectively. Substituting these energies into the set of linear equations for the atomic-orbital coefficients allows one to determine the charge distribution. This simple scheme is best carried out in an iterative way demanding one-electron self-consistency between the effective ionization potential and the charge distribution [19], although the qualitative results of the model are unchanged at the equilibrium separation.* Being a one-electron theory, this model is incapable of explaining why inert-gas atoms as represented in the periodic table should be bound at all. Both E_u and E_g go to $-I_F$ at infinite separation thus representing two F^- atoms. The model is a rather uncritical one and shows an energy-lowering from infinite separation for almost any combination of atoms. One certainly does not expect $XeNa_2$ to be bound, and the binding of XeC_2 is unlikely [20]; yet, as shown in Figure 5, both give binding-energy curves not qualitatively different from that of XeF_2. Thus it cannot be claimed that the simple, one-electron molecular-orbital model provides a discriminating explanation of the properties and place which inert-gas molecules have in the over-all scheme of inorganic

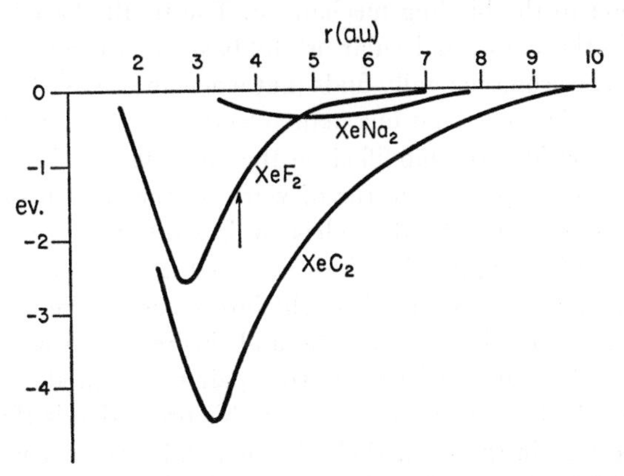

FIG. 5.—Binding energy versus internuclear separation computed from one-electron molecular-orbital theory for $XeNa_2$, XeF_2, and XeC_2.

chemistry.† This does not invalidate the usefulness of the model for some purposes, since one can calibrate the model to a known value and examine the extent to which properties such as the charge distribution and binding energy depend merely on the overlap integral and average ionization potential of the participating atoms. It appears to be adequate for explaining magnetic-resonance experiments; although, as in a many-electron molecular-orbital formula-

* The molecular charge distribution is actually quite sensitive to ω [see Jortner *et al.*, this volume, p. 369]. ω is another arbitrary parameter in the theory, and while there have been several investigations into the appropriate choice of ω for a pair of carbon atoms in π-electron systems, no investigation has been made for other types of atoms.

† Jortner *et al.* [26] have made an estimate of the left-right or charge-transfer part of the correlation energy between a xenon and a fluorine atom using a simple, semi-empirical valence-bond scheme. When the energy of the system is lowered by a charge-transfer process it generally indicates that the molecular-orbital state will yield a first approximate wave function for the bound system. However, this cannot validate or prove the adequacy of a molecular-orbital solution because it is not a sufficient criterion. Not only is the simple, semi-empirical molecular-orbital model uncritical as shown in Fig. 5, but, as noted in text, it is well known that the general molecular-orbital state does not separate into neutral atoms, and so it cannot lead to a wave function which joins the free atoms into a molecule in a continuous and well-defined manner.

tion, it is likely to place too great a weight on ionic terms even at the equilibrium separation. Further, as Jortner *et al.* have shown [this volume, p. 358], the model is capable of greatly aiding the analysis and understanding of the far-ultraviolet spectra of XeF_2 and XeF_4.

Rundle [14] and Hinze [this volume, p. 340] have directed particular attention to the small part that *d* orbitals are expected to play in the binding of XeF_2 and XeF_4. Hinze has calculated rare-gas *d*-orbital shapes and one-electron energies from a perturbed free-atom model [cf. 21, 22, and 23] and finds that such orbitals have even higher energies than atomic spectra states. This gives additional qualitative support to rejection of any suggestion that *d* orbitals make a primary contribution to the binding mechanism. The results based on including these *d* orbitals in the simple molecular-orbital basis set are not entirely conclusive, however, because one generally finds that when an extensive basis-function parameter variation is carried out in a many-electron problem, perturbed atom-basis orbitals are considerably modified by the molecular environment.

Having reviewed the present status of work on the conventional electron-pair and simple molecular-orbital models, it is worthwhile examining more closely the nature of the approximations involved and the predictions which these models make. The problem is brought into focus by the question of the symmetry of XeF_6. Both the experimental and theoretical situations are unresolved at present. The conventional electron-pair model predicts a non-octahedral XeF_6, and the fundamental reason for this prediction is that the model uses all eight electrons in the outer shell of xenon and treats them equally. The simple molecular-orbital model has given a quantitatively satisfactory explanation for a number of properties of XeF_2 and XeF_4, using only *p* orbitals with four electrons per pair of bonds, and it therefore predicts an octahedral XeF_6 using only *p* orbitals with only six electrons from xenon. Qualitatively, the conventional electron-pair model is unappealing for XeF_2 and XeF_4, because both the trigonal bipyramid and the tetragonal pyramid specified for these molecules have several protruding lone pair–electron lobes, which do not yield the smoothest charge distribution and are thus suspected of raising the energy. More quantitatively, this effect may be discussed in terms of *s* and *d* orbitals, and this has been dealt with above. On the other hand the molecule IF_7, with a pentagonal-bipyramid structure, is successfully described by the conventional electron-pair model. The difference in the ionization potential between 5*s* and 5*p* electrons is 11.3 ev. in xenon and 10.2 ev. [2] in iodine, while the spectroscopic excitation to the lowest unoccupied *d* level is 9.9 ev. and 8.4 ev. [2], respectively. These energies are not qualitatively different and certainly bond formation in IF_7 involves more than the five electrons in the outer *p* shell of iodine. The recent observation of octavalent perxenate compounds [see Malm *et al.*, this volume, p. 167] also suggests that all of the xenon outer octet may participate in chemical bonding. A satisfactory theoretical solution to this problem very likely requires the construction of an electronic wave function for XeF_6 and IF_7 by the methods described above in this chapter with the addition to the basis of *s* and *d* orbitals and the inclusion of more electrons.

The only many electron–based formulation besides that developed above has been proposed recently by Nesbet [24]. In this theory an approximate molecular-orbital state is employed in the vicinity of the equilibrium separation, and a second, single-determinant wave function (which separates into free-atom Hartree-Fock solutions) is used at large separations. In order to maintain the proper symmetry, a restriction is placed on the type of basis orbitals allowed. Many electron–perturbation theory carried to second order is used to supplement each determinant state in the joining region, but it is not clear whether the two states can be smoothly brought together or whether all contributing states have been included. At the observed internuclear separation for XeF_2 Nesbet finds that his long-range localized orbital state predominates. (This state is approximately the configuration shown in Fig. 1.) In view of the success of the simple molecular-orbital model in giving approximately correct charge distributions and far-ultraviolet spectral predictions, this seems highly unlikely and indicates that the joining region is inadequately represented. It is also impossible to get a continuous and connected binding-energy curve from the two equations Nesbet gives in his article. The binding energy ($2J$) for the molecule is the singlet-triplet separation at large internuclear separation. Physically, it seems unreasonable that this binding energy, in the case of XeF_2, should depend equally on both the first *and* second ionizational potentials of xenon, as it does in Nesbet's equations. This again suggests that an overly restrictive selection of states may have occurred. Even if one assumes that an adequate representation has been achieved, the ability to select states depends on having an inversion center in the molecule, and it is not clear that a well-defined set of contributing states can be selected when this is not true. Neither XeO_3 nor XeF_4O has an inversion center. P. W. Anderson has given a critique of this work for its application to magnetic problems in solids [25].

Note added in proof.—An interesting article on bonding in xenon fluorides has appeared while our paper was in press [R. Bersohn, *J. Chem. Phys.* **38,** 2913 (1963)]. Bersohn constructs molecular-orbital and valence-bond wave functions in which the coefficients of the components are predetermined and expressed in terms of a single parameter. This parameter specifies the number of electrons in the *p*-shell for a particular component of the total wave function. If the further approximation is made that the quadrupole coupling constant is a direct measure of this parameter, then it is possible to assign a number to the relative delocalization of charge and to state whether a simple MO or simple VB function more nearly produces this distribution. The MO model employed by Bersohn is closely related to the simple MO theory described in the section on related work. However, the valence-bond approach we have developed differs from Bersohn's because it is an in principle approximate solution of Schrödinger's equation with the weighting of components determined by energy minimization and with the quadrupole coupling constant value following as a prediction from the wave function rather than being adopted as a built in parameter of the model.

Acknowledgments.—The author wishes to thank Dr. Gordon A. Hamilton and Dr. Gordon L. Goodman for interesting discussions. He also wishes to acknowledge the past (and future) assistance of Messrs. B. Granoff, J. Harrison, and A. Lesk.

REFERENCES

1. C. D. Cornwell and R. S. Yamasaki, *J. Chem. Phys.* **27,** 1060 (1957).
2. C. E. Moore, *Atomic Energy Levels*, N.B.S. Circular 467 (1949).
3. L. C. Allen, *J. Chem. Phys.* **34,** 1156 (1961).
4. P.-O. Löwdin, *Phys. Rev.* **97,** 1474 and 1490 (1955).
5. L. C. Allen, E. Clementi, and H. M. Gladney, *Rev. Mod. Phys.*, in press.
6. H. H. Michels, *Science*, in press.
7. K. S. Pitzer, *ibid.* **139,** 414 (1963).
8. L. C. Allen, *ibid.* **138,** 892 (1962).
9. L. C. Allen and W. DeW. Horrocks, Jr., *J. Am. Chem. Soc.* **84,** 4344 (1962).
10. D. F. Smith, *J. Chem. Phys.* **38,** 270 (1963).
11. P. A. Agron *et al.*, *Science* **139,** 842 (1963).
12. G. C. Pimentel and R. D. Spratley, *J. Am. Chem. Soc.* **85,** 826 (1963).
13. G. C. Pimentel, *J. Chem. Phys.* **19,** 446 (1951).
14. R. E. Rundle, *J. Am. Chem. Soc.* **85,** 112 (1963).
15. R. S. Mulliken, *J. Chim. Phys.* **46,** 675 (1949).
16. C. J. Ballhausen, *Introduction to Ligand Field Theory*. New York: McGraw-Hill Book Co., 1962.
17. L. L. Lohr, Jr., and W. N. Lipscomb, *J. Am. Chem. Soc.* **85,** 240 (1963).
18. J. H. Waters and H. B. Gray, *ibid.* 825 (1963).
19. A. Streitweisser, Jr., *Molecular Orbital Theory for Organic Chemists*. New York: John Wiley & Sons, 1961.
20. L. C. Allen, *Nature* **197,** 897 (1963).
21. This approach has been discussed by L. C. Allen, *Phys. Rev.* **118,** 167 (1960).
22. A. Dalgarno, *Proc. Roy. Soc. A* **251,** 282 (1959).
23. R. M. Sternheimer, *Phys. Rev.* **96,** 951 (1954).
24. R. K. Nesbet, *J. Chem. Phys.* **38,** 1783 (1963).
25. P. W. Anderson in F. Seitz and D. Turnbull (eds.), *Solid State Physics*, vol. 14, p. 99. New York: Academic Press, 1963.

BINDING IN RARE-GAS COMPOUNDS

H. H. Michels
United Aircraft Corporation Research Laboratories

The extensive amount of recent experimental work on the synthesis of rare-gas compounds has stimulated several theoretical explanations of the nature of the binding in these compounds. Many of these have been based on a one-electron molecular-orbital approach. A somewhat different model for binding has been proposed by Allen [1]. The latter model suggests that radial-correlation effects between antiparallel electron pairs can produce enough separation of the outer *p* orbitals of the rare-gas atom so that sufficient overlap with the unpaired *p* orbital of the fluorine atom will occur to promote a binding situation. Allen also suggests that a considerable distortion of the outer-shell *s* electrons of the rare-gas atom will occur in molecule formation in a manner which will tend to minimize the separation of the *p* orbitals that are not involved in the binding. More recently Allen has discussed the importance of other correlation effects and the effect of mixing of various ionic and neutral valence states [this volume, p. 317]. These proposals are especially interesting, since they treat the formation of the rare-gas molecule essentially as a perturbation from its atomic constituents and, therefore, offer some explanation of the mechanism of formation.

In order to evaluate these several proposals, we must examine the relative importance of the various electron-correlation effects that occur in atomic and molecular systems. The correlation problem in quantum mechanics has been reviewed recently by Löwdin [2]. For atomic systems, there are two rather different electron-correlation effects. The first type is usually called radial or "in-out" correlation. It is, physically, a simple consequence of the fact that it is energetically highly unfavorable for two electrons to come very close to each other in space and, that if one electron at any time has a small value of r, the other will tend to have a larger value. This correlation effect exists for all electron pairs. The second type of correlation between electrons is usually called angular correlation and is a consequence of a relative localization of two electrons at slightly different angles in the charge cloud. This type of correlation can only be described using orbitals with $l > 0$.

A considerable amount of activity in developing various methods for treating these electron-correlation problems is now in progress. One method of handling radial correlation is to employ different spatial orbitals for electrons with different spin in the quantum-mechanical description of the system. Several *ab initio* calculations with such assumed open-shell structures have been performed, and all have indicated some spatial separation of antiparallel electron pairs and have resulted in substantial improvements in the calculated energies [3, 4]. Angular correlation is most often handled by the method of superposition of configurations. Although this effect must be accounted for if quantitative results are desired, calculations performed without the introduction of angular-correlation terms are similar with regard to the radial electron distribution.

In molecular systems both of these correlation effects are also present, and, in addition, other correlation factors must be considered, namely, the polarization and distortion of the original atomic charge distributions along an axis between nuclei owing to the relative attractive powers of the individual atoms. It is possible to discuss this polarization effect using an atomic viewpoint, and in this regard it is usually most convenient to employ a valence-bond framework. It is equally possible to employ a molecular-orbital approach, but here it is not as convenient to examine the changes that occur in the electronic structure as the molecule dissociates into its constituent atoms. The physical reason for these polarization effects is that it may be energetically much more favorable for the valence, or outer, electrons to shift from their normal atomic distribution to some distorted arrangement in the molecule. In this regard, we note that the really successful *ab initio* calculations on diatomic molecules have been those which have made proper allowance for this distortion from atomic symmetry, either by the direct use of polarizable molecular orbitals described by elliptical coordinates or by means of superposition of configurations composed of linear combinations of Slater orbitals. The direct use of polarizable or elliptic orbitals is especially advantageous, since it gives a clear picture of the electron distribution along any internuclear axis. This is to be contrasted with large configuration-interaction treatments, where the complex mixing of various atomic representations and ionic states is often difficult to interpret physically. It now remains to determine which, if any, of these several correlation effects is significant with regard to molecule formation in the rare gases.

There are relatively few molecular calculations that can aid our analysis of these correlation effects. One such calculation has already been discussed by Allen [5], namely, the work of Taylor and Harris [6] on the $^2\Sigma$ ground state of HeH. The orbitals used in this wave function are all of the same parametric form and can be represented in elliptic coordinates as

$$\phi_i = e^{-\delta_i \xi - \zeta_i \eta} ,$$

where δ_i and ζ_i are free parameters which are optimized using a variational treatment. The parameters of the wave function for this state are given in Table 1 as a function of internuclear separation. At large internuclear separations this wave function is essentially a molecular representation of a normal open-shell helium atom and a hydrogen atom. At close internuclear separations, however, there is a considerable polarization of the "outer" orbital of helium toward hydrogen, but this is more than compensated by a polarization of the hydrogen orbital away from the helium atom. Thus, the electron density between the nuclei is not increased during molecule formation, and the calculation indicates a repulsive state. On the other hand, some more recent calculations on the first excited $^2\Sigma$ and $^2\Pi$ states of HeH have been performed using elliptic orbitals [7]. All of these calculations show binding of the hydrogen atom with ground-state helium, and the wave functions for these states can be interpreted by the following mechanism: At large internuclear separations the charge distribution is again similar to that of the separate atoms. As the hydrogen atom approaches the helium nucleus, there is a polarization of the charge about the helium atom toward hydrogen. The hydrogen-electron density is diffuse, since we are dealing with an excited state, but there is a positive displacement of this

TABLE 1

PARAMETERS FOR THE $^2\Sigma$ GROUND-STATE WAVE FUNCTION FOR HeH

R (Bohrs)	δ_1	ζ_1	δ_2	ζ_2	δ_3	ζ_3
1.0.......	1.091	1.178	0.900	0.015	0.194	−0.625
2.0.......	2.061	2.373	1.361	0.855	0.894	−1.231
3.0.......	3.227	3.352	1.836	1.669	1.477	−1.590
4.0.......	4.345	4.403	2.395	2.333	1.991	−2.030

charge density toward helium. When these calculations were made with the restraint of atomic symmetry—even using an open-shell representation for helium—they indicated little or no binding. On the other hand, a *closed-shell* representation with allowance for polarization effects indicated all of the correct qualitative features of the molecular state. These calculations indicated that distortion polarization is probably a more important effect in these rare-gas structures than radial correlation. However, the electron distribution in the atomic states of HeH has *s*-character only, whereas mainly *p* electrons are involved in bonding of XeF_2 and XeF_4. Since *p* electrons offer a more favorable electron distribution between the nuclei for bonding, it is possible that radial-correlation effects will be more important in these systems. Our main concern, however, is not whether radial and angular correlation effects or polarization effects are most significant; it is, rather, whether or not *any* correlation effects are significant. The calculations performed on the bonding states of HeH cannot be explained using a one-electron theory, and in this regard Allen's contention that a true connection of the chemistry of the rare-gas compounds with the periodic table will require a many-electron theory is probably correct.

It is apparent that present mathematical difficulties preclude an accurate quantum-mechanical study of systems such as XeF_2 or XeF_4, including all of the electrons. However, calculations that involve interactions with only the outer electron shell for the inert gases neon, argon, krypton, and xenon, employing a central-field approximation for the core electrons, are within the realm of possibility, and we are encouraged that some calculations of this type have been undertaken by Allen and others. We believe that more useful information concerning the detailed nature of the binding in these rare-gas compounds will result from this direct approach than will result from reliance on semi-empirical theories or from analogies with isoelectronic structures that may have entirely different charge distributions.

REFERENCES

1. L. C. Allen, *Science* **138,** 892 (1962).
2. P.-O. Löwdin in *Advances in Chemical Physics*, vol. 2. New York: Interscience, 1959.
3. See, for example, G. H. Brigman and F. A. Matsen, *J. Chem. Phys.* **27,** 829 (1957).
4. H. Shull and P.-O. Löwdin, *J. Chem. Phys.* **23,** 1565 (1955); **25,** 1035 (1956).
5. L. C. Allen, *Nature* **197,** 897 (1963).
6. H. S. Taylor and F. E. Harris, *Mol. Phys.*, in press.
7. H. H. Michels and F. E. Harris, *J. Chem. Phys.*, in press.

THE NOBLE-GAS FLUORIDES, OXYFLUORIDES, AND OXIDES

PREDICTIONS OF MOLECULAR SHAPES AND BOND LENGTHS

R. J. GILLESPIE
McMaster University

There have been several recent discussions of the possible structures of the noble-gas fluorides and of the nature of the bonding involved [1–4]. Allen and Horrocks [1] have proposed that the bonds may be regarded as conventional electron-pair bonds, and, using the arguments of the valence-shell electron-pair repulsion theory [5, 6], they have predicted a linear shape for XeF_2 and a square-planar shape for XeF_4. It is the purpose of this paper to extend this discussion to show how the shapes of known and as yet unknown noble-gas compounds may be predicted and how, by comparison with the interhalogen compounds, the bond lengths in these molecules may also be estimated.

The basic assumption of the theory is that the bonds in these molecules are essentially localized electron-pair bonds whose arrangement in space is determined by the mutual interactions of all the electron pairs in the valence shell of the central atom. Lone pairs exert greater repulsions than single-bond pairs, and oxygen atoms are assumed to be bonded by double bonds containing two electron pairs, which also exert greater repulsions than single-bond electron pairs [5].

We shall first consider molecules with a total of four bonds and unshared pairs in the valence shell. These have a tetrahedral arrangement, and hence XeO_4, XeO_3, and XeO_2 are predicted to have tetrahedral, pyramidal, and angular structures, respectively (I–III).

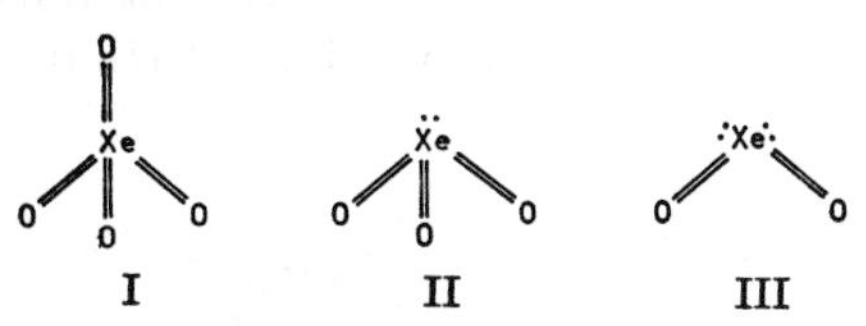

I II III

Molecules with a total of five bonds and lone pairs in their valence shells have shapes that are based on a trigonal-bipyramidal arrangement of the bonds and lone pairs. As a consequence of the tendency to minimize repulsions, lone pairs and double bonds occupy the equatorial positions rather than the axial positions

[6]. The predicted shapes of XeF_2 and some possible related oxyfluorides are shown below (IV–VII).

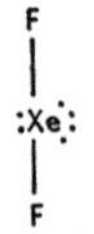

IV 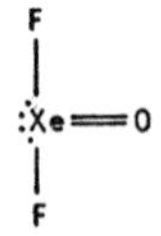 V 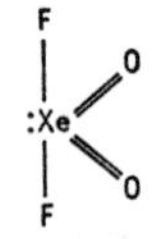 VI 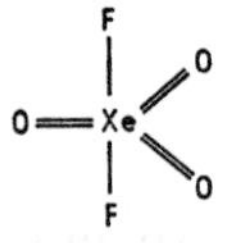 VII

The structures of a series of possible oxy-acids can be obtained by replacing F with OH in IV–VII.

Molecules with a total of six bonds and lone pairs in the valence shell of the central atom have shapes that are based on an octahedral arrangement of these bonds and lone pairs. Predicted shapes for XeF_4 and some oxyfluorides are shown below. Because of the greater repulsions exerted by lone pairs and by double-bonded oxygen atoms, they will occupy *trans* positions (VIII–X). Again the structures of possible oxy-acids can be obtained by replacing F by OH in VIII–X.

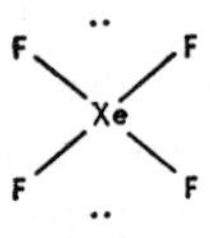

VIII 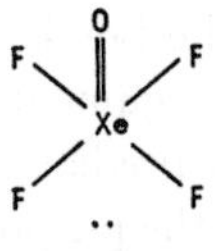 IX 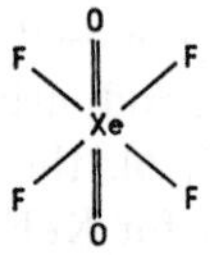 X

The arrangement of a total of seven bonds and unshared electron pairs is not quite certain, since it depends on the form of the interaction potential; whereas the arrangements of two to six electron pairs are independent of the form of the interaction potential [7]. One of the three most probable arrangements is the pentagonal bipyramid of IF_7. Xenon hexafluoride might reasonably be expected to have a closely related structure, with its lone pair in one of the equatorial positions; i.e., its structure will be that of a somewhat distorted octahedron (XI). Two alternative structures are another form of the distorted octahedron XII (lone pair located opposite a triangular face) and a distorted trigonal prism, XIII (lone pair located opposite a rectangular face). In any case it is clear that the molecule is unlikely to have a regular octahedral structure. We can also predict the pentagonal-bipyramid structure (XIV) for a possible related oxyfluoride. (Alternate structures related to XII and XIII are also possible.)

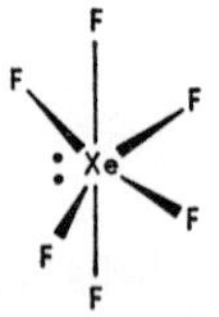

XI 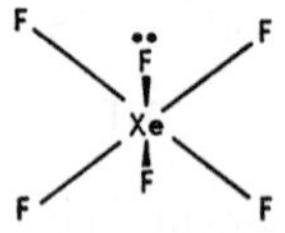 XII 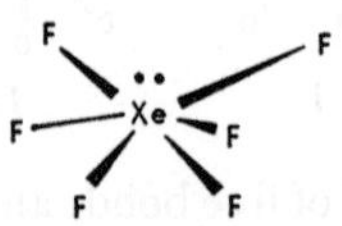 XIII 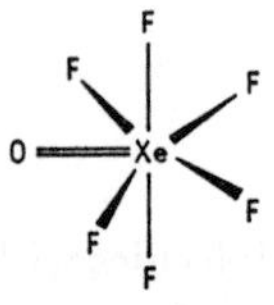 XIV

Since only elements with electronegativities greater than approximately 2.0 appear to be able to form double bonds with oxygen, and since Rundle [3] has estimated the electronegativity of xenon to be 2.25, it is not certain that all the oxyfluorides described above will necessarily exist in a simple molecular form. Some of them may polymerize, with elimination of the xenon-oxygen double bond. This is more likely to occur in the AY_5 molecules, since molecules of this type often show a tendency to adopt AY_6 coordination either by ionization or polymerization, and in those molecules containing several oxygen atoms. Thus alternative polymeric structures for IV, V, and VI are shown in XV, XVI, and XVII—in which one xenon-oxygen double bond has been eliminated in each case. The structures are based on an octahedral arrangement of bonds and lone pairs around the central xenon atom; the lone pairs and double bonds are expected to occupy *trans* positions.

XV

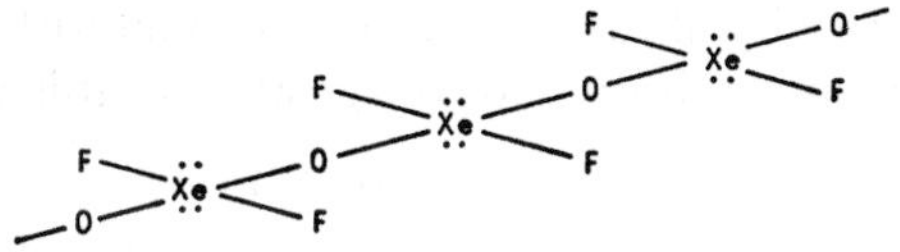

XVI

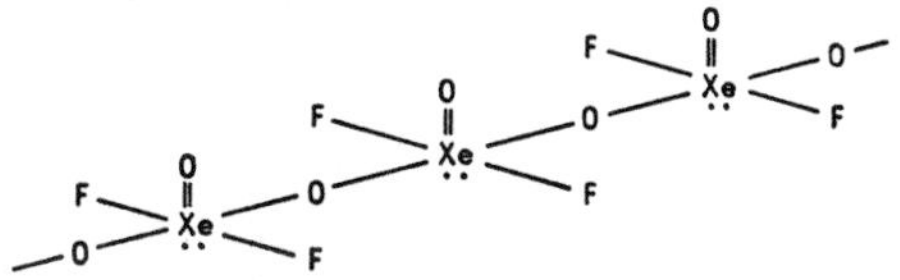

XVII

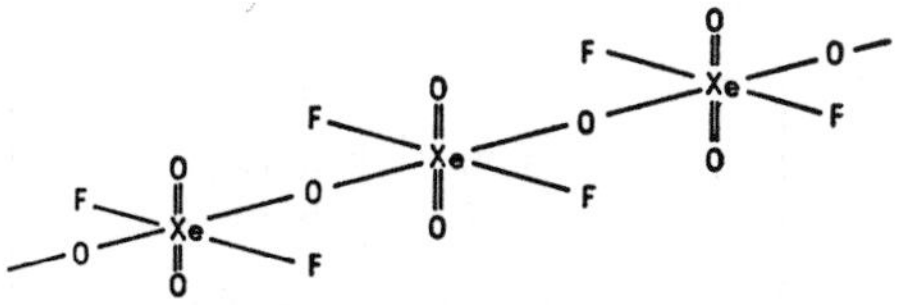

In view of the established stability of the xenon fluorides, it now seems reasonable to reformulate the compound $XePtF_6$ to which Bartlett [8] has ascribed the ionic structure $Xe^+PtF_6^-$ as a fluorine-bridged complex of XeF_2 and PtF_4 with a polymeric structure such as XVIII, possibly further polymerized by additional fluorine-bridge bonds. The recently prepared $Xe(PtF_6)_2$ [9] may also be formulated similarly as the complex $XeF_2(PtF_5)_2$ with a structure such as XIX, which may be further polymerized by additional fluorine-bridge bonds.

XVIII

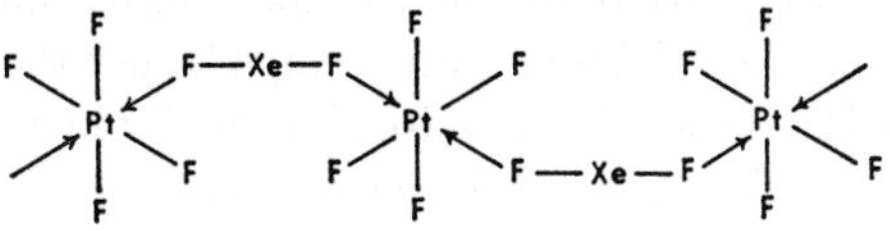

Very recent X-ray and neutron-diffraction studies [10–13] have shown that XeF_4 and XeF_2 do, in fact, have square-planar and linear structures, respectively [this volume, Part VI].

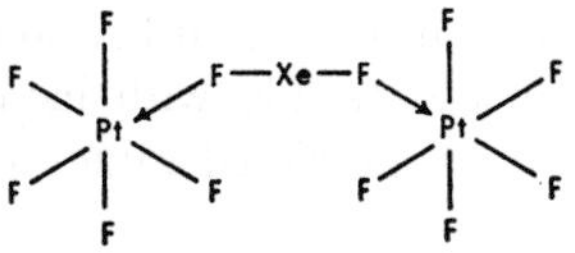

The covalent radii of argon, krypton, and xenon may be estimated from the radii of preceding elements in the periodic table as 0.95, 1.11, and 1.30 A., respectively. Taking Pauling's [14] value of 0.64 A. for the covalent radius of fluorine [15], we estimate lengths for the Ar–F bond of 1.59 A., for the Kr–F bond of 1.75 A., and for the Xe–F bond of 1.94 A. However, we would not expect the noble gas–fluorine bonds to have the same length in all the different possible compounds. Somewhat better estimates can be obtained by comparison with the halogen fluorides, in which the length of a given halogen-fluorine bond varies from compound to compound, and two different bond lengths are often found in the same molecule. In general, the bond length decreases with increasing number of fluorine ligands; this can be attributed to an increase in the effective electronegativity of the central halogen atom with increasing fluorine substitution. In addition, because of the greater repulsion exerted upon them by

TABLE 1

BOND LENGTHS IN HALOGEN FLUORIDES AND NOBLE-GAS FLUORIDES

	Normal	Long		Normal	Long
$r_{Cl}+r_F$	1.63	...	$r_{Ar}+r_F$	1.59	...
ClF	1.63	...	*ArF_2*	...	*1.68*
ClF_3	1.60	1.70	*ArF_4*	...	*1.64*
$r_{Br}+r_F$	1.78	...	$r_{Kr}+r_F$	1.75	...
BrF	1.76	...			
BrF_3	1.72	1.81	KrF_2	...	*1.81*
BrF_5	1.68	1.78	KrF_4	...	*1.77*
r_I+r_F	1.97	...	$r_{Xe}+r_F$	1.93	...
IF	*1.94*	...			
IF_3	*1.90*	*1.99*	XeF_2	...	*1.98*
IF_5	*1.86*	*1.96*	XeF_4	...	*1.94*
IF_7	1.83	1.94	XeF_6	1.80	*1.92*

neighboring electron pairs, one or more of the bonds is always abnormally long. The halogen trifluoride molecules are of the general type AX_3E_2, where A is the central atom, X a ligand, and E an unshared, or non-bonding, electron pair. They have a structure based on a trigonal-bipyramid arrangement of electron pairs, in which the axial bonds are always longer than the equatorial bonds [6]. Similarly in IF_7 the axial bonds are abnormally long. The available data on the bond lengths of the various halogen fluorides are given in Table 1 [16]. The column headed "normal" includes the ordinary or normal bond lengths together with the sum of the covalent radii. The steady decrease in this bond length with increasing number of fluorine ligands is clearly seen, particularly in the case of the bromine fluorides. The column headed "long" contains the abnormally long bond lengths; these are consistently about 0.1 A. longer than the normal bond lengths, and they also decrease steadily with increasing number of fluorine ligands. The entries in italics are predicted values for unknown compounds or for compounds for which accurate data is not available. The values for IF, IF_3, and IF_5 were obtained from the sum of the covalent radii and a consideration of the trends exhibited by the bromine-fluorine bond lengths.

From the values of the sum of the covalent radii for noble gas–fluorine bonds and a consideration of the halogen-fluorine bond lengths, one can predict Ar–F, Kr–F, and Xe–F bond lengths for various known and as yet unknown noble-gas fluorides. These values are also given in Table 1. The two electron pairs of the Xe–F bonds in XeF_2 are in the axial positions of a trigonal-bipyramid arrangement and are therefore "long" bonds. Similarly the bonds in XeF_4 are also "long" bonds.

Using Pauling's values [14] for the double-bond radii of iodine and oxygen, an I = O bond length of 1.79 A. is predicted. This agrees with the observed bond length in the IO_4^- ion [17]. Hence a bond lengthof 1.75 A. can be predicted for the Xe = O bond. In the series of molecules XeF_2, $XeOF_2$, XeO_2F_2, and XeO_3F_2

TABLE 2

PREDICTED BOND LENGTHS IN XENON OXYFLUORIDES AND OXIDES

	Xe–F	Xe=0
$XeOF_2$........	1.95	1.76
XeO_2F_2........	1.93	1.75
XeO_3F_2........	1.91	1.74
$XeOF_4$........	1.92	1.75
XeO_2F_4........	1.90	1.74
XeO_2........	....	1.76
XeO_3........	....	1.75
XeO_4........	....	1.74

and in the series XeO_2, XeO_3, XeO_4 a small decrease in both bond lengths is to be expected because of the increasing number of electronegative ligands. Some estimated values are given in Table 2. In the polymeric oxides predicted above the formally single bonds to oxygen are expected to have some double-bond character, and hence their lengths will lie between 1.75 A. and the calculated single bond length of 1.96 A.

The predicted Xe–F bond lengths of 1.98 A. for XeF_2 and of 1.94 A. for XeF_4 agree very satisfactorily with the observed values of 2.00 [10, 11] and 1.92–1.93 A. [12, 13]. This agreement lends support to the suggestion that the bonds in the noble-gas fluorides may be regarded as simple electron-pair bonds.

There have been two molecular-orbital treatments of the bonding in the noble-gas fluorides. Lohr and Lipscomb [4] have constructed molecular orbitals from the four *2s* and *2p* orbitals of each fluorine and the nine *4d*, *5s*, and *5p* orbitals of xenon. They show that the linear form of the difluoride and the square-planar form of the tetrafluoride have lower energies than other shapes and that the energy is a minimum in each case for a bond length of 2.4 A. It seems unlikely that the 4*d* electrons of xenon can contribute appreciably to the binding, as they are very tightly bound and they must surely be regarded as essentially non-bonding electrons. If this is the case this treatment does not differ substantially from that due to Rundle [3], who considers that the only xenon orbitals that are involved in bonding are the 5*p* orbitals. He constructs three molecular orbitals from a linear arrangement of a xenon 5*p* and two fluorine 2*p* orbitals. Four electrons occupy the two lowest energy orbitals of this set: a bonding three-center orbital and a non-bonding orbital that places electrons on the fluorine atoms only. In this type of three-center four-electron bonding there is effectively only one bonding pair for two bonds. This would be expected to give bonds that are weaker and longer than ordinary single bonds. Although the bonds in XeF_2 are longer than the sum of the covalent radii, the length of the XeF_4 bonds is equal to this sum (within experimental error). However, as has been pointed out above, the bond lengths in both these molecules and in the related halogen fluorides are, in fact, quite consistent with the localized electron-pair description of the bonds. It is not necessary to postulate three-center four-electron bonding in order to account for the bond lengths. Rundle's molecular-orbital treatment also predicts that XeF_2 should be linear and XeF_4 square planar, but it differs from the present theory in predicting a regular octahedral structure rather than a distorted octahedral structure for XeF_6. No experimental data are available at present to test these predictions of the structure of XeF_6. Another difference between the present treatment and the molecular-orbital treatments is that the latter make no use of the xenon 5*d* orbitals; whereas the presence of five, six, and seven electron pairs in the valence shells of xenon in XeF_2, XeF_4, and XeF_6, respectively, implies that one, two, or three 5*d* orbitals are used for bonding. Although in the free xenon atom these orbitals are of too high an energy and are too diffuse to make any effective contribution to bonding, they are very probably sufficiently lowered in energy and contracted in size by two or more fluorine ligands to be able to take part effectively in bonding [18].

REFERENCES

1. L. C. Allen and W. DeW. Horrocks, Jr., *J. Am Chem. Soc.* **84,** 4344 (1962).
2. L. C. Allen, *Science* **138,** 192 (1962).
3. R. E. Rundle, *J. Am. Chem. Soc.* **85,** 112 (1963).
4. L. L. Lohr, Jr., and W. N. Lipscomb, *ibid.* 240 (1963).
5. R. J. Gillespie and R. S. Nyholm, *Quart. Rev. Chem. Soc.* **11,** 339 (1957); R. J. Gillespie, *J. Chem. Educ.* **40,** 295 (1963).
6. R. J. Gillespie, *Can. J. Chem.* **39,** 318 (1961).
7. *Ibid.* **38,** 818 (1960).
8. N. Bartlett, *Proc. Chem. Soc.* 218 (1962).
9. ———, private communication. (See also this volume, p. 23.)
10. S. Siegel and E. Gebert, *J. Am. Chem. Soc.* **85,** 240 (1963).
11. H. A. Levy and P A. Agron, *ibid.* 241 (1963).

12. D. H. Templeton, A. Zalkin, J. D. Forrester, and S. M. Williamson, *ibid.* 242 (1963).
13. J. A. Ibers and W. C. Hamilton, *Science* **139,** 106 (1963).
14. L. Pauling, Nature of the Chemical Bond, p. 224. 3d ed.; Ithaca, N.Y.: Cornell University Press, 1962.
15. Pauling's original value for the covalent radius of fluorine is preferred to the more recent value of 0.72 A. (see reference 3, p. 228) derived from the bond length of the fluorine molecule, because it is believed that this bond is abnormally long due to strong repulsions between lone pairs, as is indeed indicated by the unexpectedly small dissociation energy of only 36 kcal.
16. E. H. Wiebenga, E. E. Havenga, and K. H. Boswijk, *Advances Inorg. Chem. Radiochem.* **3,** 133 (1961).
17. E. A. Hazelwood, *Z. Krist.* **98,** 439 (1938).
18. D. P. Craig and C. Zauli, *J. Chem. Phys.* **37,** 601 and 609 (1962).

IONIC OR *d*-HYBRID BONDS IN NOBLE-GAS HALIDES

JÜRGEN HINZE AND KENNETH S. PITZER
Rice University

Only a short time has elapsed since the preparation of true compounds of the "inert" gases was reported [1]. Nevertheless, there are already two theories advanced to explain the bonding in these systems. The essentials of these theories are:

1. Electrons on the inert gas are promoted from *p* to empty *d* orbitals, and normal covalent bonds are formed between halides and *p-d* hybrid orbitals of the inert gas [2].

2. Two singly occupied ligand orbitals interact with one doubly occupied *p* orbital of the inert gas, forming a semi-ionic delocalized bond, which may be represented by the ionic resonance structure, $FXe^+ F^- \leftrightarrow F^- {}^+XeF$ [3, 4].

It appears at first sight that these two interpretations are distinctly different; one describes the bonds as covalent involving *d* hybrids, while the other describes them as half ionic without *d* hybridization. This apparent contrast, however, is only a deviation in their initial approximation, and both theories will merge when partial ionic characters and participation of high energy *d* orbitals are considered in both schemes. Neither of the theories above is a novelty in the interpretation of chemical bonding, in as much as the phenomenon observed is no novelty but merely the expansion of the valence shell well known for many elements. Whether such an expansion of the valence shell yields predominantly ionic or *d*-hybrid bonds, is neither a new question nor limited to the inert-gas compounds. It is believed, however, that this question can be answered best by first investigating the inert-gas halides, since in these systems the effects caused by expansion of the valence shell are clear-cut and not perturbed by additional changes in bonds, which are already present.

It is noteworthy that both approximations presented predict correctly the structures of the compounds formed: XeF_2, linear; XeF_4 and KrF_4, square planar; and XeF_6, presumably octahedral. Furthermore, the simple ionic model permits a rough estimate of the stability of inert-gas compounds. For example, using the ionic model, Pitzer [3] predicted correctly that xenon-oxygen and

The research reported in this chapter was supported by a grant from the Robert A. Welch Foundation.

krypton-fluorine compounds should be on the borderline of stability. In addition, he estimated the enthalpies of the reactions

$$XeF_2 \rightarrow Xe + F_2\,,$$

and

$$XeF_4 \rightarrow Xe + 2F_2$$

to be smaller than 1.8 ev. and 4.6 ev., respectively. These values are somewhat tentative, however, because analogies have been drawn from the heats of the reactions:

$$ClF_3 \rightarrow ClF + F_2\,, \qquad BrF_3 \rightarrow BrF + F_2\,,$$

and

$$BrF_5 \rightarrow BrF + 2F_2\,,$$

assuming that the strong XF bond present is not affected by further addition of fluorines to the molecule. Such successes in predictions as well as the high heats of sublimation—12.3 kcal/mole and 15.3 kcal/mole for XeF_2 and XeF_4, respectively [5]—are strong support for the ionic model. On the other hand, it has been argued that the force constants for XeF_2—the principal stretching force constant, $k_r = 2.85$ md/A, and the interaction force constant, $k_{rr} = 0.11$ md/A—are inconsistent with a strongly ionic molecule [6]. The equivalent force constants of the isoelectronic ion ICl_2^-, which is known to be ionic, are used in this discussion, and it is pointed out that k_{rr}/k_r in XeF_2 is much smaller than in ICl_2^-. Thus it is concluded that XeF_2 can be only slightly ionic. This, however, is not convincing evidence, since the large k_{rr} in ICl_2^- can well be explained by the fact that the asymmetric mode ICl_2^- approaches the low energetic form Cl–I Cl^-. Furthermore, the ratio k_{rr}/k_r for other ionic compounds like BeF_2 or $BeCl_2$ is small too. The principal argument against the supposition that *d* hybridization is the prime source of the binding still remains. This is the high energy required to promote an outermost *p* electron of the inert gas into an empty *d* orbital. But just how high is this promotional energy?

PROMOTION ENERGY $p \rightarrow d$

The energies required to promote outermost *p* electrons into empty spectroscopic *d* orbitals of same principal quantum number can be obtained directly from atomic energy-level tables [7]. These values are 20.0, 13.8, 12.0, and 9.9 ev., respectively for neon, argon, krypton, and xenon. If these energies are compared with the 6.5 ev. required to promote carbon into its (sp^3) hybridized state [8], it appears, especially in the case of xenon, that the promotional energy is not too high to permit considerable *d* hybridization. There is one fallacy in such a comparison, however. In the case of carbon the 2*p* orbitals have approximately the same radial distribution in the region important for bonding as the 2*s* orbital, while in the inert gases the spectroscopic *n d* orbitals are much more diffuse and spread out than the *n p* orbitals. Such diffuse orbitals will yield only small overlap integrals, as well as small resonance integrals; e.g., the overlap integral of

Slater recipe $2p_z$ and $5d_{z^2}$ of fluorine and xenon, respectively, at an interatomic distance of 2.0 A. is only 0.007. Consequently these diffuse d orbitals are not suitable for bond formation, and to insure strong interactions, d orbitals that have approximately the same radial distribution in the outer region as the valence shell p orbitals are required. These latter orbitals will be called valence d orbitals.

The promotional energy to such contracted d orbitals will be higher than the promotional energy to spectroscopic d levels, since such effective valence d orbitals are not represented by a single stationary state function, giving an energy minimum. The effective promotional energy required can be obtained for the inert gases from their known diamagnetic susceptibilities and polarizabilities as outlined below. The molar diamagnetic susceptibility is given as:

$$-\chi = \frac{N e^2}{2 m c^2} \sum_k \langle r_k^2 \rangle , \tag{1}$$

where N, e, m, and c have the usual meanings—Avogadro's number, electronic charge, electronic mass, and velocity of light, respectively—and $\langle r_k^2 \rangle$ is the average radius square of orbital k. The summation extends over all occupied orbitals. It is now evident that the outermost shell will make the major contribution to the diamagnetic susceptibility. This contribution can be obtained if the small inner-shell susceptibilities are evaluated theoretically from best Hartree-Fock functions and subtracted from the experimental diamagnetic susceptibilities. Thus for the eight outer electrons it is:

$$-\chi_{\text{red}} = \frac{N e^2}{2 m c^2} \sum_i \langle r_i^2 \rangle = \frac{4 N e^2}{m c^2} \langle r_i^2 \rangle = -\chi - \frac{N e^2}{2 m c^2} \sum_j \langle r_j^2 \rangle , \tag{1a}$$

where the summation over i involves the eight outer electrons of the inert gas, while the summation over j covers all inner shell electrons.

To gain information on the promotional energy to effective d orbitals the polarizability has to be considered also. One may write in good approximation:

$$a = \frac{2 e^2}{3} \sum_{k,\, l} \frac{x_{kl}^2 + y_{kl}^2 + z_{kl}^2}{U_{kl}} , \tag{2}$$

where $x_{kl} = \int \psi_k \times \psi_l d\tau$ with ψ_k being an occupied orbital and ψ_l a vacant orbital. U_{kl} is the energy of promotion from orbital k to orbital l. Inner shells make very small contributions to the polarizability and may be ignored in the summation over k. The principal contribution to the polarizability is made by outermost p to d promotions ($n_o p \rightarrow d$), the term of interest in our considerations. However, the transitions $n_o s \rightarrow p$ as well as $n_o p \rightarrow s$ are significant and cannot be neglected. Based on Sternheimer's polarizability calculations [9] it is estimated that the $n_o s \rightarrow p$ and the $n_o p \rightarrow s$ transitions participate with about 25 per cent to the total polarizability, the remaining 75 per cent are attributed to $n_o p \rightarrow d$ transitions.

Examine x_{kl} for $k = p_z$ and $l = d_{xz}$ as a typical term of the $n_o p \rightarrow d$ transitions. Defining:

$$\psi_k = p_z = R_p(r) Y_{p_z}(\theta, \phi) \,,$$

and (3)

$$\psi_l = d_{xz} = R_d(r) Y_{d_{xz}}(\theta, \phi) \,,$$

one obtains:

$$X_{kl} = \int R_d R_p r^3 dr \iint Y_{p_z} Y_{d_{xz}} \sin^2\theta \cos\phi \, d\theta d\phi \,. \quad (4)$$

In equation 4 the double integral over the angular part of the wave functions is readily evaluated. It is $1/\sqrt{5}$ in this case and $2/\sqrt{15}$ in the only other case arising, which is of type z_{kl} for $k = p_z$ and $l = d_{z^2}$.

TABLE 1

EFFECTIVE $p \rightarrow d$ PROMOTION ENERGIES, EFFECTIVE RADII, POLARIZABILITIES, AND DIAMAGNETIC SUSCEPTIBILITIES

	χ_{mole} (10^6 cc.)	χ_{red} (10^6 cc.)	α (10^{24} cc.)	$\alpha(p \rightarrow d)$ (10^{24} cc.)	$\langle r^2 \rangle$ (10^{16} cm.^{2})	$\langle r^2 \rangle^{1/2}$ (10^8 cm.)	$U(p \rightarrow d)$ (ev.)
Ne.	7.20	7.15	0.40	0.30	0.32	0.56	40.4
Ar.	19.40	18.34	1.63	1.22	0.81	0.90	25.5
Kr.	28	22.7	2.48	1.86	1.00	1.00	20.7
Xe.	43	29.4	4.01	3.01	1.30	1.14	16.6

Now the following approximations are introduced:

$$\Sigma[\int R_k R_l r^3 dr]^2 = \Sigma \langle r_{kl} \rangle^2$$

is replaced by $\langle r_i^2 \rangle$ obtained from equation 1.

$\sum_l m_{kl} U_{kl}$ is replaced by $\bar{U}_{k,n}$, the effective promotional energy in polarization of k orbitals to n type orbitals. m_{kl} is an appropriate weighting factor determined by the size of $[x_{kl}^2 + y_{kl}^2 + z_{kl}^2]$.*

With these two approximations and equations 1 to 4 the effective promotional energy for the p to d transition becomes:

$$U_{np \rightarrow d} = \frac{8}{3} \frac{e^2 \langle r_i^2 \rangle}{\alpha(np \rightarrow d)} \,. \quad (5)$$

In Table 1 the diamagnetic susceptibilities and polarizabilities and the reduced values used in this calculation are listed together with the effective promotional energies and radii computed. The effective radii are obtained as a fringe benefit in these calculations. They can be used together with the covalent

* The first substitution would be exact if the sum over l would be over a complete set. In as much as the set is not complete, since it does not include the lower already occupied orbitals, it is an approximation. The approximation involved in the second substitution is difficult to examine due to the complexity of U_{kl}, which is discussed elsewhere [10]. Due to this necessary approximation, no high accuracy in the promotional energies obtained can be claimed.

radii of the appropriate atoms to estimate, roughly, bond distances of inert-gas compounds, suggesting, for xenon fluorides, 1.86 A. and, for krypton fluorides, 1.72 A. as lower limits for the bond distances. This is in good agreement with the values 2.00 and 1.95 A. observed for XeF_2 and XeF_4, respectively [this volume, Levy and Agron, p. 221; Burns, Agron, and Levy, p. 211].

SIMPLE MO TREATMENT

The effective energies required for $p \rightarrow d$ valence promotions, which have been obtained in the last section, can be used in a simple Hückel type MO treatment, which should give some insight in ionic and d contributions to the bonding in inert-gas halides. The details of such a calculation will be illustrated on the example of XeF_2.

First the following symmetry orbitals are formed, defining the z axis along the bonds.

$$\chi_1 = \frac{1}{\sqrt{2}}[2p_z(F_1) + 2p_z(F_2)] . \qquad (6)$$

$$\chi_2 = 5p_z(\mathrm{Xe}) . \qquad (7)$$

$$\chi_3 = \frac{1}{\sqrt{2}}[2p_z(F_1) - 2p(F_2)] . \qquad (8)$$

$$\chi_4 = d_z{}^2(\mathrm{Xe}) . \qquad (9)$$

The d orbital in χ_4 is the valence d orbital.

From these symmetry orbitals MO's are obtained as:

$$\Psi_j = \sum_i^4 c_{ij}\chi_i . \qquad (10)$$

Energy minimization with respect to the c_{ij}'s gives rise to the equation:

$$\begin{vmatrix} \alpha_1 - \epsilon & \beta_{12} - S_{12}\epsilon \\ \beta_{12} - S_{12}\epsilon & \alpha_2 - \epsilon \end{vmatrix} \times \begin{vmatrix} \alpha_3 - \epsilon & \beta_{34} - S_{34}\epsilon \\ \beta_{34} - S_{34}\epsilon & \alpha_4 - \epsilon \end{vmatrix} = 0 . \qquad (11)$$

To solve this equation, empirical values for the parameters α and β are assumed. The α's are derived from ionization potentials. However, since considerable partial charges are to be expected on the different atoms it appears unreasonable to work with α as though it were independent of the charge on the atom considered. It was thus assured that α is given as a function of q, the partial charge on the atom, as follows:

$$\alpha = I_1 + q(I_2 - I_1) \quad \text{if} \quad q > 0 , \qquad (12)$$

and

$$\alpha = I_1 + q(I_1 - E) \quad \text{if} \quad q < 0 , \qquad (13)$$

where E, I_1, and I_2 are electron affinity and first and second ionization potentials, respectively.

This dependence of α on the partial charges on the atoms requires repeated iterative computation until the system considered is self-consistent. Furthermore, expressing α directly by ionization potentials as done here, contains inherently the assumption that interatomic electron-electron repulsions are balanced by interatomic electron-core attractions. This is approximately true only if the atoms are neutral; otherwise it is necessary to add to the total energy the effective Coulomb attraction due to the partial charges on the atoms.

The resonance integral was approximated by:

$$\beta_{ij} = AS_{ij}(\alpha_i + \alpha_j) , \tag{14}$$

where the empirical constant A was taken to be 1.16, a value derived from the two long ClF bonds in ClF_3. The overlap integrals between p orbitals were evaluated using Slater type functions with ζ's obtained from diamagnetic susceptibilities and polarizabilities and the experimental interatomic separations.

TABLE 2

PARTIAL CHARGES ON CENTRAL ATOMS, δ_c, and LIGANDS, δ_l, d-ORBITAL CONTRIBUTIONS, AVERAGE BOND-DISSOCIATION ENERGIES, ENTHALPIES OF FORMATION, AND ENERGIES OF THE $V \leftarrow N$ TRANSITIONS

	δ_c	δ_l	d (%)	D_0 (kcal/mole)	ΔH(g) (kcal/mole)	$\Delta E(V \leftarrow N)$ (ev.)
XeF_2......	+.60	−.30	5	23	−10	7.5
XeF_4......	+.92	−.23	9	25	−26	9.0
KrF_2......	+.56	−.28	1	14	+ 9	7.5
KrF_4......	+.88	−.22	6	10	+34	8.9

Since the detailed radial characteristics of the effective d orbitals are not known, the p to d overlap integrals cannot be calculated. The values are assumed to be the same as those obtained for the p to p overlap integrals, an assumption not serious at all, since largely different p to d overlaps have little effect on the final results due to the high d promotional energy.

Some of the results of the computations described are exhibited in Table 2. It is seen that the principal contribution to the bonding arises from ionic character and not from d hybridization. However, it must be pointed out that large changes in these numbers, ionic and d contribution, may result if different approximations are used in the calculations. In similar computations, not including overlap, the ionic contributions were found to be lower by approximately one-fourth, while a d contribution twice as large was obtained. On the other hand, the average dissociation energies varied little in different calculations. In both computations the fluorides of argon and neon as well as the chlorides or bromides of all inert gases appeared unstable. But, as it must be expected, the model not involving overlap gave considerably lower energies for the $V \leftarrow N$ transitions.

The experimental values of the $V \leftarrow N$ transitions are 7.9 and 9.4 ev. [11], for XeF_2 and XeF_4, respectively, and ΔH of gaseous XeF_4 is -45 kcal/mole [5, 12]. These data are in good agreement when the approximate nature of the calculations is considered.

REFERENCES

1. N. Bartlett, *Proc. Chem. Soc.* 218 (1962); H. H. Claassen, H. Selig, and J. G. Malm, *J. Am. Chem. Soc.* **84**, 3593 (1962); C. Chernick *et al.*, *Science* **138**, 136 (1962); and others.
2. L. D. Allen and W. DeW. Horrocks, Jr., *J. Am. Chem. Soc.* **84**, 4344 (1962).
3. K. S. Pitzer, *Science* **139**, 414 (1963).
4. R. E. Rundle, *J. Am. Chem. Soc.* **85**, 112 (1963).
5. J. Jortner, E. G. Wilson, and S. A. Rice, *ibid.* 814 (1963).
6. P. A. Agron *et al.*, *Science* **139**, 842 (1963).
7. C. E. Moore, *Atomic Energy Levels*, N.B.S. Circular 467.
8. J. Hinze and H. H. Jaffé, *J. Am. Chem. Soc.* **84**, 540 (1962).
9. R. M. Sternheimer, *Phys. Rev.* **96**, 951 (1954); **115**, 1198 (1959).
10. K. S. Pitzer, *Advances Chem. Phys.* **2**, 59 (1959).
11. J. Jortner, E. G. Wilson, and S. A. Rice, *J. Am. Chem. Soc.* **85**, 813 and 815 (1963).
12. S. R. Gunn and S. M. Williamson, *Science* **140**, 177 (1963).

AN LCAO-MO STUDY OF RARE-GAS FLUORIDES

L. L. LOHR, JR., AND WILLIAM N. LIPSCOMB
Harvard University

ABSTRACT

The electronic structures of ArF_4, KrF_4, XeF_2, XeF_4, and XeF_6 are considered as a function of molecular geometry in terms of a programmed semi-empirical LCAO molecular-orbital theory.

INTRODUCTION

The recent preparation [1–5] of several rare-gas fluorides led us to consider the electronic structures of these compounds in terms of a programmed semi-empirical LCAO molecular-orbital theory [6–8]. Indicated molecular structures for XeF_2 and XeF_4 are linear and square planar, respectively, as have now been found [9–12]. The indicated structure for XeF_6 is probably octahedral. In addition, the bond energies for KrF_4 and ArF_4, assumed planar, were found to be approximately one-half and one-fifteenth, respectively, of the calculated XeF_4 bond energy. The LCAO-MO method, which ignores nuclear repulsions, is satisfactory for prediction of molecular symmetries but not for the calculation of bond lengths. It is concluded that the structural chemistry of xenon is not greatly different from that of iodine with fluorine, chlorine [13], or oxygen, although future studies of both xenon and iodine systems may reveal significant differences.

PROCEDURE

The LCAO-MO computer program was written primarily for studies of transition-metal complexes and requires three types of input data:

Cartesian coordinates of not more than eleven atoms.

Slater atomic orbitals for each atom, specified by n, l, and m quantum numbers and an exponent. The orbitals may be of s, p, and d types for one atom, and of s and p types for remaining atoms.

Valence-state ionization potentials (Coulomb integrals) for each atomic orbital, which are taken as the diagonal elements $\boldsymbol{H}_{ii}$ of an effective Hamiltonian matrix $\boldsymbol{H}$.

The research reported in this chapter was supported by the National Science Foundation and the Office of Naval Research.

The complete overlap matrix $\boldsymbol{S}$ is computed from the coordinates and basis atomic orbitals. Then the off-diagonal elements of $\boldsymbol{H}$ are taken to be

$$H_{ij} = -2(H_{ii} \cdot H_{jj})^{1/2} S_{ij}, \qquad i \neq j. \tag{1}$$

The eigenvalues λ_j obtained from the solution of

$$Hc = \lambda Sc \tag{2}$$

are used to define the total orbital energy E as

$$E = \sum_j n_j \lambda_j, \tag{3}$$

where n_j is the occupation number (0, 1, or 2) of the jth MO. The bond energy BE for MX_n is defined as

$$BE = \frac{1}{n}\left(\sum_i m_i H_{ii} - E\right), \tag{4}$$

where the energy of atoms at infinite separation is taken to be $\Sigma_i m_i H_{ii}$, m_i being the occupation number of the ith AO in the atomic valence state. The fraction F_{ij} of the ith AO in the jth MO is

$$F_{ij} = C_{ij} \sum_k S_{ik} C_{kj}. \tag{5}$$

Then, the charge Q_l for the lth atom is defined as

$$Q_l = Z_l' - \sum_{i \text{ on } l} \sum_j n_j F_{ij}, \tag{6}$$

where Z_l' is the core charge (nucleus plus inner electrons) for the lth atom. Finally, the overlap population OP_{lm} for atoms l, m is given by

$$OP_{lm} = \sum_{i,k} \sum_j 2 n_j C_{ij} C_{kj} S_{ik}, \tag{7}$$

where C_{ij} is the coefficient of the ith AO in the jth MO, and the summation is over AO's i on atom l and k on atom m.

The atomic orbitals are specified by a Slater exponent and a valence state ionization potential (Table 1). The $F(2s)$ and $F(2p)$ exponents are those for F^{-1}, the Xe($4d$) value is that for Xe^{+4}, while the Xe($5s$) and Xe($5p$) values are each somewhat larger than values obtained for Xe^{+4} from the standard Slater rules [14]. The reduction in the size of the Xe($5s$) and Xe($5p$) orbitals was needed to obtain reasonable Xe-F bond energies. Exponents for Ar and Kr orbitals were estimated from those of the basis functions for Ar and Kr atomic SCF calculations [15], from the xenon values used in XeF_4, and from neutral atom values given by the standard rules. The Ar($3p$), Kr($4p$), and Xe($5p$) orbitals, which are the AO's principally involved in the formation of fluorides, were assigned valence state ionization potentials which were essentially the free-atom values that had been lowered slightly to correct for the positive charge on these atoms in a molecule [7, 8]. Values for the remaining argon, krypton, and xenon orbitals were simply guessed, while those for F(2*s*) and F(2*p*) orbitals were those used previously in a treatment of $F_2N\text{-}SO_2\text{-}NF_2$ [8].

RESULTS

Table 2 gives results using F(2*s*), F(2*p*), Xe(5*s*), Xe(5*p*), and Xe(4*d*) orbitals for XeF_2 with sixteen electron pairs and seventeen MO's, and for XeF_4 with

TABLE 1

ATOMIC ORBITAL PARAMETERS

	SLATER EXPONENTS			COULOMB INTEGRALS		
	s	*p*	*d*	*s*	*p*	*d*
Fluorine.....	2.42	2.42	....	−31.4	−17.4	....
Xenon.......	2.97	2.30	3.71	−30.0	−15.0	−25.0
Krypton.....	3.43	2.65	4.28	−34.6	−17.3	−28.8
Argon.......	3.86	2.99	....	−39.0	−19.5	....

TABLE 2

THEORETICAL RESULTS FOR MF_n

Symmetry	*M-F* (A.)	−*E* (ev.)	*BE* (ev.)*	Charge *M*	*OP*†	λ (ev.) of Highest MO's		
							Occupied	
XeF_2:$D_{\infty h}$.......	2.4	707.0	3.3	1.50	0.122	−11.3	−14.9	−14.9
C_{2v}‡.......	2.4	705.5	2.6	1.52	.074	−12.1	−13.8	−14.9
C_{2v}'‡......	2.4	704.0	1.8	1.53	.030	−12.8	−13.0	−14.9
								Occupied
XeF_4:D_{4h}.......	1.7	975.5	−6.2	2.83	− .154	16.9	16.9	−11.3§
	1.9	1002.3	0.5	2.85	.088	2.3	2.3	−13.0§
	2.1	1011.6	2.8	2.88	.150	− 5.5	− 5.5	−14.0
	2.4	1013.8	3.4	2.98	.128	−11.2	−11.2	−14.8
	2.8	1011.5	2.8	3.34	.060	−14.1	−14.1	−15.1
	1.6‖	1022.2	5.5	2.39	.256	− 7.2	− 7.2	−15.1
	2.4‖	1015.3	3.8	2.76	0.135	−12.6	−12.6	−15.1
C_{4v}#.......	2.4	1009.9	2.4	3.01	0.124, 0.018	−12.1	−12.3	−12.3
C_{2v}#.......	2.4	1010.8	2.7	3.00	.130, .034	−11.2	−12.8	−12.8
T_d........	2.4	1009.6	2.4	3.04	0.156, −0.015	−12.3	−12.3	−12.3
KrF_4:D_{4h}........	2.1	1065.7	2.0	2.33	0.132	−13.3	−13.3	−16.6
ArF_4:D_{4h}........	2.1	795.5	0.2	0.78	0.040	−16.7	−16.7	−17.2

* Relative to $M + nF$.

† *M-F* overlap population per bond. The listing of two values indicates non-equivalent F atoms, arising from degenerate states for XeF_4: C_{4v} and T_d, and from low symmetry for XeF_4: C_{2v}.

‡ Assumed bond angle of 120° for C_{2v} and 90° for C_{2v}'.

§ Highest occupied MO is a_{1g}, with energy −0.1 ev. (1.7 A.) and −10.9 ev. (1.9 A.).

‖ F(2*p*σ) and Xe(5*p*σ) orbitals only.

Assumed bond angles of 70°32′ for C_{4v} and 90° for = C_{2v}.

twenty-three electron pairs and twenty-five MO's. A calculation for F_2 yielded an energy minimum at 1.45 A. (observed 1.44 A.) with a bond energy of 3.4 ev.

Calculated bond energies relative to Xe and $(n/2)F_2$ are obtained by subtracting 1.7 ev. from energies relative to Xe and nF, giving maximum values of 1.6 ev. = 37 kcal/mole for XeF_2 and 1.7 ev. = 39 kcal/mole for XeF_4 (Xe-F = 2.4 A.). A simplified calculation was made for XeF_4 using only F($2p_\sigma$) and Xe($5p_\sigma$) orbitals, yielding a slightly greater bond energy of 2.1 ev. = 48 kcal/mole at a Xe-F distance of 2.4 A. and a maximum bond energy of 3.8 ev. = 88

TABLE 3

XeF_4 ORBITAL ENERGIES (EV.) FOR XE-F = 2.4 A.

Full Calculation			$p(\sigma)$ only Calculation		
e_u	−11.21	Vacant	e_u	−12.63	Vacant
	−11.21			−12.63	
a_{2u}	−14.79		a_{2u}	−15.10	
a_{1g}	−16.92		b_{1g}	−17.40	
b_{1g}	−17.35		a_{2g}, a_{2u}, b_{2g}, b_{2u}, e_g, e_u	−17.42	
a_{2g}	−17.40			−17.42	
e_u	−17.42			−17.42	
	−17.42			−17.42	
b_{2u}	−17.42			−17.42	
e_g	−17.42			−17.42	
	−17.42			−17.42	
b_{2g}	−17.43			−17.42	
a_{2u}	−17.65		a_{1g}	−17.43	
e_u	−18.62		e_u	−18.88	
	−18.62			−18.88	
b_{1g}	−24.98		a_{1g}, b_{1g}, b_{2g}, e_g	−25.00	
a_{1g}	−24.99			−25.00	
e_g	−25.00			−25.00	
	−25.00			−25.00	
b_{2g}	−25.00			−25.00	
a_{1g}	−28.55			−30.00	
b_{1g}	−31.41		a_{1g}, a_{1g}, b_{1g}, e_u	−31.40	
e_u	−31.66			−31.40	
	−31.66			−31.40	
a_{1g}	−32.77			−31.40	

kcal/mole at 1.6 A. The simplified calculation is a six-orbital, four electron–pair problem, requiring only the solution of a 2 × 2 secular equation, as the four F($2p_\sigma$) orbitals span the irreducible representations a_{1g}, b_{1g}, and e_u of the group D_{4h}, while the Xe($5p_\sigma$) orbitals span e_u. One e_u orbital pair is unoccupied, leaving the a_{1g} and b_{1g} as occupied, non-bonding MO's. The small ligand-ligand interactions negligibly lower a_{1g} and raise b_{1g}. Orbital energies are compared in Table 3 with those for the full calculation. The simpler description differs from the more complete one primarily in its underestimation of repulsive interactions, giving rise to its excessively large bond energy, its short equilibrium Xe-F distance, and its predicted excitation energy of 7.9 ev. instead of 3.6 ev. The striking dependence of the Xe-F overlap population on the XeF_2 bond angle and the XeF_4 bond distances should also be noted.

The Slater exponents and Coulomb integrals are given in Table 1. Assumption of either smaller xenon exponents or lower Coulomb integrals gave less favorable bond energies for XeF_4. In all calculations with the Xe-F distance 2.0 A. or greater, the highest three MO's (including vacant MO's) were those consisting largely of xenon $5p$ orbitals, which span the irreducible representations $a_{2u} + e_u$ in D_{4h}, $a_1 + e$ in C_{4v}, $a_1 + b_1 + b_2$ in C_{2v}, t_2 in T_d, t_{1u} in O_h, and $\sigma_u^+ + \pi_u$ in $D_{\infty h}$. In addition, the XeF_4 calculations with assumed symmetries C_{4v} and T_d did not give closed-shell ground states, but did give a lower total orbital energy (E) than the D_{4h} structure at Xe-F = 1.7 A. None of the structures has a positive bond energy, BE, at that distance, however.

As two of the primarily Xe($5p$) MO's are vacant in XeF_4, it is seen that $XeF_6(O_h)$ would possess a vacant t_{1u} set of orbitals. Calculations were made for six assumed structures of XeF_6, all with Xe-F = 2.4 A. The largest bond energy (3.5 ev.) and smallest Xe charge (+4.45) was possessed by the octahedral (O_h) structure, the remaining structures being a pentagonal pyramid (C_{5v}), a trigonal prism (D_{3h}), a compressed octahedron (C_{3v}), a trigonal prism with Xe in a square face (C_{2v}), and a pentagonal bipyramid with an equatorial vacancy (C_{2v}). The tendency of xenon to form a higher fluoride is reflected in the XeF_2 structure [10] as a square prism of eight next-nearest fluorine atoms and in the XeF_4 structure [14] as strikingly short non-bonded Xe-F distances of 3.22 and 3.25 A. A further consequence of the relatively strong non-bonding interactions is the variation in vapor pressures (at room temperature) from approximately 3 mm. for XeF^2 [2] and XeF_4 [1] to 30 mm. for XeF_6 [3, 5].

Calculations were also made for assumed planar KrF_4 and ArF_4 over a range of M-F distances. The results in Table 2 for the E minima indicate an appreciable stability for KrF_4, but not for ArF_4. It should be noted again that nuclear repulsions are not included in the calculation, but are simulated by antibonding electronic interactions. Thus the bond distance calculations are not expected to be reliable, and, in fact, are not given.

DISCUSSION

The picture presented here of the bonding in rare-gas fluorides is not greatly different from that of the three-center, three-orbital, two electron–pair description [16], as the bonding interactions were found to involve principally the p_σ orbitals. The simplest 3-AO model for XeF_2, neglecting overlap in the MO normalizations, yields a xenon charge of +1, somewhat smaller than values given here (Table 2). The very small mixing of Xe($5s$) and Xe($4d$) orbitals in the XeF_2, XeF_4, and XeF_6 MO's corresponds to negligible s-p_σ or d_σ-p_σ hybridization and can be seen in the contributions to the total overlap population (equation 7) for $XeF_4(D_{4h})$, Xe-F = 2.4 A., which are: Xe(p_σ)-F(p_σ) + .472; Xe(p_σ)-F

(s), +.088; Xe(s)-F(p_σ), −.032; Xe(s)-F(s), −.016; and Xe(p_π)-F(p_π), +.016. The overlap populations involving the Xe$(4d_\sigma)$ and Xe$(4d_\pi)$ orbitals are all less than 0.001. The extent of s-p hybridization of the fluorine p_σ orbitals is somewhat larger than that of the xenon p_σ orbitals, but is still small relative to the "standard" one-to-one mixing.

Concerning the physical aspects of the simple molecular-orbital descriptions used, the theory is essentially that of the familiar Hückel π-electron theory of aromatic systems, with several extensions: the atoms are not confined to a plane; no nearest-neighbor approximation is made; more than one orbital per atom is considered; the complete overlap matrix is used, and the atoms are in general different (for example, xenon and fluorine), necessitating different diagonal energy matrix elements for the different types of atomic orbitals. As the atomic basis functions enter the calculation only in terms of their overlaps, the energy matrix elements have no explicit dependence on the exact form of these functions. Simple one-electron molecular orbitals defined in such a manner have proven useful in the consideration of many problems in molecular geometry [6–8, 17], with the exception of the problem of bond lengths for a given molecular symmetry.

A properly defined total energy consists [18] of the orbital energy sum (equation 3) plus the nuclear repulsions and minus the electron repulsions, the latter counted twice in the orbital energy sum. If it is assumed that the change in the nuclear repulsions upon nuclear displacement is equal to the change in the electron repulsions, the total orbital energy (equation 3) and the correctly defined energy will have their minimum values at the same geometry. Alternatively, the situation may be expressed as a cancellation of nuclear repulsions by long-range nuclear-electron attractions, each ignored by the assumption that the $\boldsymbol{H}_{ij}$ are proportional to $\boldsymbol{S}_{ij}$ (equation 1), which implies that the exponential decrease of the $\boldsymbol{S}_{ij}$ at large distances holds also for the $\boldsymbol{H}_{ij}$. The energy minima arise in the simple LCAO-MO scheme from the simulation of nuclear repulsions by the antibonding electronic interactions, this simulation being dependent upon the occupancy of at least one MO whose energy increases (antibonding) as the atoms are brought closer together.

REFERENCES

1. H. H. Claassen, H. Selig, and J. G. Malm, *J. Am. Chem. Soc.* **84,** 3593 (1962).
2. D. F. Smith, *J. Chem. Phys.* **38,** 270 (1963).
3. J. G. Malm, I. Sheft, and C. L. Chernick, *J. Am. Chem. Soc.* **85,** 110 (1963).
4. F. B. Dudley, G. Gard, and G. H. Cady, *Inorg. Chem.* **2,** 228 (1963).
5. E. E. Weaver, B. Weinstock, and C. P. Knop, *J. Am. Chem. Soc.* **85,** 111 (1963).
6. L. L. Lohr, Jr., and W. N. Lipscomb, *ibid.* 240 (1963).
7. L. L. Lohr, Jr., and W. N. Lipscomb, *J. Chem. Phys.*, **38,** 1607 (1963).
8. T. Jordan, H. W. Smith, L. L. Lohr, Jr., and W. N. Lipscomb, *J. Am. Chem. Soc.* **85,** 845 (1963).
9. S. Siegel and E. Gebert, *ibid.* 240 (1963).
10. H. A. Levy and P. A. Agron, *ibid.* 241 (1963).

11. J. A. IBERS and W. C. HAMILTON, *Science* **139**, 106 (1963).
12. D. H. TEMPLETON, A. ZALKIN, J. D. FORRESTER, and S. M. WILLIAMSON, *J. Am. Chem. Soc.* **85**, 242 (1963).
13. R. C. L. MOONEY, *Z. Krist.* **98**, 377 (1938); **100**, 519 (1939).
14. J. C. SLATER, *Phys. Rev.* **36**, 57 (1930).
15. R. E. WATSON and A. J. FREEMAN, *ibid.* **123**, 521 (1961); **124**, 1117 (1961).
16. R. E. RUNDLE, *J. Am. Chem. Soc.* **85**, 112 (1963).
17. R. HOFFMANN and W. N. LIPSCOMB, *J. Chem. Phys.* **36**, 2179, 3489 (1962); **37**, 2872 (1962).
18. J. C. SLATER, *Quantum Theory of Molecules and Solids*, vol. 1, p. 108. New York: McGraw-Hill Book Co., 1963.

A SEMI-EMPIRICAL MO TREATMENT OF THE ELECTRONIC STRUCTURE OF XENON HEXAFLUORIDE

EDWARD A. BOUDREAUX
Louisiana State University in New Orleans

INTRODUCTION

Progress is being made in the theoretical study of the structure and bonding in compounds of xenon and fluorine. A programmed molecular-orbital procedure has been carried out on XeF_2 and XeF_4 [1]; and a three-center four-electron bonding scheme has been proposed [2*a*, *b*]. Additional calculations have been accomplished on XeF_4 using a semi-empirical method [3], and more refined computations are now in progress [4, 5].

Since the recent discovery of xenon hexafluoride, it is equally interesting to perform semi-empirical calculations regarding the electronic structure of this molecule. Hence, the electronic structure of XeF_6 will be treated according to Wolfsberg-Helmholz procedure [6*a*], with the hope that a reasonable scheme for the electronic distribution can be given, and the lowest electronic excited state identified.

PROCEDURE

The molecular-orbital distribution will be derived by the LCAO method using Slater type orbitals (STO) as basis functions, which are considered orthogonal and normalized. The molecule is represented with a total of thirty-eight electrons, eight from the xenon atom and five from each of the fluorine atoms. The total number of electrons is placed in σ and π orbitals derivable from the $5d$, $5p$, and $5s$ orbitals of xenon and the $2p$ orbitals of fluorine.

The molecule is considered to be symmetrical and is therefore treated in the octahedral, O_h, point group. Standard group theoretical methods [6*a*, *b*] are employed in deriving the various molecular orbital combinations. The various orbital energies were obtained as solutions to the determinants.

$$|H_{ij} - G_{ij}E| = 0 , \tag{1}$$

where H_{ij}, the resonance integrals, are approximated from the expression

$$H_{ij} = FG_{ij}\frac{H_{ii} + H_{jj}}{2} . \tag{2}$$

The Coulomb integrals, H_{ii} and H_{jj}, required in equation 2, were deduced from atomic spectral data [7] and are presented in Table 1 and Figure 1. The values employed correspond closely to those of Lohr and Lipscomb [1] for the xenon atom but somewhat lower values are assigned to the fluorine atoms. These values were derived by contrasting the valence state ionization energies of the fluorine atoms as a function of charge distribution in several fluoride molecules, including XeF_4 [3].

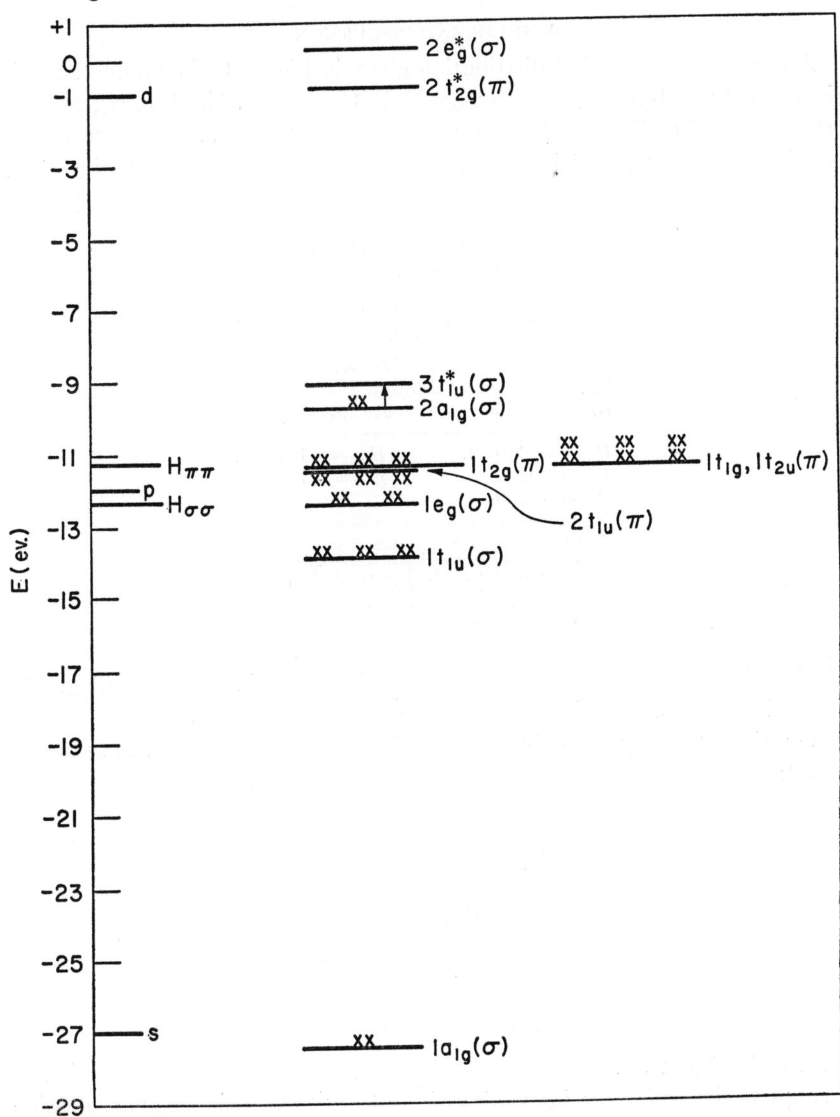

Fig. 1.—The calculated relative molecular-orbital energies and the lowest singlet spectral excitation for XeF_6.

It should also be noticed that the Coulomb integral for the $5d$ shell of xenon is assigned a value of -1.0 ev., which is derived from the ionization potential of the neutral atom minus the energy of the $5p^6 \rightarrow 5p^5 5d$ excitation. The proportionality parameter, F, is assigned values of 1.7 and 1.9 for σ and π orbitals respectively, based on arguments presented elsewhere for analogous molecular systems [6a]. The various group overlap integrals, G_{ij}, required in the computations were obtained from the literature [8, 9, 10] and are given in Table 1.

RESULTS AND DISCUSSION

According to the orbital distribution given in Figure 1, the ground state is given as: $(1a_{1g})^2(1e_g)^4(1t_{1u})^6(2t_{1u})^6(1t_{2g})^6(1t_{1g})^6(1t_{2u})^6(2a_{1g})^2$. The $1a_{1g}$, $1e_g$, and $1t_{1u}$ orbitals are σ bonding, the $1t_{2g}$ is π bonding. The $3t_{1u}$, $2a_{1g}$, and $2e_g$ are σ antibonding, and the $1t_{2u}$ and $1t_{1g}$ are non-bonding π orbitals on the fluorines. The

TABLE 1

GROUP OVERLAPS* AND COULOMB INTEGRALS FOR XeF_6

$G_{a_{1g}}(s, \sigma)$	=	$\sqrt{6}\, S(5s, 2p_\sigma)$	=	0.282
$G_{e_g}(d, \sigma)$	=	$-\sqrt{3}\, S(5d_\sigma, 2p_\sigma)$	=	$-$ .338
$G_{t_{1u}}(p, \sigma)$	=	$\sqrt{2}\, S(5p_\sigma, 2p_\sigma)$	=	.239
$G_{t_{1u}}(p_\pi, \pi)$	=	$2\, S(5p_\pi, 2p_\pi)$	=	.080
$G_{t_{2g}}(d, \pi)$	=	$2\, S(5d_\pi, 2p_\pi)$	=	0.146

$H_{dd} = -1.0$ ev.	$H_{\sigma\sigma} = -12.3$ ev.
$H_{pp} = -12.0$	$H_{\pi\pi} = -11.3$
$H_{ss} = -27.0$	

* These overlaps were evaluated at a Xe-F distance of 2 A., and do not change significantly within the range of 1.9–2.1 A. The required effective charges were approximated from charge distributions calculated previously in XeF_4, where about 30 per cent ionicity was the computed result.

$2t_{1u}$ is only very slightly π bonding and is thus essentially non-bonding. The $1t_{2u}$ and $1t_{1g}$ orbitals are degenerate and lie at the energy $H_{\pi\pi}$.

For the symmetry O_h, only the following electric dipole transitions are allowed:

$$A_{1g} \leftrightarrow T_{1u}$$

$$A_{2g} \leftrightarrow T_{2u}$$

$$E_g \leftrightarrow T_{1u},\ T_{2u}$$

$$T_{1g} \leftrightarrow A_{1u},\ E_u,\ T_{1u},\ T_{2u}$$

$$T_{2g} \leftrightarrow A_{2u},\ E_u,\ T_{1u},\ T_{2u}\ .$$

On neglecting any states that may arise via vibronic coupling, and making no attempt to determine the intensities of the transitions, it is possible to assign the lowest singlet-singlet transition from Figure 1. Of the several lowest energy allowed transitions, the one most likely to give rise to a strong spectroscopic absorption maximum should involve the promotion of an electron from the filled ground state orbital $2a_{1g}$ to the vacant $3t_{1u}^*$ orbital. The calculated energy for this allowed singlet transition is about 0.6 ev. This result is apparently in disagreement with experimental data [11], which indicate a single absorption maximum at 3300 A. and more intense absorption at wave lengths shorter than 2750 A. for XeF_6 vapor. The discrepancy probably lies in the fact that the $2a_{1g}$ orbital is calculated to be too highly antibonding; whereas, in reality, it is probably very nearly non-bonding. Consequently the energy of the $2a_{1g}$ orbital should be much closer to that of the xenon $5p$ shell, which will result in a $2a_{1g} \rightarrow 3t_{1u}^*$ transition energy of about 3 ev.

Finally it may be concluded that the electronic structure of XeF_6 should be at least qualitatively correct if the molecule is in fact octahedrally symmetric.

Acknowledgment.—The author wishes to extend his gratitude to Dr. Gordon L. Goodman of Argonne National Laboratory for invaluable discussions and for citing some very critical points in the course of these calculations.

REFERENCES

1. L. L. Lohr, Jr., and W. N. Lipscomb, *J. Am. Chem. Soc.* **85,** 240 (1963).
2a. G. C. Pimentel, *J. Chem. Phys.* **19,** 446 (1951).
2b. R. E. Rundle, *J. Am. Chem. Soc.* **85,** 112 (1963).
3. E. A. Boudreaux, submitted for publication.
4. L. C. Allen, private communication.
5. C. Cusacs and E. A. Boudreaux, submitted for publication.
6a. M. Wolfsberg and L. Helmholz, *J. Chem. Phys.* **20,** 837 (1952).
6b. C. J. Ballhausen, *Introduction to Ligand Field Theory.* New York: McGraw-Hill Book Co., 1962.
7. C. E. Moore, *Atomic Energy Levels,* N.B.S. Circular 467.
8. R. S. Mulliken, C. A. Rieke, D. Orloff, and H. Orloff, *J. Chem. Soc.* **17,** 1248 (1949).
9. H. H. Jaffé and G. O. Doak, *J. Chem. Phys.* **21,** 196 (1953).
10. H. H. Jaffé, *ibid.* 258 (1953).
11. J. G. Malm, I. Sheft, and C. L. Chernick, *J. Am. Chem. Soc.* **85,** 110 (1963).

THEORETICAL AND EXPERIMENTAL STUDIES OF THE ELECTRONIC STRUCTURE OF THE XENON FLUORIDES

JOSHUA JORTNER, E. GUY WILSON, and STUART A. RICE
University of Chicago

ABSTRACT

A semi-empirical molecular-orbital model for XeF_2 and XeF_4 is presented. The results of experiments to determine the absorption spectrum in the range 1100–2500 A. and determinations of the heat of sublimation of these compounds are described. The proposed model is shown to be in good agreement with the physical and chemical properties of these compounds.

INTRODUCTION

The recent discovery of stable compounds of the rare gases xenon, radon, and krypton is an event of great current interest, since the stability of these compounds seems to violate one of the oldest and most widely accepted rules of valence theory. Extensive information concerning the chemical and physical properties of these compounds is rapidly accumulating, and the current state of knowledge is summarized in the proceedings of this conference. The most extensive of the available data concern the xenon fluorides, and it is these compounds that will be considered in the present work. Of course, the development of rare-gas chemistry serves to refocus attention on problems important to theoretical inorganic chemistry. Several possible models of the xenon fluorides have recently been proposed to account for the binding in these compounds. These proposals will now be briefly reviewed.

We may first attempt to apply to the xenon fluorides a conventional bonding scheme involving sp^n hybridization. If a bonding scheme employing sp hybridization is applied to XeF_2, this molecule is predicted to be linear. However, the promotion energy of xenon is expected to be of the order of the observed first singlet-singlet excitation energy ${}^1S(5s^25p^6) \rightarrow {}^1P_1(5s^25p^56s)$, ($\Delta E = 9.57$ ev.) the gain in the bonding energy of the resultant molecule is likely to be less than this large promotion energy, and the formation of stable compounds is unlikely.

Allen has recently proposed [1] that binding in these compounds is determined by correlation effects, in the sense that there is a difference in overlap between the xenon orbitals occupied separately by electrons of different spin and the fluorine filled orbitals. [See this volume, p. 319.] It is not evident to us that this scheme will give any bonding at all. (The interaction energy may be positive.) Moreover, this model is not specific and does not seem to establish why there are not more compounds of xenon.

We believe that a bonding scheme must be provided that involves decoupling of electrons in a closed-shell system (i.e., the xenon atom) upon molecule formation. Such cases are not new in valence theory; they have arisen previously in the molecular-orbital formulation of ligand field theory and in the theory of the structure of the polyhalide ions [2, 3, 4].

The close analogy between the polyhalide ions and the case under consideration was independently pointed out by Jortner, Rice, and Wilson [5], Lohr and Lipscomb [6], Pimentel [7], Pitzer [8], and Rundle [9]. The polyhalide anions provide an interesting analogy for the xenon fluorides, since XeF_2 and XeF_4 are isoelectronic with IF_2^- and IF_4^-, respectively. The preparation of IF_4^- has been reported [10], although its structure is unknown. A bonding scheme consistent with the geometry of the polyhalide ions and with the observed nuclear quadrupole coupling constants [11] was constructed by using delocalized orbitals to provide accommodation for excess electrons over the number required to fill the outermost occupied p orbitals of a halogen molecule and a halide ion.

In the present work the structure of the xenon fluorides will be discussed on the basis of the delocalized molecular-orbital model. We shall also present the results of an experimental far ultraviolet–spectroscopic study of XeF_2 and XeF_4. A preliminary report of these results has already been published [12]. Electronic spectroscopic data are an obvious source of information concerning the excited electronic levels of these molecules and provide a crucial test for the adequacy of any theory concerning the nature of binding. Finally, the adequacy of the binding scheme in accounting for the intermolecular interactions in the solid compounds will be considered.

LONG-RANGE XeF INTERACTIONS

The model we wish to discuss in this paper uses delocalized bonding orbitals. The use of delocalized bonding orbitals must be viewed with caution, since the naïve delocalized molecular-orbital scheme leads to a stabilization of H_3 relative to H_2 and H [13, 14]. In the $H_2 + H$ system, intermolecular-dispersion forces are weak, and repulsive forces keep the H and H_2 apart. In order for the delocalization model to be applicable and lead to binding, it is necessary that long-range attractive forces be operative, i.e., the electrostatic polarization and dispersion forces in $I_2 + I^-$.

In order to validate the use of delocalized molecular orbitals to account for the binding in the xenon fluorides, long-range interactions between xenon and fluorine atoms have to be considered. It is well known that the molecular-orbital treatment breaks down for distances larger than about one and one-half times the equilibrium intramolecular separation, and the valence-bond method has to be applied [15]. These weak long-range interactions can be adequately described in terms of Mulliken's charge-transfer theory. We shall demonstrate these arguments for the XeF diatomic molecule at large Xe-F separation ($R > 3$ A.). In this system the xenon atom acts as an electron donor, while the fluorine atom acts as an acceptor. It will be shown that charge-transfer interactions are quite strong because of the relatively large overlap between the $5p\sigma$ orbital of xenon and the vacant $2p\sigma$ fluorine atom orbital. The long-range stabilization of the Xe-F pair is assigned to σ-type interaction. The application of Mulliken's charge-transfer theory [16] is straightforward. The system is described in terms of resonance between the "no bond structure" (which may of course be stabilized by dispersion forces) and the ionic dative structure, $Xe^+ F^-$. The valence bond wave function is

$$\Psi = \Psi_0(\mathrm{XeF}) + a\Psi_1(\mathrm{Xe^+F^-}) , \tag{1}$$

where a is a mixing coefficient. The charge transfer–stabilization energy is given by the second-order perturbation-theory expression:

$$\Delta E^{C.T.} = -\frac{(H_{01} - S_{01}H_{00})^2}{H_{00} - H_{11}} , \tag{2}$$

where the H_{ij} are the matrix elements of the total Hamiltonian for the system, and S_{01} is the overlap integral between the states Ψ_0 and Ψ_1. Application of a one-electron treatment similar to that given by Murrel [17] and approximation of the exchange integrals using the Mulliken magic formula [18] lead to the result:

$$\Delta E^{C.T.} = -\frac{2S^2}{1+S^2} \cdot \frac{(E_F + e^2/2R)^2}{(I_{Xe} - E_F - e^2/R)} , \tag{3}$$

where E_F is the electron affinity of a fluorine atom ($E_F = 3.57$ ev.) [19], I_{Xe} is the ionization potential of xenon (12.1 ev.), and R is the Xe-F separation. S is the overlap integral between the filled Xe $5p\sigma$ and the empty F $2p\sigma$ orbitals. For the estimation of S, the forms of the atomic wave functions at large distances are required. The use of the usual Slater-type orbital is inadequate, as it seriously underestimates the magnitude of the tails of the wave functions. Clearly, the effective nuclear charges "seen" by an electron at large distances in a neutral atom and a negative ion tend toward unity and zero, respectively. The main contribution to the overlap integral in this calculation involves the large R portion of the wave function. One expects that the SCF Hartree-Fock wave function would be better everywhere and particularly in the tail. We have fitted a $2p$ single Slater-type function—i.e., $Nre^{-ar}Y_{1m}(\theta, \phi)$—to the $2p$ SCF wave function of the F^- ion (20) and found that the orbital exponent $a =$ 0.78 a.u. provides a reasonable fit in the region 5–10 a.u. The SCF wave function for the xenon atom is unknown. To overcome this difficulty the $4p$ SCF wave functions of Rb^+ [21] and Kr [22] were compared. Using the Hartree scaling method [23], the Rb^+ $4p$ wave function is transformed to the Kr $4p$ wave function in the region 5–10 a.u. by multiplying by the scale factor 1.07. The same scale factor was employed to derive from the known Cs^+ $5p$ orbital [23] the Xe $5p$ orbital. This pseudo xenon wave function can be adequately represented in the region 5–10 a.u. by a $5p$ single Slater function—i.e., $Nr^3e^{-ar}Y_{1m}(\theta, \phi)$—characterized by the orbital exponent $a = 1.44$ a.u. The scaling procedure is displayed in Figure 1. It should be pointed out that the SCF-like Xe $5p$ orbital thus obtained may be useful for a complete calculation of the binding energy at equilibrium interatomic separation in the xenon compounds. The overlap integral, S, was then calculated, and hence an estimate of the long-range charge-transfer stabilization was obtained.

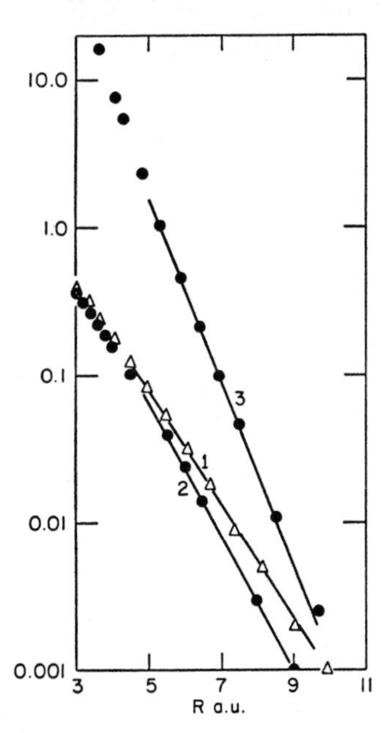

FIG. 1.—Scaling of wave functions: (1) ru versus r for Kr $4p$ orbital, (2) ru versus r for Cs^+ $4p$ orbital, and (3) 10^3 $(ru)/r^4$ versus r for the pseudo xenon $5p$ orbital.

In addition to the charge-transfer interaction, there is a sizable contribution to the binding energy from the (second-order) dispersion forces ΔE^{dis}. These can be represented in the conventional form:

$$\Delta E^{\text{dis}} = -\frac{3}{2}\cdot\frac{I_{\text{Xe}}I_{\text{F}}}{I_{\text{Xe}}+I_{\text{F}}}\cdot\frac{\alpha_{\text{Xe}}\alpha_{\text{F}}}{R^6}. \tag{4}$$

Taking the polarizabilities as $\alpha_{\text{Xe}} = 3.99 \times 10^{-24}$ cc. and $\alpha_{\text{F}} = 1 \times 10^{-24}$ cc. [25] and the known ionization potentials of xenon and fluorine, the contribution of the dispersion forces was estimated. In Table 1 we present the results of the calculation for long-range interactions. These numbers are probably accurate within a factor of 2. These results demonstrate unambiguously that, in the case of Xe + F, dispersion and charge-transfer forces are predominant at large separation, thus validating the use of delocalized molecular orbitals at smaller separations. At smaller distances the second-order perturbation theory used for the estimation of charge transfer forces breaks down and the MO method based on the variation method must be applied, although essentially the same trial wave functions are used.

PRESENTATION OF THE MODEL

It is proposed to describe the bonding in compounds of xenon and fluorine in terms of delocalized molecular orbitals formed mainly by combination of $p\sigma$-type xenon and fluorine orbitals. Thus, in XeF_2, one doubly filled Xe $5p\sigma$ atomic orbital and two F $2p\sigma$ atomic orbitals, each containing one electron, are considered. Similar considerations apply to XeF_4, where we start with four F $2p\sigma$ and two Xe $5p\sigma$ orbitals. In a zero*th* approximation, xenon and fluorine

TABLE 1

LONG-RANGE INTERACTIONS BETWEEN XENON AND FLUORINE

R (a.u.)	$-E^{C.T.}$ (ev.)	$-E^{\mathrm{dis}}$ (ev.)
6	0.42	0.032
7	.14	.013
8	.045	.006
9	0.014	0.003

$np\pi$ and ns orbitals, and xenon d-type orbitals can be considered to be non-bonding. Later we shall show that the effects of π-bonding are relevant to the interpretation of the electronic spectra and the ionization potentials of the xenon fluorides.

The semi-empirical treatment is reduced to a simple LCAO theory equivalent to the treatment of hetero-atomic, π-electron systems, except for the different symmetries of the orbitals involved. In the standard LCAO scheme the molecular orbitals, Ψ_r, are represented in the form

$$\Psi_r = \sum_i c_{ri}\mu_i \,, \tag{5}$$

where the μ_i are atomic orbitals. The secular equations are

$$(\alpha_i - E_r)\, c_{ri} + \sum_{j\neq i} (\beta_{ij} - E_r S_{ij})\, c_{rj} = 0 \,. \tag{6}$$

The matrix elements may be written in terms of the effective one-electron Hamiltonian, h (not very clearly defined in this scheme), so that the Coulomb integral is given by

$$\alpha_i = \int \mu_i h \mu_i d\tau \,; \tag{7}$$

the exchange integral is

$$\beta_{ij} = \int \mu_i h \mu_j d\tau \; ; \tag{8}$$

and the overlap integral is given by

$$S_{ij} = \int \mu_i \mu_j d\tau \; . \tag{9}$$

The total orbital energy is then

$$\epsilon = \sum_r m_r E_r , \tag{10}$$

where m_r is the occupation number of the rth molecular orbital. The energy of the atoms at infinite separation is taken as

$$I\epsilon = \sum_i m_i a_i , \tag{11}$$

with m_i being the occupation number of the ith atomic orbital in the atoms in their ground state. The binding energy per bond (considering N bonds in the molecular system) is easily seen to be

$$\Delta E = \frac{(I\epsilon - \epsilon)}{N} . \tag{12}$$

The approximations involved in the derivation of equation 12 have to be considered. It is well known that the sum of the orbital energies is not equal to the total energy of an atomic or molecular system in the Hartree self-consistent scheme, as the Coulomb repulsions between electron pairs have to be properly included. To the extent that the one electron effective Hamiltonian h represents a SCF Hamiltonian, equation 10 includes the interelectronic Coulomb repulsion energy, $\sum_{r>s} J_{rs}$, twice. On the other hand, the internuclear repulsion term V_{nn} was omitted from the molecular-energy expression. When these effects are taken into account, equation 12 becomes

$$\Delta E = \frac{1}{N}\left(I\epsilon - \epsilon + \sum_{r>s} J_{rs} - V_{nn} \right) . \tag{12a}$$

In the derivation of equation 12 it is assumed that the last two terms in equation 12*a* cancel out. However, this cancellation is by no means exact even for simple molecular systems [14]. The absolute values of binding energies derived from equation 12 are usually overestimated by a factor of 2. The semi-empirical LCAO method is not expected to lead to reliable values of bond energies, which are not easily obtained even by much more elaborate SCF computations. Instead, the semi-empirical treatment will prove to be extremely useful for the computation of energy differences between different nuclear or electronic configurations.

The eigenvectors of the secular matrix yield information concerning the charge distribution in the molecule. This calculation is to be carried out, taking account of the non-orthogonality of the atomic orbitals [25]. The normalization condition for Ψ_r is given by

$$\sum_{ij} S_{ij} c_{ri} c_{rj} = 1 . \tag{13}$$

Now, the overlap integrals may be regarded as the components of a metric tensor in the space defined by the MO coefficients. Consider the set:

$$d_{ri} = \sum_{j} S_{ij} c_{rj} . \tag{14}$$

Then the charge on atom i, q_i, can be expressed in the form [25]:

$$q_i = \sum_{r} m_r c_{ri} d_{ri} . \tag{15}$$

THE SEMI-EMPIRICAL TREATMENT OF XeF_2 AND XeF_4

In this section the semi-empirical LCAO scheme is applied to the xenon fluorides. In Table 2 we have listed the symmetry orbitals proper to XeF_2 and XeF_4 constructed by standard group theoretical methods. The phase convention used is given in Figure 2. The effect of Xe 5*s*, 6*s*, 4*d*, and 5*d* and the fluorine 2*s* and 3*s* orbitals will be neglected for reasons which will be considered later (energy considerations). The three appropriate σ-type molecular orbitals for the linear XeF_2 are given by

$$\Psi(a_{2u}^{-}) = \frac{a_-}{\sqrt{2}}(\mathrm{F}p\sigma 1 - \mathrm{F}p\sigma 2) + b_-\mathrm{Xe}p\sigma ,$$

$$\Psi(a_{1g}) = \frac{1}{\sqrt{2}}(\mathrm{F}p\sigma 1 + \mathrm{F}p\sigma 2), \tag{16}$$

and,

$$\Psi(a_{2u}^{+}) = \frac{a_+}{\sqrt{2}}(\mathrm{F}p\sigma 1 - \mathrm{F}p\sigma 2) + b_+\mathrm{Xe}p\sigma ,$$

while the π-type orbitals are given by

$$\Psi(e_{1u}) = \beta \begin{pmatrix} \mathrm{Xe}p\pi_x \\ \mathrm{Xe}p\pi_y \end{pmatrix} + a \begin{pmatrix} \mathrm{F}p\pi 1_x + \mathrm{F}p\pi 2_x \\ \mathrm{F}p\pi 1_y + \mathrm{F}p\pi 2_y \end{pmatrix},$$

and

$$\Psi(e_{1g}) = \frac{1}{\sqrt{2}} \begin{pmatrix} \mathrm{F}p\pi 1_x - \mathrm{F}p\pi 2_x \\ \mathrm{F}p\pi 1_y - \mathrm{F}p\pi 2_y \end{pmatrix}. \tag{17}$$

The secular determinant is readily factorized leading to the type energy levels,

$$E_{\pm}(a_{2u}) = \frac{A \pm \sqrt{(A^2 - B)}}{2C},$$

TABLE 2

SYMMETRY ORBITALS FOR LINEAR XeF_2 AND SQUARE-PLANAR XeF_4

	Symmetry Group	Orbital Symmetry	Xenon Orbital	Fluorine Orbitals σ	Fluorine Orbitals π
XeF_2	$D_{\infty h}$	a_{1g}	s, d_{z^2}	$\sigma_1-\sigma_2$	…
		a_{2u}	p_z	$\sigma_1+\sigma_2$	…
		e_{1u}	$p\pi_x$	…	$p\pi 1_x+p\pi 2_x$
			$p\pi_y$	…	$p\pi 1_y+p\pi 2_y$
		e_{1g}	d_{xz}	…	$p\pi 1_x-p\pi 2_x$
			d_{yz}	…	$p\pi 1_y-p\pi 2_y$
		e_{2g}	$d_{x^2-y^2}$	…	…
			d_{xy}	…	…
XeF_4	D_{4h}	a_{1g}	s, d_{z^2}	$\sigma_1+\sigma_2+\sigma_3+\sigma_4$	…
		e_u	p_x	$\sigma_1-\sigma_3$	$p\pi 2_y-p\pi 4_x$
			p_y	$\sigma_2-\sigma_4$	$p\pi 1_x-p\pi 3_y$
		a_{2u}	$p\pi_z$	…	$p\pi 1_y+p\pi 2_x-p\pi 2_x-p\pi 4_y$
		b_{1g}	$d_{x^2-y^2}$	$\sigma_1-\sigma_2+\sigma_3-\sigma_4$	…
		b_{2g}	d_{xy}	…	$p\pi 1_x+p\pi 2_y+p\pi 3_y+p\pi 4_x$
		e_g	d_{zx}	…	$p\pi 1_y+p\pi 3_x$
			d_{zy}	…	$p\pi 2_x+p\pi 4_y$
		b_{2u}	…	…	$p\pi 2_x-p\pi 1_y+p\pi 3_x-p\pi 4_y$
		a_{2g}	…	…	$p\pi 1_x-p\pi 2_y-p\pi 4_x+p\pi 2_x$

The phase convention is that given in Figure 2.

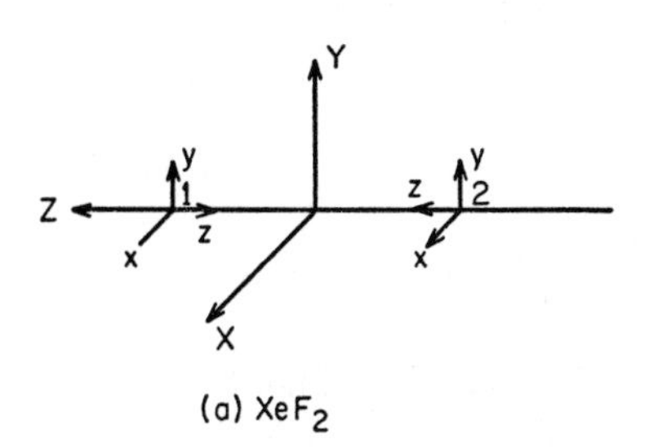

(a) XeF_2

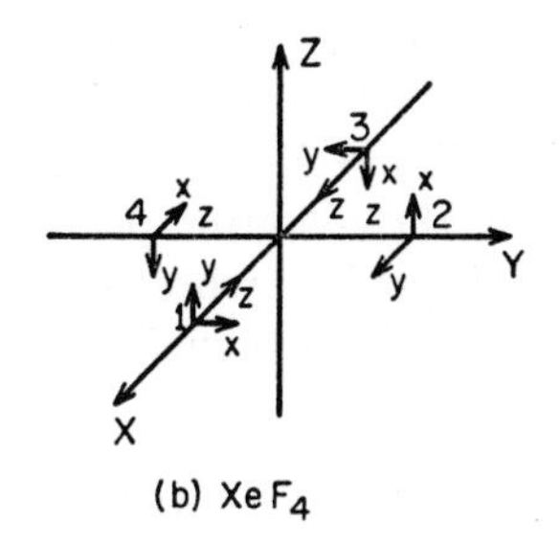

(b) XeF_4

FIG. 2.—Phase convention and notation for wave functions

where

$$A = \alpha(\mathrm{Xe}) + \alpha(\mathrm{F}) - 4\beta(\mathrm{Xe}p\sigma, \mathrm{F}p\sigma)S(\mathrm{Xe}p\sigma, \mathrm{F}p\sigma)\,,$$

$$B = 4C[\alpha(\mathrm{Xe})\alpha(\mathrm{F}) - 2\beta^2(\mathrm{Xe}p\sigma, \mathrm{F}p\sigma)]\,, \tag{18}$$

and,

$$C = 1 - 2S^2(\mathrm{Xe}p\sigma, \mathrm{F}p\sigma)\,;$$

and where

$$E(a_{1g}) = \alpha(\mathrm{F})\,. \tag{18a}$$

The energy levels for the π-type orbitals are of a similar form. The two energies $E_\pm(e_{1u})$ are obtained from $E_\pm(a_{2u})$ by substituting $\beta(\mathrm{Xe}\pi, \mathrm{F}\pi)$ and $S(\mathrm{Xe}\pi, \mathrm{F}\pi)$ for $\beta(\mathrm{Xe}\sigma, \mathrm{F}\sigma)$ and $S(\mathrm{Xe}\sigma, \mathrm{F}\sigma)$, respectively. Since the σ orbitals are responsible for bonding in these compounds, they will be considered first.

In the ground state of XeF_2 the bonding $\Psi\ (a_{2u}^-)$ and the non-bonding $\Psi(a_{1g})$ orbitals are doubly occupied while the antibonding $\Psi(a_{2u}^+)$ orbital is empty.

An alternative representation of the ground-state wave function of XeF_2 is of some interest. It can be easily shown that the ground-state wave function described by a single Slater determinant for this system reduces to a form characteristic of two localized Xe-F bonds. As in the usual equivalent-orbitals treatment [27] we may for the sake of this argument consider only orbitals with one spin function. Since the determinantal function is invariant under a unitary transformation, it is apparent that

$$\begin{vmatrix} \Psi(a_{2u}^-)(1)\Psi(a_{2u}^-)(2) \\ \Psi(a_{1g})(1)\Psi(a_{1g})(2) \end{vmatrix}$$

$$= \begin{vmatrix} \left(\sqrt{a_-}\mathrm{F}p\sigma 1 + \frac{b_-}{\sqrt{2a_-}}\mathrm{Xe}p\sigma\right)(1) & \left(\sqrt{a_-}\mathrm{F}p\sigma 1 + \frac{b_-}{\sqrt{2a_-}}\mathrm{Xe}p\sigma\right)(2) \\ \left(\sqrt{a_-}\mathrm{F}p\sigma 2 - \frac{b_-}{\sqrt{2a_-}}\mathrm{Xe}p\sigma\right)(1) & \left(\sqrt{a_-}\mathrm{F}p\sigma 2 - \frac{b_-}{\sqrt{2a_-}}\mathrm{Xe}p\sigma\right)(2) \end{vmatrix}. \tag{19}$$

Thus, the term "delocalization" should not be interpreted in a literal sense.

The simple treatment sketched above leads to some interesting conclusions, independent of the numerical values assigned to the molecular integrals. The comparison of the bond energies of XeF_2 and the radical XeF [see Morton and Falconer, p. 245, this volume] will be now considered. The bond energy per Xe-F bond in linear XeF_2 is $\alpha(\mathrm{Xe}) - E_-$ per bond. The XeF diatomic radical can be treated as a two-center three-electron problem. The secular equation leads to the two energy levels E_+ and E_- represented by equation 18, with the level E_- doubly occupied and E_+ singly occupied. The σ-orbital energy of the $^2\Sigma$ XeF ground state is thus $2E_- + E_+$, while the bond energy is

$$2\alpha(\mathrm{Xe}) + \alpha(\mathrm{F}) - (2E_- + E_+) = [\alpha(\mathrm{Xe}) + \alpha(\mathrm{F}) - (E_+ + E_-)] + [\alpha(\mathrm{Xe}) - E_-]\,. \tag{20}$$

Because of the non-orthogonality correction (note that β is negative) in equation 18, $\alpha(\mathrm{Xe}) + \alpha(\mathrm{F}) - (E_+ + E_-) < 0$; hence the bond energy in the XeF radical is expected to be less than the bond energy (per bond) in XeF_2.

At this point it is convenient to consider the hypothetical bent structure (bond angle 90°) of XeF_2. In this problem, six electrons located in four σ-type molecular orbitals have to be considered (two Xe $p\sigma$ and two F orbitals). The four energy levels are E_- and E_+, each double degenerate. The ground state of this molecule is therefore a triplet state. The bond energy is

$$\tfrac{1}{2}[4\alpha(\mathrm{Xe}) + 2\alpha(\mathrm{F}) - 4E_- - 2E_+]\,. \tag{21}$$

But this is just the bond energy of the XeF radical (equation 20). Hence, we may conclude that the bent structure (bond angle 90°) of XeF_2 is less stable than the corresponding linear structure. The conclusions thus obtained are consistent with the experimental data.

Considering now the structure of the square-planar XeF_4 (symmetry group D_{4h}), the molecular orbitals can be readily constructed from the symmetry orbitals presented in Table 2. It should be noted that the doubly degenerate $e_u\sigma$-type orbital involves now a contribution from the fluorine π-type orbitals. This molecular orbital is given in the form:

$$\Psi(e_u) = \begin{cases} A\,\mathrm{Xe}p\sigma_x + \dfrac{a}{\sqrt{2}}(\mathrm{F}p\sigma 1_z - \mathrm{F}p\sigma 3_z) + \dfrac{a}{\sqrt{2}}(\mathrm{F}p\pi 2_y - \mathrm{F}p\pi 4_x) \\ A\,\mathrm{Xe}p\sigma_y + \dfrac{a}{\sqrt{2}}(\mathrm{F}p\sigma 2_z - \mathrm{F}p\sigma 4_z) + \dfrac{a}{\sqrt{2}}(\mathrm{F}p\pi 1_x - \mathrm{F}p\pi 3_y)\,. \end{cases} \tag{22}$$

The three e_u-type energy levels (i.e., two doubly degenerate σ levels and one doubly degenerate π level) are obtained from the solution of the secular equation:

$$\begin{vmatrix} \alpha(\mathrm{Xe}) - E & \sqrt{2}\beta(\mathrm{Xe}p\sigma_x, \mathrm{F}p\sigma 1_z) - E\sqrt{2}S(\mathrm{Xe}p\sigma_x, \mathrm{F}p\sigma 1_z) & \sqrt{2}\beta(\mathrm{Xe}p\sigma_x, \mathrm{F}p\pi 2_y) - E\sqrt{2}S(\mathrm{Xe}p\sigma_x, \mathrm{F}p\pi 2_y) \\ \sqrt{2}\beta(\mathrm{Xe}p\sigma_x, \mathrm{F}p\sigma 1_z) - E\sqrt{2}S(\mathrm{Xe}p\sigma_x, \mathrm{F}p\sigma 1_z) & \alpha(\mathrm{F}) - E & 2\beta(\mathrm{F}p\sigma 1_z, \mathrm{F}p\pi 2_y) - 2ES(\mathrm{F}p\sigma 1_z, \mathrm{F}p\pi 2_y) \\ \sqrt{2}\beta(\mathrm{Xe}p\sigma_x, \mathrm{F}p\pi 2_y) - E\sqrt{2}S(\mathrm{Xe}p\sigma_x, \mathrm{F}p\pi 2_y) & 2\beta(\mathrm{F}p\sigma 1_z, \mathrm{F}p\pi 2_y) - 2ES(\mathrm{F}p\sigma 1_z, \mathrm{F}p\pi 2_y) & \alpha(\mathrm{F}) - E \end{vmatrix} = 0 \tag{23}$$

Here the interaction between F atoms pairs 1–3 and 2–4 separated by 3.9 A. was neglected.

The other two σ-type orbitals can be presented in the form

$$\Psi(b_{1g}) = \frac{1}{2[1 - 2S(\mathrm{F}p\sigma 1_z, \mathrm{F}p\sigma 2_z)]^{1/2}} \times (\mathrm{F}p\sigma 1_z - \mathrm{F}p\sigma 2_z + \mathrm{F}p\sigma 3_z - \mathrm{F}p\sigma 4_z) .$$

$$\Psi(a_{1g}) = \frac{1}{2[1 + 2S(\mathrm{F}p\sigma 1_z, \mathrm{F}p\sigma 2_z)]^{1/2}} \times (\mathrm{F}p\sigma 1_z + \mathrm{F}p\sigma 2_z + \mathrm{F}p\sigma 3_z + \mathrm{F}p\sigma 4_z) . \qquad (24)$$

It should be noted that the interaction between the adjacent fluorine atoms (separated by 2.82 A.) cannot now be neglected. These orbitals energies are given in the form:

$$E(b_{1g}) = \frac{\alpha(\mathrm{F}) - 2\beta(\mathrm{F}p\sigma 1_z, \mathrm{F}p\sigma 2_z)}{1 - 2S(\mathrm{F}p\sigma 1_z, \mathrm{F}p\sigma 2_z)} ,$$

$$E(a_{1g}) = \frac{\alpha(\mathrm{F}) + 2\beta(\mathrm{F}p\sigma 1_z, \mathrm{F}p\sigma 2_z)}{1 + 2S(\mathrm{F}p\sigma 1_z, \mathrm{F}p\sigma 2_z)} . \qquad (25)$$

Thus the interaction between adjacent $\mathrm{F}p\sigma$ orbitals which are perpendicular in respect to each other leads to the splitting of the orbital energies. The energy difference is given by

$$E(b_{1g}) - E(a_{1g}) = \frac{-4\beta(\mathrm{F}p\sigma 1_z, \mathrm{F}p\sigma 2_z) + 2\alpha(\mathrm{F})S(\mathrm{F}p\sigma 1_z, \mathrm{F}p\sigma 2_z)}{1 - 4S^2(\mathrm{F}p\sigma 1_z, \mathrm{F}p\sigma 2_z)} . \qquad (25a)$$

As we shall see in the ground state of XeF_4, the e_u^+ orbital is empty, and the ground state of the molecule is $^1A_{1g}$.

NUMERICAL CALCULATIONS

The first step in performing numerical calculations for the xenon fluorides in the semi-empirical molecular-orbital scheme is the selection of suitable atomic wave functions for the construction of the molecular orbitals. As SCF type atomic orbitals for xenon are not available, Slater-type atomic orbitals were used to account for the Xe-F interactions at relatively small separations of the order of the equilibrium interatomic distances. These orbitals were taken to be

$$\Psi_{2p}(\mathrm{F}) = Nre^{-2.600r}Y_{1m}(\theta, \phi) ,$$

and,

$$\Psi_{5p}(\mathrm{Xe}) = N'r^3e^{-2.063r}Y_{1m}(\theta, \phi) . \qquad (26)$$

The orbital exponents were chosen using the well-known Slater rules. In order to account for the interaction between adjacent F atoms in XeF_4 the use of a single Slater-type orbital seems inappropriate because of the relatively large distances (2.8 A.). We used in our calculation the best available fluorine atom SCF wave function represented in the form of a linear combination of four Slater-type wave functions [28]

$$\Psi_{2p}(\mathrm{F}) = r \sum_{i=1}^{4} a_i \alpha_i^{5/2} e^{-\alpha_i r} Y_{1m}(\theta, \phi) . \tag{27}$$

The coefficients, a_i, and the orbital exponents, α_i, are those given by Clementi and Roothaan [28].

The choice of the Coulomb and exchange integrals is subject to uncertainty. The Coulomb integrals were taken as the appropriate atomic ionization potentials. In Table 3 we present the Coulomb integrals for xenon and fluorine estimated from the atomic energy levels [29] and from the valence state energies compiled by Pritchard and Skinner [30]. As is well known, the necessary condition for effective combination of atomic orbitals in simple MO theory is that the atomic orbital energies should be of comparable magnitude. It is apparent from the data presented in Table 3 that the orbitals leading to effective binding are the p-type xenon and fluorine orbitals.

TABLE 3

ORBITAL IONIZATION POTENTIALS FOR XENON AND FLUORINE

Orbital	I ev.
Xe 6s	2.5
Xe 5p	12.1
Xe 5s	23.0
Xe 5d	2.0
Xe 4d	25.0
F 2p	17.4
F 2s	40.0

An independent calculation was carried out using the iterative ω technique [31]. The Coulomb integral is assumed to depend on the net charge on the given atom i:

$$\alpha_i^{\mu} = \alpha_i^{\mu-1} - (m_i - q_i^{\mu-1})\omega . \tag{28}$$

Here $q_i^{\mu-1}$ is the negative charge associated with atom i obtained from the $(\mu - 1)th$ iteration. The iterative process is repeated until self-consistency for q is achieved. Usually ten iterations were sufficient to reach self-consistency. Values of ω in the region 0–4 ev. were used in our calculations. This method is supposed to account semi-empirically for interelectronic repulsions.

The exchange integrals β were taken to be proportional to the overlap integrals,

$$\beta(i, j) = K(i, j)S(i, j) , \tag{29}$$

a procedure which is common in the semi-empirical treatment of aromatic molecules [32]. The proportionality parameter was chosen using the Wolfsberg-Helmholtz [32] recipe,

$$K(i, j) = g\left(\frac{\alpha(i) + \alpha(j)}{2}\right), \tag{30}$$

where g is an empirical parameter chosen as $g = 2.0$ [32]. This recipe is based essentially on the Mulliken approximation [18], i.e.,

$$\phi(i)\phi(j) = \tfrac{1}{2}S(i,j)[\phi^2(i) + \phi^2(j)] . \tag{31}$$

The Xe-F overlap integrals were evaluated using Slater-type orbitals (equation 26), applying the master formulas given by Mulliken *et al.* [33]. The F-F overlap integrals required for the calculation of the XeF_4 molecule were evaluated using the SCF orbital (equation 27). This procedure seems to us legitimate because of the relatively large internuclear F-F separation in this case. These integrals were calculated by numerical integration using 10,000 integration points. This calculation requires approximately one minute on an IBM 7090 computer. The integrals calculated using the phase convention of Mulliken *et al.* [33] are presented in Table 4. The F-F overlap integrals were expanded

TABLE 4

OVERLAP INTEGRALS

S(F 2p, Xe 5p) CALCULATED FROM SLATER ORBITALS

R (a.u.)	S(F 2$p\pi$, Xe 5$p\pi$)	S(F 2$p\sigma$, Xe 5$p\sigma$)
2.00	0.27681	−0.12984
2.50	.18375	− .22943
3.00	.11310	− .23895
3.50	.06536	− .19825
3.78	.04696	− .16749
4.00	.03584	− .14310
4.50	.01881	− .09365
5.00	.00951	− .05692
5.50	.00466	− .03264
6.00	0.00222	−0.01786

S(F 2p, F 2p) CALCULATED FROM THE CLEMENTI-ROOTHAAN SCF WAVE FUNCTION

R (a.u.)	S(F 2$p\sigma$, F2$p\sigma$)	S(F 2$p\pi$, F 2$p\pi$)
1.00	0.0975	0.6011
1.50	− .1731	.3811
2.00	− .2476	.2304
2.50	− .2294	.1357
3.00	− .1808	.0788
3.50	− .1308	.0453
4.00	− .0899	.0260
4.50	− .0576	.0149
5.00	− .0389	.0085
5.50	− .0249	.0049
6.00	− .0158	.0028
6.50	− .0100	.0016
7.00	− .0062	.0009
7.50	− .0039	.0005
8.00	−0.0024	0.0003

in the form of a linear combination of $S(\pi\pi)$- and $S(\sigma\sigma)$- type integrals

$$-S_{12}=\frac{(\boldsymbol{n}_1\cdot\boldsymbol{R})(\boldsymbol{n}_2\cdot\boldsymbol{R})}{R^2}\,S(\sigma\sigma)-\left[\boldsymbol{n}_1\cdot\boldsymbol{n}_2-\frac{(\boldsymbol{n}_1\cdot\boldsymbol{R})(\boldsymbol{n}_2\cdot\boldsymbol{R})}{R^2}\right]S(\pi\pi)\,. \quad (32)$$

Here $\boldsymbol{n}_1$ and $\boldsymbol{n}_2$ are the unit vectors defining the direction of the orbitals 1 and 2, and R is the vector connecting the two atoms. From equation 32 it readily follows that for the case under consideration

$$\begin{aligned} S(\mathrm{F}p\sigma 1_z,\ \mathrm{F}p\sigma 2_z) &= \tfrac{1}{2}[S(\mathrm{F}\sigma,\ \mathrm{F}\sigma)+S(\mathrm{F}\pi,\ \mathrm{F}\pi)] \\ S(\mathrm{F}p\sigma 2_z,\ \mathrm{F}p\pi 1_z) &= \tfrac{1}{2}[S(\mathrm{F}\sigma,\ \mathrm{F}\sigma)-S(\mathrm{F}\pi,\ \mathrm{F}\pi)]\,. \end{aligned} \quad (33)$$

The integrals relevant to our calculations are presented in Table 5.

ENERGY LEVELS AND CHARGE DISTRIBUTION

Some elevations of the energy levels and the charge distributions for XeF_2 at the equilibrium nuclear distance (R = 2.0 A.) were performed in order to test the choice of the empirical parameters. The results displayed in Table 6 were obtained changing g in the region 0.66–2.2. The value of g = 2.0 was chosen for further calculations. This choice seems to be consistent with previous experience [32]. Thus, the values of the proportionality factors $K(i, j)$ used in our calculations of the exchange integrals were taken to be K(Xe, F) = −29.5 ev.

TABLE 5

OVERLAP INTEGRALS FOR MOLECULAR CALCULATIONS

Integral	R (a.u.)	S
S (Xe $p\sigma_z$, F $p\sigma 1_z$)	3.78	0.1675
S (Xe $p\pi_z$, F $p\pi 1_y$).......	3.78	.04696
S (F $p\sigma 1_z$, F $p\sigma 2_y$).......	5.34	.01157
S (F $p\sigma 1_z$, F $p\sigma 2_z$).......	5.34	0.01771

The phase convention is that given in Figure 2.

TABLE 6

SEMI-EMPIRICAL LCAO CALCULATIONS FOR LINEAR XeF_2

α (Xe) = −12.12 ev.; α (F) = −17.4 ev.

K	E_-	E_+	q_F	q_{Xe}
−10.......	−17.98	−12.05	0.967	0.063
−22.......	−17.54	−10.92	.920	.160
−26.......	−17.98	− 9.96	.846	.308
−30.......	−18.56	− 8.86	.787	.427
−32.......	−18.88	− 8.28	0.762	0.475

All values quoted are in units of ev. The quantities tabulated are the energies of the bonding and antibonding states E_-, E_+; the charges q_F and q_{Xe} calculated from the bonding MO coefficients represent the negative charge associated with the fluorine; and the residual charge on the xenon atom. Note that $2q_F + q_{Xe} = 2$.

and $K(F, F) = -34.8$ ev. In Tables 7 and 8 we have displayed the results of some MO calculations for the σ and π orbitals of XeF_2 and for the e_u orbitals of XeF_4, using the ω technique. The starting values for the Coulomb integrals are taken from Table 3.

An attempt was made to estimate the equilibrium separation and the binding energy for XeF_2. In this calculation the effects of π-bonding were included. The calculations were performed for a number of internuclear distances. The contributions to the binding energy in XeF_2 are summarized in Figure 3. At small internuclear separations the binding energy decreases because of the

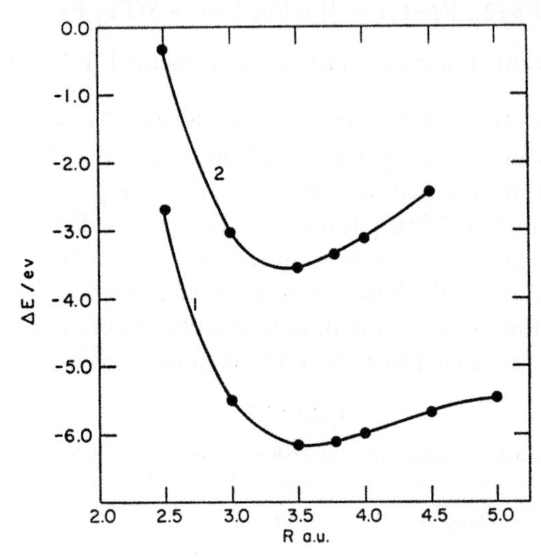

FIG. 3.—Binding energy per bond for XeF_2 versus xenon-fluorine internuclear distance. 1. $W = 0$; 2. $W = 2$ ev.

effect of the antibonding e_{2u} π-type orbital. The electronic contribution to the binding energies attains a maximum value at $R_{Xe-F} = 1.85$ A., which compares quite favorably with the experimental value of 2.00 A. [34]. The binding energy estimated for this theoretical equilibrium distance is higher than the experimental value (1.4 ev.). [Compare Gunn and Williamson, p. 133; Stein and Plurien, p. 144; Weinstock, Weaver, and Knop, p. 50—this volume.] The electronic contributions to the binding energies per bond for XeF_2 and XeF_4 calculated at the experimental internuclear separation of 2.0 A. turn out to be equal.

Our calculations indicate that in the ground state of the xenon fluorides there is substantial migration of negative charge from the xenon to the fluorine. It should be noted that the π orbitals do not contribute to the charge migration as both the bonding and the antibonding orbitals are filled. In Table 9 we have presented the charge distributions predicted for XeF_2 and XeF_4 obtained by

TABLE 7

ENERGY LEVELS AND CHARGE DISTRIBUTION FOR XeF_2 AS CALCULATED BY THE ω METHOD

R(Xe-F) = 2.00 A.

ω (ev.)	$E(a_{2u}^+)$ (ev.)	$E(a_{2u}^-)$ (ev.)	$E(e_{1u}^+)$ (ev.)	$E(e_{1u}^-)$ (ev.)	ΔE (ev.)	b_+	b_-	a_+	a_-	β_+	β_-	α_+	α_-	q_F	q_{Xe}
0.0...	− 9.27	−18.47	−11.87	−17.51	6.10	0.967	0.351	−0.570	0.856	0.991	0.144	−0.210	0.979	0.805	0.389
2.0...	−10.88	−18.16	−13.81	−17.58	3.53	.876	.539	− .731	.723	.977	.219	− .284	.961	.629	.742
3.0...	−11.22	−18.23	−14.06	−16.54	2.75	0.831	0.607	−0.786	0.663	0.923	0.390	−0.522	0.895	0.536	0.928

TABLE 8

THE e_u ENERGY LEVELS OF XeF_4

R(Xe-F) = 2.0 A.

ω (ev.)	α(Xe) (ev.)	α(F) (ev.)	E_1 (ev.)	E_2 (ev.)	E_3 (ev.)	A_1	a_1	α_1	A_2	a_2	α_2	A_3	a_3	α_3
0.0......	−12.10	−17.40	−18.72	−17.14	− 9.16	0.336	0.760	0.396	0.090	0.411	−0.913	0.971	−0.559	−0.119
0.5......	−13.57	−17.03	−18.67	−16.78	−10.37	.415	.729	.348	.098	.365	− .931	.939	− .628	− .126
1.0......	−14.72	−16.74	−18.73	−16.50	−11.22	.492	.688	.306	.105	.326	− .946	.900	− .692	− .129
2.0......	−16.53	−16.45	−19.08	−16.07	−12.28	0.631	0.587	0.232	0.114	0.254	−0.966	0.807	−0.805	−0.126

The three e_u energy levels are labeled as E_1, E_2, and E_3, the first two being filled. The LCAO coefficients are labeled as in equation 22.

the ω technique. In view of the charge migration the xenon fluorides should be considered as *semi-ionic compounds.*

In keeping with the large charge transfer some comments are in order. The charge distribution predicted is confirmed by the observed fluorine NMR chemical shifts obtained for the xenon fluorides in the solid and in HF solution [compare Hindman and Svirmickas, this volume, p. 260]. The comparison between the theoretical results and the charge distributions calculated from the NMR data is gratifying.

The implication of these results for the interpretation of the infrared data of the xenon fluorides should be briefly considered. The ratio between the interaction and stretching force constants for XeF_2 is $f_{rr}/f_r = 0.05$. It was recently argued by Smith [35] that the difference of the force constants ratio, f_{rr}/f_r, for

TABLE 9

CHARGE DISTRIBUTIONS IN XeF_2 AND XeF_4

ω (ev.)	q_F*	
	XeF_2	XeF_4
0.0	0.805	0.806
0.5	.761	.734
1.0	.721	.655
2.0	.629	.491
Experimental [36]	0.70	0.55

* q_F represents the excess negative charge on one fluorine atom.

XeF_2 and the trihalide ions contradicts the molecular-orbital description [see also this volume, p. 298]. However, it should be noted that the charge distribution is entirely different in these two cases. In the case of the xenon fluorides the central xenon atom carries a considerable positive charge, while in the trihalide ions (say, $I\ Cl_2^-$) the central atom is almost neutral (or slightly negatively charged). The interaction force constant, f_{rr}, is expected to be reduced in the case of the xenon fluorides due to electrostatic interaction. The ratio f_{rr}/f_r for xenon difluoride should not be compared to that for the trihalide ions but to the corresponding ratio for ClF_3 ($f_{rr}/f_r = 0.08$).

It is interesting to speculate on the solution chemistry of the xenon fluorides in inert solvents. Chemical exchange between the bound fluorine and fluoride ions in solution is expected to be efficient, increasing with the extent of charge transfer (i.e., $XeF_2 > XeF_4 > XeF_6$). In solvents with high dielectric constants or where special solvent bonds may be formed (e.g., HF_2^- in HF), ionic dissociation of the xenon fluorides XeF_n into XeF_{n-1}^+ and F^- ions is feasible. [Compare Hindman and Svirmickas, this volume, p. 254, and Hyman and Quarterman, *ibid.*, p. 277.]

EXCITED ELECTRONIC STATES

Group theory has been used to classify allowed energy levels by their symmetry. The model presented has also given estimates of the energy of the various levels. A combination of these results allows a description of the excited electronic states. In particular, it allows an estimate of the types and nature of the optical transitions from the ground state to excited electronic states.

The energy levels for XeF_2 and XeF_4 are presented in Figure 4. The ground state configuration of XeF_2 is expected to be

$$(a_{2u}^{-})^2(e_{1u})^4(e_{1g})^4(a_{1g})^2(e_{1u})^4$$

and that of XeF_4,

$$(e_u)^4(a_{1g})^2(b_{2g})^2(a_{2u})^2(b_{2u})^2(e_g)^4(e_u)^4(a_{2g})^2(b_{1g})^2(a_{2u})^2 .$$

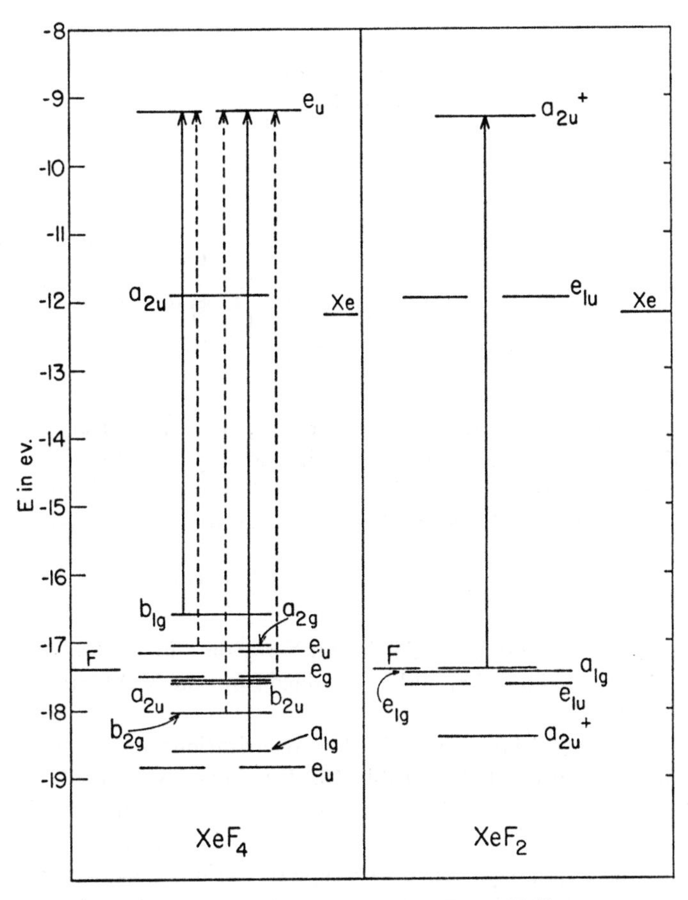

FIG. 4.—Energy levels of XeF_2 and XeF_4

Thus the ground state of both molecules is expected to be a totally symmetric singlet state $^1A_{1g}$. The highest filled orbital in both cases is a π-type antibonding orbital.

Consider first the linear XeF_2 (symmetry group $D_{\infty h}$). The first singlet-singlet allowed transition is from the non-bonding σ_g to the bonding $\sigma_u{}^+$, i.e., $a_{1g} \rightarrow a^+{}_{2u}$ (or $^1A_{1g} \rightarrow {}^1A_{2u}$). The estimated transition energy is 8.1 ev. This transition is polarized along the molecular axis (z). In order to evaluate the transition probability the explicit form of the calculated molecular orbitals is required. The transition dipole moment Q is

$$Q = \sqrt{2}\left\langle \frac{1}{\sqrt{2}}(\mathrm{F}p\sigma 1+\mathrm{F}p\sigma 2) \middle| z \middle| \frac{a_+}{\sqrt{2}}(\mathrm{F}p\sigma 1-\mathrm{F}p\sigma 2)+b_+\mathrm{Xe}p\sigma \right\rangle$$

$$= \frac{a_+}{\sqrt{2}}[\langle \mathrm{F}p\sigma 1 \mid z \mid \mathrm{F}p\sigma 1\rangle - \langle \mathrm{F}p\sigma 2 \mid z \mid \mathrm{F}p\sigma 2\rangle] + 2b_+\langle \mathrm{F}p\sigma \mid z \mid \mathrm{Xe}p\sigma\rangle. \quad (34)$$

TABLE 10

EXCITED STATES OF XeF_4

Transition	Polarization	$h\nu$ Calculated (ev.)	$h\nu$ Experimental (ev.)
$b_{1g} \rightarrow e_u$	x, y	7.4	6.8
$a_{2g} \rightarrow e_u$	x, y	7.8 }	
$e_g \rightarrow e_u$	z	8.2 }	8.3?(shoulder)
$b_{2g} \rightarrow e_u$	x, y	8.6 }	
$a_{1g} \rightarrow e_u$	x, y	9.1	9.4

The matrix elements elements $\langle \mathrm{F}p\sigma i|z|\mathrm{F}p\sigma i\rangle$, ($i$ = 1, 2), are the mean distances of an electron measured in the z direction from the fluorine atoms i = 1 and 2, respectively. The matrix element $\langle \mathrm{F}p\sigma|z|\mathrm{Xe}p\sigma\rangle$ is approximated by $RS(\mathrm{F}p\sigma, \mathrm{Xe}p\sigma)/4$. Hence Q is related to the internuclear F-F separation R by

$$Q = a_+R/\sqrt{2} + b_+\frac{RS(\mathrm{F}p\sigma, \mathrm{Xe}p\sigma)}{2}. \quad (35)$$

Using the value R = 4.0 A. obtained from X-ray data in crystalline XeF_2 [34], and applying the values of a_+ and b_+ obtained from the MO treatment, f (estimated) = 1.1.

The allowed transitions for XeF_4, their polarizations and respective transition energies are presented in Table 10. Two strongly allowed $\sigma \rightarrow \sigma$ type transitions are expected, i.e., $b_{1g} \rightarrow e_u$ and $a_{1g} \rightarrow e_u$ separated approximately by $4\beta(p\sigma\mathrm{F1}, p\sigma\mathrm{F2})$. The symmetry allowed $\pi \rightarrow \sigma$ transitions $a_{2g} \rightarrow e_u$ and $b_{2g} \rightarrow e_u$ should overlap the two $\sigma \rightarrow \sigma$ type transitions. The $e_g \rightarrow e_u$ transition is expected to be located between the two strong $\sigma \rightarrow \sigma$ transitions. The predicted allowed optical transitions for XeF_4 are displayed in Figure 4.

We shall now show how these predictions are confirmed by the spectroscopic experimental data.

EXPERIMENTAL METHOD

The experiments were carried out, at 5 A. resolution, using a McPherson Seya-Namioka Vacuum U.V. Monochromator. The light source was a windowless hydrogen gas discharge; a sodium salicylate–coated E.M.I. 9514B photomultiplier tube was used as a detector. The optical cell, 5 cm. in length, was of stainless steel, with LiF end windows attached by epoxy resin. The sample gas in the cell was in equilibrium with solid sample contained in a Monel bottle. By changing the temperature of the solid sample, the amount of gas in the cell could be varied. Experiments were carried out down to ~1100 A., the LiF transmission cutoff.

The intensity of the light leaving the cell was measured as a function of wave length first with the cell empty and then containing the sample. The light intensity incident on the cell was stable to ~1 per cent over the time of the experiment. The optical density against wave length obtained in this way was reproducible in shape from sample to sample, although its absolute magnitude varied by ~10 per cent.

The optical density was also measured at fixed wave length as a function of the temperature of the solid sample. The proportionality of the optical density to the vapor pressure allows the heat of sublimation of the solid to be deduced.

THE ABSORPTION SPECTRA OF XeF_2 AND XeF_4

The absorption spectra obtained are displayed in Figures 5–7. Those of Figures 5 and 6 were obtained with a solid sample temperature of 0° C. and the absorptions of Figure 7 with a temperature of 22° C. The molar extinction coefficients were estimated assuming the vapor pressure at 25° C. is 3.8 mm. Hg [36] for XeF_2 and 2 mm. Hg for XeF_4. The heats of sublimation of XeF_2 and XeF_4, which were also determined (presented below, pp. 384–86), were used in the analysis of these data. The experimental oscillator strengths were evaluated from the expression

$$f = 4.59 \times 10^{-9} \epsilon_{max} \Delta\nu \,, \qquad (36)$$

where ϵ_{max} is the molar extinction coefficient at the band maximum (in units of liter cm.$^{-2}$ mole^{-1}) and $\Delta\nu$ is the half-band width (in cm.$^{-1}$). Equation 36 is based on a gaussian approximation for the shape of the absorption bands. The data relevant to the interpretation of the spectra are summarized in Table 11. The experimental oscillator strengths are estimated to be reliable to a factor of 2.

The absorption spectrum of XeF_2 is characterized by a weak band followed by a strong absorption band located at 1580 A., accompanied by a series of sharp bands. The absorption spectrum of XeF_4 is characterized by a weak band followed by two strong bands located at 1840 A. and 1325 A. No vibrational fine structure of the bands was observed. This may indicate excitation to dissociative states. However, high-resolution experiments are required to clarify this point.

It should be noted that no absorptions characteristic of atomic xenon or molecular fluorine are present in the results. The ${}^1S \rightarrow {}^3P_1$ and ${}^1S \rightarrow {}^1P_1$ transitions of xenon are located at 1470 A. and 1295 A., respectively [29], while the

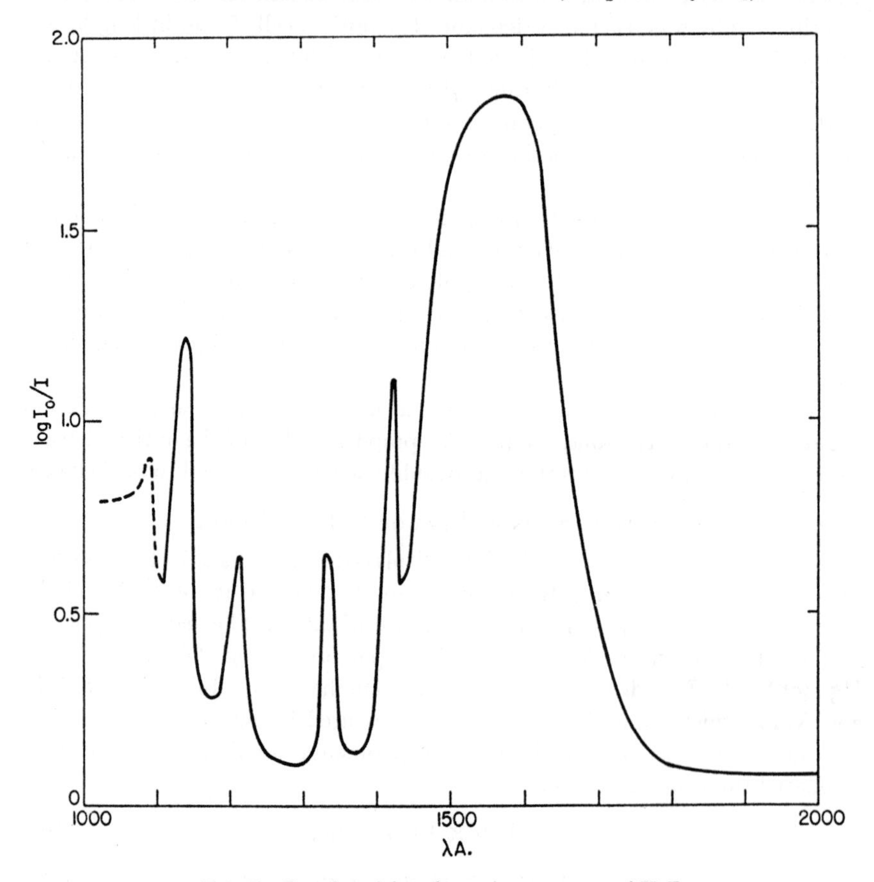

FIG. 5.—Far ultraviolet-absorption spectrum of XeF_2

F_2 molecule is characterized by a weak $\pi_g{}^* \rightarrow \sigma_u{}^*$ transition at 2800 A. ($\epsilon_{max} \approx$ 6 liter/cm^{-2}/mole^{-1}) followed by the $N \rightarrow V(\sigma_g \rightarrow \sigma_u{}^*)$ transition originating at 1090 A. [35]. This allows some confidence that decomposition products are not being observed.

We turn now to the interpretation of these spectroscopic data. The strong absorption in XeF_2 observed at 1580 A. is assigned to the singlet-singlet $a_{1g} \rightarrow a_{2u}{}^-$ transition. The calculated transition energy (8.1 ev.) compares quite favorably with the observed value of 7.9 ev. The calculated oscillator strength

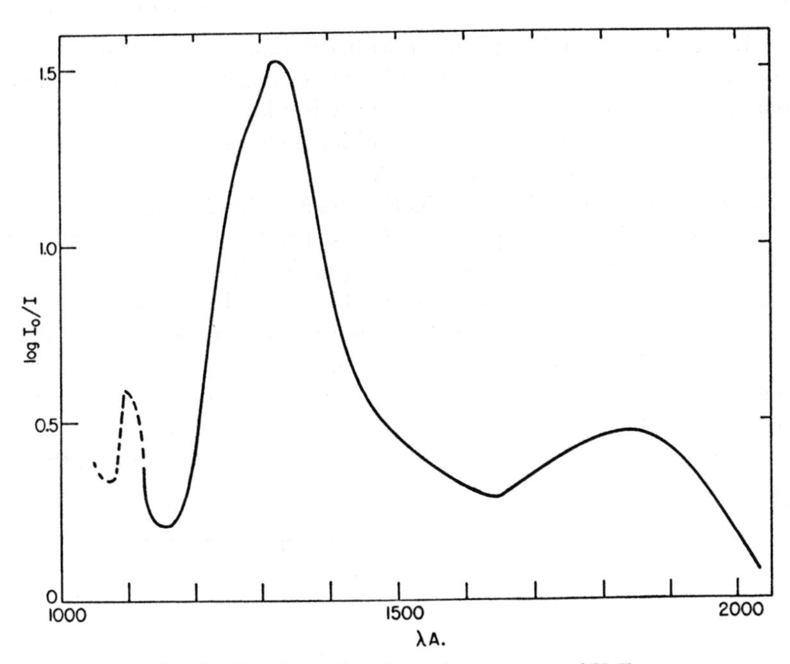

Fig. 6.—Far ultraviolet–absorption spectrum of XeF_4

TABLE 11

Far-Ultraviolet Spectroscopic Data for the Xenon Fluorides

XeF_2

max (A.)	$\Delta\nu$ (cm.$^{-1}$)	D_{max}	ϵ_{max} (liter/mole/cm^2)	f
2300.......	8249	0.014	86	0.0033
1580.......	8060	1.84	1.12×10^4	.42
1425.......	(1000)	0.70	$0.4\ \times10^4$	.02
1335.......	(1290)	0.67	$0.4\ \times10^4$	.02
1215.......	(2070)	0.67	$0.4\ \times10^4$	.03
1145.......	(2730)	1.0	$0.6\ \times10^4$	0.06

XeF_4

max (A.)	$\Delta\nu$ (cm.$^{-1}$)	D_{max}	ϵ_{max} (liter/mole/cm^2)	f
2265.......	7000	0.044	440	0.014
1840.......	10000	0.475	4.75×10^3	.22
1325.......	11200	1.53	$1.5\ \times10^4$	0.80
(1090).....	....	....	....	....

($f = 1.1$) is in reasonable agreement with the experimental value ($f = 0.4$). This transition involves a considerable charge migration from the fluorine atoms to the central xenon atoms. This strongly allowed transition from a non-bonding to an antibonding molecular orbital may be regarded as an intramolecular charge-transfer transition [37].

The two allowed transitions in XeF_4 separated by 2.6 ev. are assigned to $\sigma \rightarrow \sigma$ type transitions. The presence of these two bands in XeF_4 confirms our hypothesis of the splitting of the a_{1g} and b_{1g} orbitals because of interaction between adjacent F atoms. The 1850 A. band is assigned to the $b_{1g} \rightarrow e_u$ and the $a_{2g} \rightarrow e_u$ transitions, while the 1325 A. is assigned to the $a_{1g} \rightarrow e_u$ and the $b_{2g} \rightarrow e_u$ transitions. The symmetry allowed $\pi \rightarrow \sigma$ type transition, $e_g \rightarrow e_u$, should be located between the two $\sigma \rightarrow \sigma$ type transitions. This transition is probably hidden in the asymmetric onset of the $a_{1g} \rightarrow e_u$ transition. In Table 10 the experimental transition energies are compared with the results of the semi-empirical calculations.

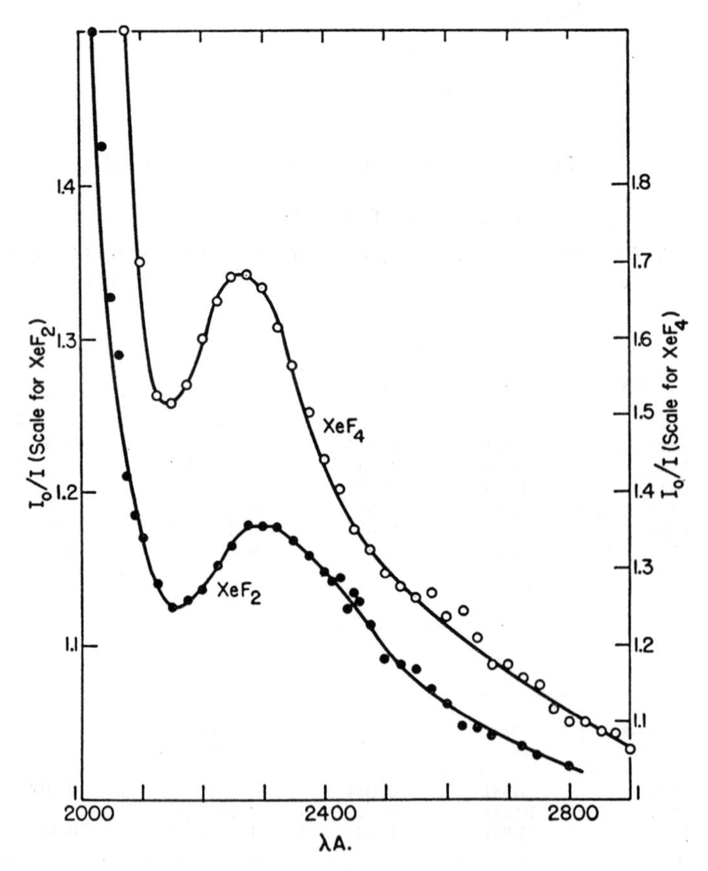

FIG. 7.—Weak absorption bands of XeF_2 and XeF_4. Note that the ordinate scale is linear

We may thus conclude that both for XeF_2 and XeF_4 the agreement between theory and experiment is gratifying, providing strong support for the binding scheme proposed for these compounds.

Another potentially interesting feature of the absorption spectrum of XeF_4 should be noted. Since the excited MO is doubly degenerate (e_u type), a Jahn-Teller configurational distortion in the excited state is expected. This may lead to absorption bands exhibiting a doublet peak. The shape of the $a_{1g} \rightarrow e_u$ band may perhaps be caused by such an effect. Similar effects may rise in the case of XeF_6 where the vacant level is expected to be a triply degenerate t_{1u} MO. Further experimental and theoretical studies are required to investigate these interesting aspects of the electronic spectra of the xenon fluorides.

The possibility of observing singlet-triplet transitions, corresponding to the $a_{1g} \rightarrow a_{2u}^+$ and $b_{1g} \rightarrow e_u$ transitions in XeF_2 and XeF_4, respectively, are of interest. These spin forbidden transitions may be enhanced by an intramolecular heavy atom effect [38]. We do not assign the 2300 A. band in XeF_2 and the 2265 A. band in XeF_4 (Fig. 7) to singlet-triplet transitions for the following reasons: *a*) The bands intensities seem to be too high when compared with the intermolecular spin-orbit induced transitions in iodo aromatic compounds [38]. *b*) The energy difference between the 2300 A. and 1580 A. bands in XeF_2 is 2.5 ev., leading to quite a large value for the appropriate exchange integral. On the other hand, the splitting between the weak 2265 A. and the allowed 1840 A. bands in XeF_4 is 1.0 ev. It is not clear why the exchange integral should be drastically reduced for XeF_4 relative to XeF_2.

We thus propose to assign these weak bands in XeF_2 and XeF_4, not to spin forbidden transitions, but to symmetry forbidden transitions.

VIBRONICALLY INDUCED TRANSITIONS IN THE XENON FLUORIDES

We suggest that the weak 2300 A. band in XeF_2 arises from a singlet-singlet transition from the highest filled π orbital to the σ_u^+ orbital, i.e., $e_{1u} \rightarrow a_{2u}^+$ (or ${}^1A_{1g} \rightarrow {}^1E_{1g}$). As seen from Table 7 the highest filled πe_{1u} orbital consists predominantly of the $5p\pi$ Xe orbital, and thus this is essentially a Xe $5\pi \rightarrow a_{2u}^+$ transition. The $D_{\infty h}$ symmetry of the linear XeF_2 molecule forbids this transition, but it becomes allowed because of vibronic coupling with the $E_u(\nu_2)$ out of axis vibration. In a similar way, the weak transition of XeF_4 at 2265 A. is assigned to the symmetry forbidden $a_{2u} \rightarrow e_u$ transition, which is vibronically induced by the E_u-type vibration [39].

Vibronically induced transitions have been studied in relation to the calculation of the intensities of the absorption bands of transition-metal ions, where they have been treated by the methods of ligand field theory [40]. The linear triatomic molecule CO_2 is characterized by a weak transition beginning at 1712 A., which may serve as an interesting analogy to the vibronically induced transition in XeF_2. The CO_2 band has been tentatively assigned [41] to the symmetry forbidden transition $\pi_g \rightarrow \sigma_g$.

In order to obtain further support for the assignment of the weak XeF_2 band to a vibronically induced transition, a calculation of the oscillator strength was carried out. As is well known [40], a symmetry forbidden transition is due to mixing of states of different symmetry. The zero-order wave functions are written as a product of electronic (Ψ_N) and vibrational (χ_n) wave functions

$$\Psi_{Nn} = \Psi_N \chi_n \, . \tag{37}$$

States are labeled by specifying both electronic and vibrational quantum numbers. The perturbation Hamiltonian for the molecular system is

$$H' = \sum_i Q_i \frac{\delta H}{\delta Q_i}, \tag{38}$$

where Q_i is the amplitude of the ith normal mode of vibration. It should be noted that as H' transforms like a_{1g}, and $\delta H/\delta Q_i$ has to mix in electronic states of different parity, then Q_i must have u symmetry.

The most important contribution to the vibronic coupling in XeF_2 rises from mixing in of the e_{1u} and a_{1g} wave functions by the E_u vibration. First-order–perturbation theory leads to the following expression for the perturbed e_{1u} state:

$$| e_{1u}'\chi_n\rangle' = | e_{1u}\chi_n\rangle + \sum_m | a_{1g}\chi_m\rangle \frac{\langle a_{1g}\chi_m | H' | e_{1u}\chi_m\rangle}{E(e_{1u}) - E(a_{1g})}. \tag{39}$$

The dipole moment for the $e_{1u}' \rightarrow a_{2u}$ transition takes the form

$$\langle R\rangle = \sum_{n,\,m} \frac{\langle a_{2u} | Q | a_{1g}\rangle \langle a_{1g}\chi_m | H' | e_{1u}\chi_n\rangle}{E(e_{1u}) - E(a_{1g})}. \tag{40}$$

The perturbation matrix element reduces to a product of electronic and vibrational parts,

$$\langle a_{1g}\chi_m | H' | e_{1u}\chi_n\rangle = \sum_i \Big\langle \chi_m \Big| Q_i \Big| \chi_n \Big\rangle \Big\langle a_{1g} \Big| \frac{\delta H}{\delta Q_i} \Big| e_{1u} \Big\rangle. \tag{41}$$

The selection rule for the vibrational parts in the harmonic approximation is

$$\begin{aligned} \langle \chi_m | Q_i | \chi_n\rangle &\neq 0 \quad \text{if } n - m = \pm 1 \\ &= 0 \quad \text{if } n - m \neq \pm 1 \, . \end{aligned} \tag{42}$$

A straightforward, though lengthy, calculation leads to the following expression for the oscillator strength of the forbidden transition:

$$\begin{aligned} f(e_{1u} \rightarrow a_{2u}) &= f(a_{1g} \rightarrow a_{2u}) \times \mathrm{N} \\ &\times \frac{E(e_{1u}) - E(a_{2u})}{[E(e_{1u}) - E(a_{1g})]^2 [E(a_{2u}) - E(a_{1g})]} \times \Big\langle a_{1g} \Big| \frac{\delta H}{\delta X} \Big| e_{1u} \Big\rangle \\ &\times \frac{h}{8\pi^2 \left(m_{\mathrm{Xe}} + \dfrac{m_{\mathrm{Xe}}^2}{2 m_{\mathrm{F}}} \right) \nu} \coth\left(\frac{h\nu}{2kT} \right), \end{aligned} \tag{43}$$

where $f(a_{1g} \rightarrow a_{2u})$ is the oscillator strength for the allowed charge-transfer transition, N is a numerical factor (N = 8), E is the orbital energy, $h/\{8\pi^2[m_{Xe} + (m^2_{Xe}/2m_F)]\nu\}$ is the square of the matrix element for a harmonic oscillator $0 \rightarrow 1$ transition, ν is the E_u vibration frequency, k is the Boltzman constant and T is the absolute temperature. The matrix element $\langle a_{1g} | \delta H/\delta X | e_{1u} \rangle$ was approximated by a method similar to that used in our semi-empirical molecular-orbital calculations

$$v = \left\langle a_{1g} \left| \frac{H}{\delta X} \right| e_{1u} \right\rangle \approx \sqrt{2} \left[\frac{\delta}{\delta X} \langle Xep\pi \,|\, h \,|\, \mathrm{F}p\sigma \rangle \right]_{X=0} . \quad (44)$$

This approximation leads to a value of $v^2 = 3.6 \times 10^{-7}$ C.G.S. which is of the same order of magnitude as obtained for transition-metal ions [42].

It is apparent from equation 43 that the vibronic-coupling mechanism will be effective if a suitable allowed transition of high intensity is located in the vicinity of the forbidden band. In this case the vibronically induced band may effectively "steal" intensity from the allowed transition. Another factor which enhances the forbidden transition is a low frequency of the "scrambling" u-type vibration. Both these conditions are fulfilled in the case of XeF_2, where the forbidden and allowed bands are separated by 2.5 ev., and the E_u-type vibration frequency is relatively low (ν_2 = 213 cm.$^{-1}$) [38]. A numerical estimate of the oscillator strength for the forbidden transition, from equations 43 and 45, leads to

$$f_{(e_{1u} \rightarrow a_{2u})} = 5.0 \times 10^{-3} f_{(a_{1g} \rightarrow a_{2u})} . \quad (45)$$

From the experimental value of $f_{(a_{1g} \rightarrow a_{2u})} = 0.4$ it follows that $f_{(e_{1u} \rightarrow a_{2u})}$ (estimated) = 0.002 which is in good agreement with the experimental value of 0.003. This result strongly supports the assignment of this transition. A crucial test of this assignment will involve the investigation of the temperature dependence of the intensity of this forbidden band. From equation 43 it follows that the band intensity should increase by about 0.3 per cent per degree at temperatures around 300° K. Experiments are now in progress to establish this point.

It is likely that the first singlet-triplet transition for xenon difluoride is located on the onset of the 1580 A. band. The 2265 A. band onset in XeF_4 shows a marked asymmetry in the region of 2600 A. where the first singlet-triplet transition for this molecule may be hidden.

RYDBERG STATES OF XeF_2

The set of sharp bands observed on the high-energy side of the 1580 A. band of XeF_2 are assigned to Rydberg states. The highest filled orbital in XeF_2 is the e_{1u} π-type orbital involving mainly the Xe $5p\pi$ AO. The sharp bands of XeF_2 located at 1425 A., 1335 A., 1215 A., and 1145 A. are attributed to one-electron excitation from the e_{1u} orbital. The first two bands are due to excitation into s-type Rydberg states. It should be noted that the $^3P_1$ excited states in the xenon atom are located at 1469.6 A. and 1295.6 A., respectively [29]. The splitting between the 1425 A. and 1335 A. bands in XeF_2 is due to the p electron spin–orbit coupling of the xenon atom. The bands located at 1215 A. and 1145 A. may correspond to d-type Rydberg states. In the case of the xenon atom the three allowed $5p^6 \rightarrow 5p^5 5d$ transitions are located at $5d_{1/2,\ 1} =$ 1250.2 A., $5d_{3/2,\,1} =$ 1192.0 A., and $5d_{3/2,\,1'} =$ 1068.2 A. [29, 42]. Alternatively, all the four Rydberg bands observed for XeF_2 may belong to the same series. In the latter case, two series of Rydberg states are expected, split approximately by the p electron spin–orbit coupling of xenon. The observed bands can be fitted by the two series

$$\nu_1 = 92{,}000 - \frac{\mathrm{Ry}}{(n+0.2)^2}\ \mathrm{cm.}^{-1}, \tag{46a}$$

and,

$$\nu_2 = 98{,}000 - \frac{\mathrm{Ry}}{(n+0.2)^2}\ \mathrm{cm.}^{-1} \tag{46b}$$

for n = 2, 3. The 1425 A. and 1215 A. bands are then assigned to the series (46*a*), while the transitions at 1335 A. and 1145 A. correspond to (46*b*).

The energy difference between the two sets is 0.75 ev. The spin-orbit coupling in atomic xenon is $\frac{3}{2}\zeta$ = 1.12 ev. [29]. The reduction of the spin-orbit–coupling constant upon molecule formation has previously been encountered in the cases of transition-metal ions [40].

The first ionization potential of XeF_2 is thereby estimated to be 11.5 ± 0.1 ev., compared to the value 12.12 ev. for the ionization potential of xenon. Some of the difference between these values may be due to the effect of π-bonding. The result should be compared with the energy of the highest filled orbital, viz., the antibonding π orbital (e_{1u}). Taking the Coulomb integrals equal to the atomic ionization potentials (i.e., ω = 0), the calculated e_{1u} orbital energy is 11.87 ev., in adequate agreement with experiment. However, more refined calculations of the ionization potentials, similar to those applied to aromatic hydrocarbons [43], are desirable.

THE HEATS OF SUBLIMATION OF XeF_2 AND XeF_4

The observation that XeF_2 and XeF_4 are crystalline solids at room temperature is somewhat surprising. If the stability of these solids were due primarily to dispersion forces, the expected heats of sublimation would be similar to those of the rare gases, and the compounds would be gaseous at NTP. This anomalous behavior suggested the study of the heats of sublimation and the stability of the solids.

In the case of XeF_4 the absorptions at 1330 A., 1586 A., and 2010 A. were investigated, while for XeF_2 the absorption at 1750 A. was studied. As can be seen from Figure 8, the logarithm of the optical density at a constant wave length was found to be a linear function of T^{-1} over the temperature region $-15°$ C. to 22° C. It was found that

$$\Delta H_{sub}\ (XeF_2) = 12.3 \pm 0.2\ \text{kcal/mole}\,,$$

and that,

$$\Delta H_{sub}\ (XeF_4) = 15.3 \pm 0.2\ \text{kcal/mole}\,.$$

We now turn to the interpretation of these very large heats of sublimation. We focus attention on XeF_2 to illustrate our considerations. The crystal structure of XeF_2 has been established [34] to be a body-centered tetragonal with $c = 6.995$ A. and $a = 4.315$ A. The xenon atoms are located at the corners and at the body center, and the molecular axis lies along the c axis of the unit cell. All molecular axes are parallel. Thus, with an Xe-F bond length of 2.0 A., the

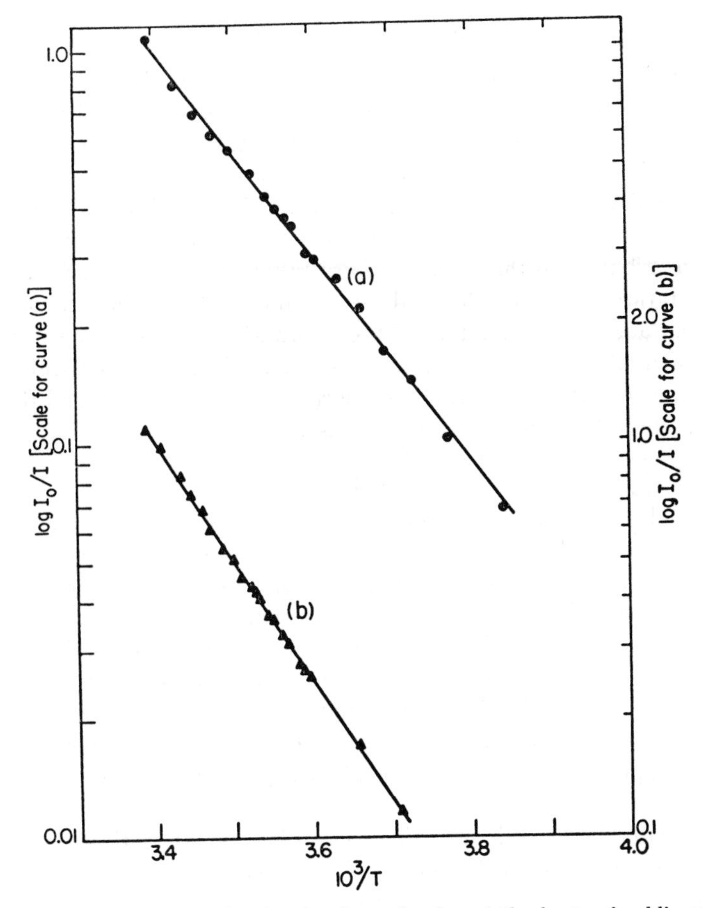

FIG. 8.—Clausius-Clapeyron plot for the determination of the heats of sublimation of crystalline XeF_2 and XeF_4: (*a*) XeF_2 at 1750 A, (*b*) XeF_4 at 2010 A.

F-F separation along the *c* axes is about 3.0 A., and the shortest intermolecular X-F distance is about 3.4 A.

We would not expect dispersion and repulsive overlap forces alone to lead to these large heats of sublimation; therefore, other contributions to ΔH_{sub} must be considered. It has been seen that the proposed model for the binding of the xenon fluorides shows a substantial charge migration from the xenon to the fluorine, such that the effect of electrostatic interactions on the heat of sublimation has to be considered. Long-range intermolecular interactions in the XeF_2 crystal can be adequately described by (weak) quadrupole-quadrupole forces, but the interaction between near-neighbors is better described by the interaction between point charges located at the xenon and fluorine atoms. The computed electrostatic stabilization of the solid is found to be

$$\Delta H_{sub}{}^{electros} = 45.2\ q_F{}^2\ \text{kcal/mole}\,, \tag{47}$$

where q_F is the charge on the fluorine atom. This sum is due to electrostatic interaction of a molecule, in the point-charge approximation, with the eighteen nearest neighbors whose centers lie within 7 A. of the center of the central molecule; interaction with neighbors further away than this distance was neglected. Using the value of q_F obtained from the model of XeF_2, the electrostatic stabilization energy is 11.31 kcal/mole. Thus we conclude that the dominant contribution to the stability of crystalline XeF_2 (and XeF_4) arises from electrostatic interactions. Our model for the intramolecular interactions in XeF_2 yields an adequate description of the intermolecular interaction in the solid. XeF_6 should have a lower heat of vaporization than XeF_2 and XeF_4, because the charge migration from xenon to fluorine should be smaller, and the electrostatic stabilization energy should decrease.

CONCLUSIONS

It has been seen that the ground state of the xenon fluorides is semi-ionic; that is, substantial charge transfer occurs from the xenon to the fluorine. Thus, it is expected that a low rare-gas (R) ionization potential and a high halogen (X) ionization potential are conducive to formation of a rare gas halide (RX_n). In this sense it is not surprising that krypton, xenon, and radon fluorides have been the first of these compounds to be discovered. We would expect that argon halides and rare-gas chlorides, etc., would be progressively more weakly bound, if bound at all. Of course a further energetic restriction for compound formation in the gas phase is that the R-X bond energy be greater than the X-X bond energy.

A valuable source of information concerning the interaction of halogen and rare-gas atoms is obtained from experiments on the catalysis of halogen atom recombination by rare gases. It is fairly well established that the recombination involves a diatomic radical formed by the interaction of a rare gas and a halogen atom. From the study of the negative energies of activation involved in these reactions, Porter and Smith [44] concluded that the heat of formation of the complex is greater than that which would result from normal dispersion forces; the estimated He-I and A-I interaction energies are ≈ 1 kcal. mole^{-1} and 2 kcal. mole^{-1}, respectively. Charge-transfer interactions were invoked to explain the binding. It will be interesting to apply the semi-empirical MO scheme to these reactions. Further experiments on the recombination of halogen atoms by xenon and krypton, etc., will be of interest.

The present semi-empirical MO treatment of the xenon fluorides is surprisingly successful in providing an adequate interpretation of a variety of chemical and physical properties. This may be regarded as one of the successes of MO theory as applied to *inorganic* compounds.

In 1952 Wolfsberg and Helmholtz applied semi-empirical MO theory, similar to that presented in this work, to account for the properties of tetrahedral transition metal complexes [32]. This treatment applied to complexes such as MnO_4^- and $CrO_4^=$ leads to a wrong order for some of the energy levels [45]. This failure of the semi-empirical MO treatment indicates the difficulties involved in applying a deductive approach to the structure of inorganic compounds. Clearly, any a priori theoretical treatment based on a semi-empirical scheme should be viewed with caution. It is emphasized that we regard the present treatment as essentially semi-empirical. *Ab initio* calculations are now required for these interesting molecules.

Acknowledgments.—We wish to thank the following staff members of the Chemistry Division, Argonne National Laboratory, for pre-publication information and discussion: C. L. Chernick and J. G. Malm (preparation of samples of XeF_2 and XeF_4), S. Siegel and E. Gebert (X-ray structure analysis), H. H. Claassen (IR and Raman spectral data), H. H. Hyman and L. Quarterman (information relevant to solutions of XeF_2 in HF), J. C. Hindman and A. Svirmickas (NMR measurements and interpretation), and G. L. Goodman (UV electronic spectral data). We are also indebted to Professor M. H. Cohen and Professor J. Halpern for several provocative discussions. We are grateful to Dr. G. Goodman for his comments on the manuscript. This research was supported by the AFOSR (Grant 61-52) and the USPHS and has benefited from the nonspecific support of material research at the University of Chicago by the AEC and ARPA.

REFERENCES

1. L. C. Allen, *Science* **138,** 892 (1962).
2. R. J. Hach and R. E. Rundle, *J. Am. Chem. Soc.* **73,** 4321 (1951).
3. G. C. Pimentel, *J. Chem. Phys.* **19,** 446 (1951).
4. E. E. Havenga and E. H. Wiebenga, *Rec. Trav. Chim.* **78,** 724 (1959).
5. J. Jortner, S. A. Rice, and E. G. Wilson, *J. Chem. Phys.* **38,** 2302 (1963).
6. L. L. Lohr, Jr., and W. N. Lipscomb, *J. Am. Chem. Soc.* **85,** 240 (1963).
7. G. C. Pimentel and R. D. Spratley, *ibid.* 826 (1963).
8. K. S. Pitzer, *Science* **139,** 414 (1963).

9. R. E. Rundle, *J. Am. Chem. Soc.* **85**, 112 (1963).
10. G. B. Hargreaves and R. D. Peacock, *J. Chem. Soc.* 2373 (1960).
11. C D. Cornwell and R. Yamasaki, *J. Chem. Phys.* **27**, 1060 (1957).
12. J. Jortner, E. G. Wilson, and S. A. Rice, *J. Am. Chem. Soc.* **85**, 813 (1963).
13. J. H. Van Vleck and A. Sherman, *Rev. Mod. Phys.* **7**, 174 (1935).
14. J. C. Slater, *Acta Cryst.* **12**, 197 (1959).
15. C. A. Coulson, *Valence.* New York: Oxford University Press, 1952.
16. R. S. Mulliken, *J. Am. Chem. Soc.* **74**, 811 (1962).
17. J. N. Murrel, *ibid.* **81**, 5037 (1958).
18. R. S. Mulliken, *J. Chim. Phys.* **46**, 497 (1949).
19. H. O. Pritchard, *Chem. Rev.* **52**, 529 (1953).
20. D. R. Hartree, *Proc. Roy. Soc. A* **151**, 96 (1935).
21. *Ibid.* 101 (1935).
22. R. S. Knox, *Solid State Phys.* **4**, 413 (1957).
23. D. R. Hartree, *The Calculation of Atomic Structures.* New York: John Wiley & Sons, 1957.
24. ———, *Proc. Roy. Soc., A* **143**, 506 (1934).
25. C. Kittel, *Introduction to Solid State Physics.* New York: John Wiley & Sons, 1956.
26. C. A. Coulson and M. Chirgwin, *Proc. Roy. Soc. A* **201**, 196 (1950).
27. R. Daudel, R. Lefebvre, and C. Moser, *Quantum Chemistry.* New York: Interscience, 1956.
28. E. Clementi and C. C. J. Roothaan, Phys. Rev. **127**, 1618 (1962).
29. C. E. Moore, *Atomic Energy Tables,* N.B.S. Circular 467.
30. H. O. Pritchard and H. A. Skinner, *Chem. Rev.* **33**, 745 (1958).
31. A. Streitweisser, Jr., *Molecular Orbital Theory.* New York: John Wiley & Sons, 1961.
32. M. Wolfsberg and L. Helmholtz, *J. Chem. Phys.* **20**, 837 (1952).
33. R. S. Mulliken, C. A. Rieke, D. Orloff, and H. Orloff, *ibid.* **17**, 1248 (1949).
34. S. Siegel and E. Gebert, *J. Am. Chem. Soc.* **85**, 240 (1963).
35. D. F. Smith, *J. Chem. Phys.* **38**, 270 (1963).
36. P. A. Agron, *et al.*, *Science* **139**, 842 (1963).
37. R. S. Mulliken, *J. Chem. Phys.* **7**, 20 (1939).
38. S. P. McGlyn, R. Sunseri, and N. Christodouleas, *J. Chem. Phys.* **37**, 1818 (1962).
39. G. Herzberg, *Infrared and Raman Spectra.* New York: D. Van Nostrand Co., 1952.
40. C. J. Ballhausen, *Introduction to Ligand Field Theory.* New York: McGraw-Hill Book Co., 1962.
41. R. S. Mulliken, *J. Chem. Phys.* **3**, 720 (1935).
42. The notation used is that of G. Racah, *Phys. Rev.* **61**, 537 (1942).
43. J. R. Hoyland and L. Goodman, *J. Chem. Phys.* **36**, 12 (1962).
44. G. Porter and M. Smith, *Proc. Roy. Soc. A* **261**, 28 (1961).
45. A. Carrington and M. C. R. Symons, *J. Chem. Soc.* **889** (1960).

APPENDIX

SELECTED PROPERTIES OF THE NOBLE GASES*

Vapor Pressures of Noble Gases

T° K	Gas					
	He^3	He^4	Ne	Ar	Kr	Xe
1	8.56	0.12				
2	150.55	23.77				
3	619.92	182.07				
5		1478.535				
20			28.5			
25			375			
30			*2.2*			
40			*14.5*			
70				58		
80				305	3.0	
90				1003	18.5	
100				*3.21*	82	
110				*6.59*	295	3.0
120				*11.98*	770	12.0
130				*19.99*	*2.05*	39
140				*31.30*	*3.8*	107
150				*46.8*	*6.4*	255
160					*10.0*	560
180					*22.0*	*2.15*
200					*42.0*	*5.0*
220						*10.0*
240						*18.5*
260						*30.5*
280						*47.0*

Vapor pressures are given in mm. Hg (roman) and atm. (*italics*).

* From G. A. Cook (ed.), *Argon, Helium and the Rare Gases* (New York: Interscience, 1961).

Some Physical Properties of the Noble Gases

Property	Gas						
	He	He^3	Ne	Ar	Kr	Xe	Rn
Triple point (° K).....			24.55	83.78	115.95	161.3	202
Pressure at triple point (mm. Hg)..........			324	516	548	612	~500
Heat of fusion at triple point (cal. $mole^{-1}$)..	44.5*	43.0*	80.1	281	390.7	548.5	
Boiling point (° K)....	4.215	3.19	27.07	87.27	119.8	165.05	211
Heat of vaporization at boiling point (cal. $mole^{-1}$)...........	19.4	6.09	414	1557.5	2158	3020	
Critical temperature (° C).............	−267.9	−269.82	−228.7	−122.3	−63.8	16.59	105
Critical pressure (Atm.)	2.26	1.15	26.9	48.3	54.3	57.64	62
Critical density (gm. cm^{-3}).............	0.0693		0.484	0.536	0.908	1.100	
Solubility in water at 0° C [cc(STP)/1000 gm]...............	9.78		14.0	52.4	99.1	203.2	510

* Value given is taken at 25° K.

NAME INDEX

Participants in the Conference on Noble-Gas Compounds (Argonne National Laboratory, April 22–23, 1963) and authors of chapters are cited in capitals.
Chapter references are given in boldface type.
Addresses are given for participants in a code; the key follows.

KEY

ANL	Argonne National Laboratory, Argonne, Ill.
APC	Air Products and Chemicals, Inc., Allentown, Pa.
APLC	Division of Applied Chemistry
ARF	Armour Research Foundation, 10 W. 35th Street, Chicago, Ill.
BIM	Division of Biological and Medical Research
BNL	Brookhaven National Laboratory, Upton, Long Island, N.Y.
CCC WDO	Callery Chemical Co., Western District Office, Sherman Oaks, Calif.
CEN	Chemical Engineering Division
CENS	Centre d'Études Nucléaire de Saclay, Boite Postale No. 2, Gif-sur-Yvette (Seine-et-Oise), France
CHEM	Department of Chemistry
CHM	Chemistry Division
CIT	California Institute of Technology, Pasadena, Calif.
Coe Coll.	Coe College, Cedar Rapids, Iowa
Cor. U.	Cornell University, Ithaca, N.Y.
DuP	E. I. Dupont de Nemours Experimental Station, Wilmington, Del.
FMCSL	Ford Motor Co. Scientific Laboratory, Box 2053, Dearborn, Mich.
Gen. Chem.	General Chemical Division, Allied Chemical Corp., P.O. Box 405, Morristown, N.J.
GIT, EES	Georgia Institute of Technology, Engineering Experimental Station, Atlanta, Ga.
Har. U.	Harvard University, Cambridge, Mass.
HIS	Department of History of Science
ISM-UC	Institute for the Study of Metals, University of Chicago, Chicago 37, Ill.
ISU	Iowa State University, Ames, Iowa
JSI	Nuclear Institute Jožef Stefan, Ljubljana, Slovenia, Yugoslavia
LASL	Los Alamos Scientific Laboratory, P.O. Box 1663, Los Alamos, N.M.
LIV	University of California, Lawrence Radiation Laboratory, P.O. Box 808, Livermore, Calif.
LSU NO	Louisiana State University in New Orleans, New Orleans 22, La.
McM. U.	McMaster University, Hamilton College, Hamilton, Ontario, Canada
MRC	Monsanto Research Corp., Mound Lab., Miamisburg, Ohio
MSTR	Anorganisch–Chemisches Inst. der Universitat, 44 Mün-

SUBJECT INDEX

Pages given in italics are the initial pages of chapters largely devoted to the subject indexed.